STUDENT'S SOLUTIONS MANUAL

to accompany

Fundamentals of Differential Equations

FIFTH EDITION

AND

Fundamentals of Differential Equations and Boundary Value Problems

THIRD EDITION

STUDENT'S SOLUTIONS MANUAL

V. Maymeskul
E. B. Saff
Anne M. Kusmierczyk

to accompany

Fundamentals of Differential Equations

FIFTH EDITION

AND

Fundamentals of Differential Equations and Boundary Value Problems

THIRD EDITION

R. Kent Nagle

E. B. Saff
University of South Florida

Arthur David Snider
University of South Florida

Boston San Francisco New York
London Toronto Sydney Tokyo Singapore Madrid
Mexico City Munich Paris Cape Town Hong Kong Montreal

Reprinted with corrections, October 2000

Copyright © 2000 Addison Wesley Longman.

All rights reserved. No part of this publication may be reproduced, stored in a retrieval system, or transmitted, in any form or by any means, electronic, mechanical, photocopying, recording, or otherwise, without the prior written permission of the publisher. Printed in the United States of America.

ISBN 0-201-33869-6

2 3 4 5 6 7 8 9 10 PHTH 020100

Contents

CHAPTER 1: Introduction

EXERCISES 1.1: Background, page 5

3. This equation is an ODE because it contains no partial derivatives. Since the highest order derivative is dy/dx, the equation is a first order equation. This same term also shows us that the independent variable is x and the dependent variable is y. This equation is nonlinear because of the y in the denominator of the term $\dfrac{y(2-3x)}{x(1-3y)}$.

5. This equation is an ODE because it contains only ordinary derivatives. The term $\dfrac{dp}{dt}$ is the highest order derivative and thus shows us that this is a first order equation. This term also shows us that the independent variable is t and the dependent variable is p. This equation is nonlinear since in the term $kp(P-p)=kPp-kp^2$ the dependent variable p is squared (compare with equation (7) on page 5 of the text).

11. This equation contains partial derivatives, thus it is a PDE. Because the highest order derivative is a second order partial derivative, the equation is a second order equation. The terms $\partial N/\partial t$ and $\partial N/\partial r$ show that the independent variables are t and r and the dependent variable is N.

17. In classical physics, the instantaneous acceleration, a, of an object moving in a straight line is given by the second derivative of distance, x, with respect to time, t; that is

$$\frac{d^2x}{dt^2}=a.$$

Integrating both sides with respect to t and using the given fact that a is constant we obtain

$$\frac{dx}{dt}=at+C. \tag{1}$$

The instantaneous velocity, v, of an object is given by the first derivative of distance, x, with respect to time, t. At the beginning of the race, $t=0$, both racers have zero velocity. Therefore we have $C=0$. Integrating equation (1) with respect to t we obtain

$$x=\frac{1}{2}at^2+C_1.$$

For this problem we will use the starting position for both competitors to be $x=0$ at $t=0$. Therefore, we have $C_1=0$. This gives us a general equation used for both racers as

$$x=\frac{1}{2}at^2$$

or

$$t = \sqrt{\frac{2x}{a}},$$

where the acceleration constant a has different values for Kevin and for Alison.

Kevin covers the last $\frac{1}{4}$ of the full distance, L, in 3 seconds. This means Kevin's acceleration, a_K, is determined by:

$$t_K - t_{3/4} = 3 = \sqrt{\frac{2L}{a_K}} - \sqrt{\frac{2(3L/4)}{a_K}},$$

where t_K is the time it takes for Kevin to finish the race. Solving the above equation for a_K gives,

$$a_K = \frac{\left(\sqrt{2} - \sqrt{3/2}\right)^2}{9} L.$$

Therefore the time required for Kevin to finish the race is given by:

$$t_K = \sqrt{\frac{2L}{\left(\sqrt{2} - \sqrt{3/2}\right)^2 L/9}} = \frac{3}{\sqrt{2} - \sqrt{3/2}}\sqrt{2} = 12 + 6\sqrt{3} \approx 22.39 \text{ sec}.$$

Alison covers the last $\frac{1}{3}$ of the distance, L, in 4 seconds. This means Alison's acceleration, a_A, is found by:

$$t_A - t_{2/3} = 4 = \sqrt{\frac{2L}{a_A}} - \sqrt{\frac{2(2L/3)}{a_A}},$$

where t_A is the time required for Alison to finish the race. Solving the above equation for a_A gives

$$a_A = \frac{\left(\sqrt{2} - \sqrt{3/2}\right)^2}{16} L.$$

Therefore the time required for Alison to finish the race is given by:

$$t_A = \sqrt{\frac{2L}{\left(\sqrt{2} - \sqrt{4/3}\right)^2 L/16}} = \frac{4}{\sqrt{2} - \sqrt{4/3}}\sqrt{2} = 12 + 4\sqrt{6} \approx 21.80 \text{ sec}.$$

The time required for Alison to finish the race is less than Kevin; therefore Alison wins the race by $6\sqrt{3} - 4\sqrt{6} \approx 0.594$ seconds.

EXERCISES 1.2: Solutions and Initial Value Problems, page 14

3. Since $y = \sin x + x^2$, we have $y' = \cos x + 2x$ and $y'' = -\sin x + 2$. These functions are defined on $(-\infty, \infty)$. Substituting these expressions into the differential equation $\frac{d^2y}{dx^2} + y = x^2 + 2$ gives

$$y'' + y' = -\sin x + 2 + \sin x + x^2 = 2 + x^2 = x^2 + 2 \qquad \text{for all } x \text{ in } (-\infty, \infty).$$

Therefore, $y = \sin x + x^2$ is a solution to the differential equation on the interval $(-\infty, \infty)$.

9. Differentiating the equation $x^2 + y^2 = 6$ implicitly, we obtain

$$2x + 2yy' = 0 \qquad \Rightarrow \qquad y' = -\frac{x}{y}.$$

Since there can be no function $y = f(x)$ that satisfies the differential equation $y' = \frac{x}{y}$ *and* the differential equation $y' = -\frac{x}{y}$ *on the same interval*, we see that $x^2 + y^2 = 6$ does *not* define an implicit solution to the differential equation.

13. Differentiating the equation $\sin y + xy - x^3 = 2$ implicitly with respect to *x*, we obtain

$$y' \cos y + xy' + y - 3x^2 = 0$$

$$\Rightarrow \qquad (\cos y + x)y' = 3x^2 - y$$

$$\Rightarrow \qquad y' = \frac{3x^2 - y}{\cos y + x}.$$

Differentiating the second equation above again, we obtain

$$(-y' \sin y + 1)y' + (\cos y + x)y'' = 6x - y'$$

$$\Rightarrow \qquad (\cos y + x)y'' = 6x - y' + (y')^2 \sin y - y' = 6x - 2y' + (y')^2 \sin y$$

$$\Rightarrow \qquad y'' = \frac{6x - 2y' + (y')^2 \sin y}{\cos y + x}.$$

Multiplying the right-hand side of this last equation by $\frac{y'}{y'} = 1$ and using the fact that

$$y' = \frac{3x^2 - y}{\cos y + x},$$

we get

$$y'' = \frac{6x - 2y' + (y')^2 \sin y}{\cos y + x} \cdot \frac{y'}{\dfrac{3x^2 - y}{\cos y + x}}$$

$$\Rightarrow \qquad y'' = \frac{6xy' - 2(y')^2 + (y')^3 \sin y}{3x^2 - y}.$$

Thus y is an implicit solution to the differential equation.

19. Squaring and adding the terms $\frac{dy}{dx}$ and y in the equation $\left(\frac{dy}{dx}\right)^2 + y^2 + 3 = 0$ gives a nonnegative number. Therefore when these two terms are added to 3, the left-hand side will always be greater than or equal to three and hence can never equal the right-hand side which is zero.

21. For $\phi(x) = x^m$, we have $\phi'(x) = mx^{m-1}$ and $\phi''(x) = m(m-1)x^{m-2}$.

(a) Substituting these expressions into the differential equation, $3x^2 \frac{d^2y}{dx^2} + 11x\frac{dy}{dx} - 3y = 0$, gives

$$3x^2\left[m(m-1)x^{m-2}\right] + 11x\left[mx^{m-1}\right] - 3x^m = 0$$

$$\Rightarrow \qquad 3m(m-1)x^m + 11mx^m - 3x^m = 0$$

$$\Rightarrow \qquad \left[3m(m-1) + 11m - 3\right]x^m = 0$$

$$\Rightarrow \qquad \left[3m^2 + 8m - 3\right]x^m = 0.$$

For the last equation to hold on an interval for x, we must have

$$3m^2 + 8m - 3 = (3m-1)(m+3) = 0.$$

Thus either $(3m - 1) = 0$ or $(m + 3) = 0$, which gives $m = \frac{1}{3}, -3$.

(b) Substituting the above expressions for ϕ, ϕ' and ϕ'' into the differential equation, $x^2 \frac{d^2y}{dx^2} - x\frac{dy}{dx} - 5y = 0$, gives

$$x^2\left[m(m-1)x^{m-2}\right] - x\left[mx^{m-1}\right] - 5x^m = 0$$

$$\Rightarrow \qquad \left[m^2 - 2m - 5\right]x^m = 0.$$

For the last equation to hold on an interval for x, we must have

$$m^2 - 2m - 5 = 0.$$

To solve for m we use the quadratic formula:

$$m = \frac{2 \pm \sqrt{4+20}}{2} = 1 \pm \sqrt{6}.$$

26. Here $f(x, y) = 3x - \sqrt[3]{y-1}$ and $\frac{\partial f}{\partial y} = -\frac{1}{3}(y-1)^{-2/3}$. Unfortunately, $\frac{\partial f}{\partial y}$ is not continuous or defined when $y = 1$. So there is no rectangle containing (2,1) in which both f and $\frac{\partial f}{\partial y}$ are continuous. Therefore, we are not guaranteed a unique solution to this initial value problem.

31. **(a)** To try to apply Theorem 1 we must first write the equation in the form $\frac{dy}{dx} = f(x, y)$. Here $f(x, y) = 4\frac{x}{y}$ and $\frac{\partial f}{\partial y} = -4\frac{x}{y^2}$. Neither f nor $\frac{\partial f}{\partial y}$ are continuous or defined when $y = 0$. Therefore there is no rectangle containing $(x_0, 0)$ in which both f and $\frac{\partial f}{\partial y}$ are continuous, so Theorem 1 cannot be applied.

(b) Suppose for the moment that there is such a solution $y(x)$ with $y(x_0) = 0$ and $x_0 \neq 0$. Substituting into the differential equation we get

$$y(x_0)y'(x_0) - 4x_0 = 0 \qquad (2)$$

or

$$0 \cdot y'(x_0) - 4x_0 = 0 \qquad \Rightarrow \qquad 4x_0 = 0.$$

Thus $x_0 = 0$, which is a contradiction.

(c) Taking $C = 0$ in the implicit solution $4x^2 - y^2 = C$ given in Example 5 on page 9 gives $4x^2 - y^2 = 0$ or $y = \pm 2x$. Both solutions $y = 2x$ and $y = -2x$ satisfy $y(0) = 0$.

EXERCISES 1.3: Direction Fields, page 22

5. **(a)** The graph of the directional field is shown in Figure B.4 in the answers section of the text.

(b), (c) The direction field indicates that all solution curves (other than $p(t) \equiv 0$) will approach the horizontal line (asymptote) $p = 1.5$ as $t \to +\infty$. Thus $\lim_{t \to +\infty} p(t) = 1.5$.

(d) No. The direction field shows that populations greater than 1500 will steadily decrease but can never reach 1500 or any smaller value, i.e., the solution curves cannot cross the line $p = 1.5$. Indeed, the constant function $p(t) \equiv 1.5$ is a solution to the given logistic equation, and the uniqueness part of Theorem 1, page 12, prevents intersections of solution curves.

6. **(a)** The slope of a solution to the differential equation $\frac{dy}{dx} = x + \sin y$ is given by $\frac{dy}{dx}$. Therefore the slope at $\left(1, \frac{\pi}{2}\right)$ is equal to

$$\frac{dy}{dx} = 1 + \sin\frac{\pi}{2} = 2.$$

(b) The solution curve is increasing if the slope of the curve is greater than zero. From part (a) we know the slope to be $x + \sin y$. The function $\sin y$ has values ranging from -1 to 1; therefore if x is greater than 1 then the slope will always have a value greater than zero. This tells us that the solution curve is increasing.

(c) The second derivative of every solution can be determined by finding the derivative of the differential equation $\frac{dy}{dx} = x + \sin y$. Thus

$$\frac{d}{dx}\left(\frac{dy}{dx}\right) = \frac{d}{dx}(x + \sin y)$$

$$\frac{d^2y}{dx^2} = 1 + (\cos y)\frac{dy}{dx} \qquad \text{(chain rule)}$$

$$= 1 + (\cos y)[x + \sin y]$$

$$= 1 + x\cos y + \sin y \cos y$$

$$\frac{d^2y}{dx^2} = 1 + x\cos y + \frac{1}{2}\sin 2y.$$

(d) Relative minima occur when the first derivative, $\frac{dy}{dx}$, is equal to zero and the second derivative, $\frac{d^2y}{dx^2}$, is greater than zero. The value of the first derivative at the point (0,0) is given by

$$\frac{dy}{dx} = 0 + \sin 0 = 0.$$

This tells us that the solution has a critical point at the point (0,0). Using the second derivative found in part (c) we have

$$\frac{d^2y}{dx^2} = 1 + 0 \cdot \cos 0 + \frac{1}{2}\sin 0 = 1.$$

This tells us the point (0, 0) is a relative minimum.

7. **(a)** The graph of the directional field is shown in Figure B.5 in the answers section of the text.

(b) The direction field indicates that all solution curves with $p(0)>1$ will approach the horizontal line (asymptote) $p=2$ as $t\to+\infty$. Thus $\lim_{t\to+\infty} p(t)=2$ when $p(0)=3$.

(c) The direction field shows that a population between 1000 and 2000 (that is $1<p(t)<2$) will approach the horizontal line $p=2$ as $t\to+\infty$.

(d) The direction field shows that an initial population less than 1000 (that is $0\le p(t)<1$) will approach zero as $t\to+\infty$.

(e) As noted in part (d), the line $p=1$ is an asymptote. The direction field indicates that a population of 900 $p=0.9$ steadily decreases with time and therefore cannot increase to 1100.

13. For the equation $\dfrac{\partial y}{\partial x}=-\dfrac{x}{y}$, the isoclines are the curves $-\dfrac{x}{y}=c$. These are lines that pass through the origin and have equations of the form $y=mx$, where $m=-\dfrac{1}{c}$, $c\neq 0$. If we let $c=0$ in $-\dfrac{x}{y}=c$, we see that the *y*-axis ($x=0$) is also an isocline. Each element of the direction field associated with a point on an isocline has slope *c* and is, therefore, perpendicular to that isocline. Since circles have the property that at any point on the circle the tangent at that point is perpendicular to a line from that point to the center of the circle, we see that the solution curves will be circles with their centers at the origin. But since we cannot have $y=0$ (since $-\dfrac{x}{y}$ would then have a zero in the denominator) the solutions will not be defined on the *x*-axis. (Note however that a related form of this differential equation is $y\dfrac{dy}{dx}+x=0$. This equation has implicit solutions given by the equations $y^2+x^2=C$. These solutions will be circles.) The graph of $\phi(x)$, the solution to the equation satisfying the initial condition $y(0)=4$, is the upper semicircle with center at the origin and passing through the point (0, 4) (see Figure B.8 in the answers of the text).

EXERCISES 1.4: The Phase Line, page 31

3. **(a)** First we rewrite the equation in the equivalent form $y'=1-y^2$. The function $f(y)=1-y^2$ has zeros $y=\pm1$. Since $f(y)<0$ for $|y|>1$ and $f(y)>0$ for $|y|<1$, the phase line for the given equation is the one shown in Figure B.11 in the answers of the text.

(b) The equilibrium solutions are $y(t)\equiv-1$ and $y(t)\equiv 1$.

(c) From the arguments above and the criteria given on page 26 we conclude that the equilibrium point $y=-1$ is a source while $y=1$ is a sink.

7. **(a)** Writing the equation in the equivalent form $y'=y\cos y$, we observe that the function $f(y)=y\cos y$ has zeros

$$y=0 \qquad \text{and} \qquad \pm y_k=\pm(\pi/2+\pi k), \quad k=0,1,2,\ldots.$$

For $0<y<\pi/2=y_0$ we have $y\cos y>0$; for $\pi/2<y<3\pi/2=y_1$ the $\cos y$ term is negative, so $y\cos y<0$; for $3\pi/2<y<5\pi/2=y_2$, the cosine term is positive again, so $y\cos y>0$; and so on. Thus the sign of $f(y)$ changes at each y_k. Since $f(-y)=-f(y)$ (f is an odd function) the opposite sign changes occur at the points $-y_k$. Thus we conclude that the phase line for the given equation is the one shown in Figure B.13 in the answers of the text.

(b) The equilibrium solutions $y(t)\equiv 0$ and $y(t)\equiv\pm y_k$, $k=0,1,2,\ldots$, are also shown in Figure B.13.

(c) From the part (a) we find that the equilibrium points $y=\pm y_{2j}=\pm\left(\frac{\pi}{2}+2\pi j\right)$, $j=0,1,2,\ldots$, are sinks while $y=0$ and $y=\pm y_{2j+1}=\pm\left(\frac{\pi}{2}+(2j+1)\pi\right)$, $j=0,1,2,\ldots$, are sources.

11. Solving the equation $f(y)=y^2-7y+10=0$, we find two equilibrium points $y=2$ and $y=5$. Since $f(y)<0$ for $2<y<5$ and $f(y)>0$ otherwise, from the criteria on page 26 of the text we conclude that $y=2$ is a sink (or a stable point) while $y=5$ is a source. The initial point $y(0)=3\in(2,5)$ which gives $y(t)\to 2$ as $t\to+\infty$.

15. In this problem, the function $f(y)=\sin^2 y$ has zeros $y_k=\pi k$, $k=0,\pm1,\pm2,\ldots$, and $f(y)\geq 0$ for all y. Therefore all equilibrium points y_k are nodes and all the solutions are increasing functions. Since $y(0)=\pi/2\in(0,\pi)$ we have $y(t)\to\pi$ as $t\to+\infty$.

19. Assuming that s rats are killed per week, we get a modified equation

$$y'=-y(y-100)/50-s.$$

The zeros of the quadratic function $f(y)=-y(y-100)/50-s$, which are the equilibrium points, are $50\pm\sqrt{2500-50s}$. The bifurcation diagram for this problem is similar to that for Example 4 on page 30 of the text. Solving $2500-50s=0$ we conclude that the coalescence of equilibria occurs when $s=50$. So, more than 50 rats should be killed weekly to eradicate their population.

EXERCISES 1.5: The Approximation Method of Euler, page 35

2. In this problem, $x_0 = 0,\ y_0 = 4,\ h = 0.1$, and $f(x,y) = -\dfrac{x}{y}$. Thus, the recursive formulas given in equations (2) and (3) on page 32 of the text become

$$x_{n+1} = x_n + h = x_n + 0.1,$$

$$y_{n+1} = y_n + hf(x_n, y_n) = y_n + 0.1\cdot\left(-\frac{x_n}{y_n}\right), \qquad n = 0, 1, 2, \ldots .$$

To find an approximation for the solution at the point $x_1 = x_0 + 0.1 = 0.1$, we let $n = 0$ in the last recursive formula to find

$$y_1 = y_0 + 0.1\cdot\left(-\frac{x_0}{y_0}\right) = 4 + 0.1\cdot(0) = 4.$$

To approximate the value of the solution at the point $x_2 = x_1 + 0.1 = 0.2$, we let $n = 1$ in the last recursive formula to obtain

$$y_2 = y_1 + 0.1\cdot\left(-\frac{x_1}{y_1}\right) = 4 + 0.1\cdot\left(-\frac{0.1}{4}\right) = 4 - \left(\frac{1}{400}\right) = 3.9975 \approx 3.998.$$

Continuing in this way we find

$$x_3 = x_2 + 0.1 = 0.3, \qquad y_3 = y_2 + 0.1\cdot\left(-\frac{x_2}{y_2}\right) = 3.9975 + 0.1\cdot\left(-\frac{0.2}{3.9975}\right) = 3.992,$$

$$x_4 = 0.4, \qquad y_4 = 3.985,$$

$$x_5 = 0.5, \qquad y_5 = 3.975,$$

where all of the answers have been rounded off to three decimal places.

7. For this problem notice that the independent variable is t and the dependent variable is x. Hence, the recursive formulas given in equations (2) and (3) on page 32 of the text become

$$t_{n+1} = t_n + h \quad \text{and} \quad \phi(t_{n+1}) \approx x_{n+1} = x_n + hf(t_n, x_n), \qquad n = 0, 1, 2, \ldots .$$

For this problem, $f(t,x) = 1 + t\sin(tx)$, $t_0 = 0$ and $x_0 = 0$. Thus the second recursive formula above becomes

$$x_{n+1} = x_n + h[1 + t_n \sin(t_n x_n)], \qquad n = 0, 1, 2, \ldots .$$

For the case $N = 1$, we have $h = \dfrac{1-0}{1} = 1$ which gives us

$$t_1 = 0 + 1 = 1 \quad \text{and} \quad \phi(1) \approx x_1 = 0 + 1\cdot(1 + 0\cdot\sin 0) = 1.$$

For the case $N = 2$, we have $h = \frac{1}{2} = 0.5$. Thus we have

$$t_1 = 0 + 0.5 = 0.5, \qquad x_1 = 0 + 0.5 \cdot (1 + 0 \cdot \sin 0) = 0.5,$$

and

$$t_2 = 0.5 + 0.5 = 1, \qquad \phi(1) \approx x_2 = 0.5 + 0.5 \cdot [1 + 0.5\sin(0.25)] = 1.06185.$$

For the case $N = 4$, we have $h = \frac{1}{4} = 0.25$, and so the recursive formulas become

$$t_{n+1} = t_n + 0.25 \quad \text{and} \quad x_{n+1} = x_n + 0.25 \cdot [1 + t_n \sin(t_n x_n)].$$

Therefore, we have

$$t_1 = 0 + 0.25 \quad \text{and} \quad x_1 = 0 + 0.25 \cdot [1 + 0 \cdot \sin(0)].$$

Plugging these values into the recursive equations above yields

$$t_2 = 0.25 + 0.25 = 0.5 \quad \text{and} \quad x_2 = 0.25 + 0.25 \cdot [1 + 0.25\sin(0.0625)] = 0.503904.$$

Continuing in this way gives

$$t_3 = 0.75 \quad \text{and} \quad x_3 = 0.503904 + 0.25 \cdot [1 + 0.5\sin(0.251952)] = 0.785066,$$

$$t_4 = 1.00 \quad \text{and} \quad \phi(1) \approx x_4 = 1.13920.$$

For $N = 8$, we have $h = \frac{1}{8} = 0.125$. Thus, the recursive formulas become

$$t_{n+1} = t_n + 0.125 \quad \text{and} \quad x_{n+1} = x_n + 0.125 \cdot [1 + t_n \sin(t_n x_n)].$$

Using these formulas and starting with $t_0 = 0$ and $x_0 = 0$, we can fill in Table 1-A below. From this we see that $\phi(1) \approx x_8 = 1.19157$, which is rounded to five decimal places.

Table 1-A. Euler's method approximations for the solutions of $x' = 1 + t\sin(tx)$, $x(0) = 0$ at $t = 1$ with 8 steps ($h = 1/8$)

n	t_n	x_n
1	0.125	0.125
2	0.250	0.250244
3	0.375	0.377198
4	0.500	0.508806
5	0.625	0.649535
6	0.750	0.805387
7	0.875	0.983634
8	1.000	1.191572

---

13. From Problem 12, $y_n = \left(1+\frac{1}{n}\right)^n$ and so $\lim_{n\to\infty} \frac{e - y_n}{1/n}$ is a 0/0 indeterminant. If we let $h = 1/n$ in y_n and use L'Hôpital's rule, we get

$$\lim_{n\to\infty} \frac{e - y_n}{1/n} = \lim_{h\to\infty} \frac{\left[e - (1+h)^{1/h}\right]}{h} = \lim_{h\to 0} \frac{g(h)}{h} = \lim_{h\to 0} \frac{g'(h)}{1},$$

where $g(h) = e - (1+h)^{1/h}$. Writing $(1+h)^{1/h}$ as $e^{\ln(1+h)/h}$ the function $g(h)$ becomes

$$g(h) = e - e^{\ln(1+h)/h}.$$

The first derivative is given by

$$g'(h) = 0 - \frac{d}{dh}\left[e^{\ln(1+h)/h}\right] = -e^{\ln(1+h)/h} \frac{d}{dh}\left[\frac{1}{h}\ln(1+h)\right].$$

Substituting Maclaurin's series for $\ln(1+h)$ we obtain

$$g'(h) = -(1+h)^{1/h} \frac{d}{dh}\left[\frac{1}{h}\left(h - \frac{1}{2}h^2 + \frac{1}{3}h^3 - \frac{1}{4}h^4 + \cdots\right)\right]$$

$$= -(1+h)^{1/h} \frac{d}{dh}\left[1 - \frac{1}{2}h + \frac{1}{3}h^2 - \frac{1}{4}h^3 + \cdots\right];$$

$$g'(h) = -(1+h)^{1/h}\left[-\frac{1}{2} + \frac{2}{3}h - \frac{3}{4}h^2 + \cdots\right].$$

Hence

$$\lim_{h\to 0} g'(h) = \lim_{h\to 0}\left\{-(1+h)^{1/h}\left[-\frac{1}{2} + \frac{2}{3}h - \frac{3}{4}h^2 + \cdots\right]\right\}$$

$$= \left[-\lim_{h\to 0}(1+h)^{1/h}\right]\left[\lim_{h\to 0}\left\{-\frac{1}{2} + \frac{2}{3}h - \frac{3}{4}h^2 + \cdots\right\}\right].$$

Form calculus we know that $e = \lim_{h\to 0}(1+h)^{1/h}$, which gives

$$\lim_{h\to 0} g'(h) = -e\left(-\frac{1}{2}\right) = \frac{e}{2}.$$

So we have

$$\lim_{n\to\infty} = \frac{e - y_n}{1/n} = \frac{e}{2}.$$

16. For this problem notice that the independent variable is t and the dependent variable is T. Hence, in the recursive formulas for Euler's method, the t will take the place of the x and the T will take the place of the y. Also we see that $h=0.1$ and $f(t,T)=K(M^4-T^4)$ where $K=40^{-4}$ and $M=70$. Therefore, the recursive formulas given in equations (2) and (3) on page 32 of the text become

$$t_{n+1}=t_n+h=t_n+0.1,$$

$$T_{n+1}=T_n+hf(t_n,T_n)=T_n+0.1(40^{-4})(70^4-T_n^4),\quad n=0,1,2,\ldots.$$

From the initial condition $T(0)=100$, we see that $t_0=0$ and $T_0=100$. Therefore, for $n=0$, we have

$$t_1=t_0+0.1=0+0.1=0.1,$$

$$T_1=T_0+0.1(40^{-4})(70^4-T_0^4)=100+0.1(40^{-4})(70^4-100^4)=97.0316,$$

where we have rounded off to four decimal places. For $n=1$, we have

$$t_2=t_1+0.1=0.1+0.1=0.2,$$

$$T_2=T_1+0.1(40^{-4})(70^4-T_1^4)=97.0316+0.1(40^{-4})(70^4-97.0316^4)=94.5068.$$

By continuing this way, we fill in Table 1-B below.

Table 1-B: Euler's method approximations for the solutions of $T=K(M^4-T^4)$, with $K=40^{-4}, M=70, T(0)=100$, and $h=0.1$.

n	t_n	T_n	n	t_n	T_n
0	0	100	11	1.1	81.8049
1	0.1	97.0316	12	1.2	80.9934
2	0.2	94.5068	13	1.3	80.2504
3	0.3	92.3286	14	1.4	79.5681
4	0.4	90.4279	15	1.5	78.9403
5	0.5	88.7538	16	1.6	78.3613
6	0.6	87.2678	17	1.7	77.8263
7	0.7	85.9402	18	1.8	77.3311
8	0.8	84.7472	19	1.9	76.8721
9	0.9	83.6702	20	2.0	76.4459
10	1.0	82.6936			

From this table we see that

$$T(1) = T(t_{10}) \approx T_{10} = 82.694,$$

and

$$T(2) = T(t_{20}) \approx T_{20} = 76.446,$$

where we have rounded to three decimal places.

CHAPTER 2: First Order Differential Equations

EXERCISES 2.2: Separable Equations, page 51

3. This equation is separable because: $\frac{dy}{dx}=\frac{ye^{x+y}}{x^2+2}=\left(\frac{e^x}{x^2+2}\right)ye^y=g(x)p(y)$.

13. Writing the given equation in the form $\frac{dx}{dt}=x-x^2$. We separate the variables to obtain $\left(\frac{1}{x-x^2}\right)dx=dt$. Integrate (the left side is integrated by partial fractions, with $\frac{1}{x-x^2}=\frac{1}{x}+\frac{1}{1-x}$) to obtain:

$$\ln|x|-\ln|1-x|=t+c$$

$$\Rightarrow \quad \ln\left|\frac{x}{1-x}\right|=t+c$$

$$\Rightarrow \quad \frac{x}{1-x}=\pm e^{t+c}=Ce^t, \qquad \text{where } C=e^c$$

$$\Rightarrow \quad x=Ce^t-xCe^t$$

$$\Rightarrow \quad x+xCe^t=Ce^t$$

$$\Rightarrow \quad x\left(1+Ce^t\right)=Ce^t$$

$$\Rightarrow \quad x=\frac{Ce^t}{1+Ce^t}.$$

Note: When C is replaced by $-K$, this answer can also be written as $x=\frac{Ke^t}{Ke^t-1}$. Further we observe that since we divide by $x-x^2=x(1-x)$, then $x\equiv 0$ and $x\equiv 1$ are also solutions. Allowing K to be zero gives $x\equiv 0$, but no choice for K will give $x\equiv 1$, so we list this as a separate solution.

21. Separate variables to obtain $\frac{1}{2}(y+1)^{-1/2}dy=\cos x\,dx$. Integrating we have

$$(y+1)^{1/2}=\sin x+C.$$

Using the fact that $y(\pi)=0$, we find

$$1 = \sin\pi + C \qquad \Rightarrow \qquad C = 1.$$

Thus

$$(y+1)^{1/2} = \sin x + 1$$

and so

$$y = (\sin x + 1)^2 - 1 = \sin^2 x + 2\sin x.$$

25. By separating variables we obtain $(1+y)^{-1}\,dy = x^2\,dx$. Integrating yields

$$\ln|1+y| = \frac{x^3}{3} + C. \qquad (1)$$

Substituting $y=3$ and $x=0$ from the initial condition, we get $\ln 4 = 0 + C$, which implies that $C = \ln 4$. By substituting this value for C into equation (1) above, we have

$$\ln|1+y| = \frac{x^3}{3} + \ln 4.$$

Hence,

$$e^{\ln|1+y|} = e^{(x^3/3)+\ln 4} = e^{x^3/3}e^{\ln 4} = 4e^{x^3/3}$$

$$\Rightarrow \qquad 1 + y = 4e^{x^3/3}$$

$$\Rightarrow \qquad y = 4e^{x^3/3} - 1.$$

We can drop the absolute signs above because we are assuming from the initial condition that y is close to 3 and therefore $1+y$ is positive.

27. **(a)** The differential equation $\dfrac{dy}{dx} = e^{x^2}$ separates if we multiply by dx. We integrate the separated equation from $x=0$ to $x=x_1$ to obtain

$$\int_0^{x_1} e^{x^2}\,dx = \int_{x=0}^{x=x_1} dy = y\Big|_{x=0}^{x=x_1} = y(x_1) - y(0).$$

If we let t be the variable of integration of integration and replace x_1 by x and $y(0)$ by 0, then we can express the solution to the initial value problem as

$$y(x) = \int_0^x e^{t^2}\,dt.$$

(b) The differential equation $\dfrac{dy}{dx} = e^{x^2}y^{-2}$ is separable if we multiply by y^2 and dx. We integrate the separated equation from $x=0$ to $x=x_1$ to obtain

$$\int_0^{x_1} e^{x^2}\,dx = \int_{x=0}^{x=x_1} y^2\,dx = \frac{1}{3}y^3\Big|_{x=0}^{x=x_1} = \frac{1}{3}\left[y(x_1)^3 - y(0)^3\right].$$

If we let t be the variable of integration and replace x_1 by x and $y(0)$ by 1 in the above equation, then we can express the initial value problem as

$$\int_0^{x} e^{t^2}\,dt = \frac{1}{3}\left[y(x)^3 - 1\right].$$

Solving for $y(x)$ we arrive at

$$y(x) = \left[1 + 3\int_0^{x} e^{t^2}\,dt\right]^{1/3}. \tag{2}$$

(c) The differential equation $\frac{dy}{dx} = \sqrt{1+\sin x}\left(1+y^2\right)$ separates if we divide by $\left(1+y^2\right)$ and multiply by dx. We integrate the separated equation from $x = 0$ to $x = x_1$ and find

$$\int_0^{x_1} \sqrt{1+\sin x}\,dx = \int_{x=0}^{x=x_1} \left(1+y^2\right)^{-1} dy = \tan^{-1} y(x_1) - \tan^{-1} y(0).$$

If we let t be the variable of integration and replace x_1 by x and $y(0)$ by 1 then we can express the solution to the initial value problem by

$$y(x) = \tan\left\{\int_0^{x} \sqrt{1+\sin t}\,dt + \frac{\pi}{4}\right\}.$$

(d) We will use Simpson's rule (Appendix B) to approximate the definite integral found in part (b). (Simpson's rule is implemented on the website for the text.) Simpson's rule requires an even number of intervals, but we don't know how many are required to obtain the desired three-place accuracy. Rather than make an error analysis, we will compute the approximate value of $y(0.5)$ using 2, 4, 6, ... intervals for Simpson's rule until the approximate values for $y(0.5)$ change by less than five in the fourth place.
For $n = 2$, we divide [0, 0.5] into 4 equal subintervals. Thus each interval will be of length $\frac{0.5-0}{4} = \frac{0.5}{4} = \frac{1}{8}$. Therefore, the integral is approximated by

$$\int_0^{0.5} e^{x^2}\,dx = \frac{1}{24}\left[e^0 + 4e^{(0.125)^2} + 2e^{(0.25)^2} + 4e^{(0.375)^2} + e^{(0.5)^2}\right] \approx 0.544999003.$$

Substituting this value into equation (2) from part (b) yields

$$y(0.5) \approx \left[1 + 3(0.544999003)\right]^{1/3} \approx 1.38121.$$

Repeating these calculations for $n = 3, 4,$ and 5 yields

Table 2-A. Successive approximations for $y(0.5)$ using Simpson's rule.

Number of Intervals	$y(0.5)$
6	1.38120606
8	1.38120520
10	1.38120497

Since these values do not change by more than 5 in the fourth place, we can conclude that the first three places are accurate and that we have obtained an approximate solution $y(0.5) \approx 1.381$.

33. Let $A(t)$ be the number of kilograms of salt in the tank at t minutes after the process begins. Then we have

$$\frac{dA(t)}{dt} = \text{rate of salt in} - \text{rate of salt out}.$$

Rate of salt in $= 10 \text{ L/min} \times 3 \text{ kg/L} = 30 \text{ kg/min}.$

Since the tank is kept uniformly mixed, $\frac{A(t)}{400}$ is the mass of salt per liter that is flowing out of the tank at time t. Thus we have

Rate of salt out $= 10 \text{ L/min} \times \frac{A(t)}{400} \text{ kg/L} = \frac{A(t)}{40} \text{ kg/min}.$

Therefore,

$$\frac{dA}{dt} = 30 - \frac{A}{40} = \frac{1200 - A}{40}.$$

Separating this differential equation yields $\frac{40}{1200 - A} dA = dt$. Integrating gives

$$-40 \ln|1200 - A| = t + C$$

$$\Rightarrow \quad \ln|1200 - A| = -\frac{t}{40} + C, \text{ where } -\frac{C}{40} \text{ is replaced by } C$$

$$\Rightarrow \quad 1200 - A = Ce^{-t/40}, \text{ where } C \text{ can now be positive or negative}$$

$$\Rightarrow \quad A = 1200 - Ce^{-t/40}.$$

There are 20 kg of salt in the tank initially, thus $A(0) = 20$. Using this initial condition, we find

$$20 = 1200 - C \quad \Rightarrow \quad C = 1180.$$

Substituting this value of C into the solution, we have

$$A = 1200 - 1180e^{-t/40}$$

Thus

$$A(10) = 1200 - 1180e^{-10/40} \approx 281 \text{ kg}.$$

Note: There is a detailed discussion of mixture problems in Section 3.2.

35. In Problem 34 we saw that the differential equation $\frac{dT}{dt} = k(M - T)$ can be solved by separation of variables to yield

$$T = Ce^{-kt} + M.$$

When the oven temperature is 120°, we have $M = 120$. Also $T(0) = 40$. Thus

$$40 = C + 120 \quad \Rightarrow \quad C = -80.$$

Because $T(45) = 90$, we have

$$90 = -80e^{-45k} + 120$$

$$\Rightarrow \quad \frac{3}{8} = e^{-45k} \quad \Rightarrow \quad -45k = \ln\left(\frac{3}{8}\right).$$

Thus $k = \dfrac{\ln(8/3)}{45} \approx 0.02180$. This k is independent of M. Therefore, we have the general equation

$$T(t) = Ce^{-0.02180t} + M.$$

(a) We are given that $M = 100$. To find C we must solve the equation $T(0) = 40 = C + 100$. This gives $C = -60$. Thus the equation becomes

$$T(t) = -60e^{-0.02180t} + 100.$$

We want to solve for t when $T(t) = 90$. This gives us

$$90 = -60e^{-0.02180t} + 100$$

$$\Rightarrow \quad \frac{1}{6} = e^{-0.02180t}$$

$$\Rightarrow \quad -0.0218t = \ln\left(\frac{1}{6}\right)$$

$$\Rightarrow \quad 0.02180t = \ln 6.$$

Therefore $t = \dfrac{\ln 6}{0.02180} \approx 82.2$ min.

(b) Here $M = 140$, so we solve

$$T(0) = 40 = C + 140 \qquad \Rightarrow \qquad C = -100.$$

As above, solving for t in the equation

$$T(t) = -100e^{-0.02180t} + 140 = 90$$

$$\Rightarrow \qquad t \approx 31.8.$$

(c) With $M = 80$, we solve

$$40 = C + 80,$$

yielding $C = -40$. Setting

$$T(t) = -40e^{-02180t} + 80 = 90$$

$$\Rightarrow \qquad -\frac{1}{4} = e^{0.02180t}.$$

This last equation is impossible because an exponential function is never negative. Hence it never attains desired temperature. The physical nature of this problem would lead us to expect this result. A further discussion of Newton's law of cooling is given in Section 3.3.

37. The differential equation $\dfrac{dP}{dt} = \dfrac{r}{100}P$ separates if we divide by P and multiply by dt.

$$\int \frac{1}{P}\,dP = \frac{r}{100}\int dt \qquad \Rightarrow \qquad \ln P = \frac{r}{100}t + C$$

$$\Rightarrow \qquad P(t) = Ke^{rt/100},$$

where K is the initial amount of money in the savings account, $K = \$1000$, and $r\%$ is the interest rate, $r = 5$. This results in

$$P(t) = 1000e^{5t/100}. \tag{3}$$

(a) To determine the amount of money in the account after 2 years we substitute $t = 2$ into equation (3), which gives

$$P(2) = 1000e^{10/100} = \$1105.17.$$

(b) To determine when the account will reach \$4000 we solve equation (3) for t with $P = \$4000$ and $r = 5$:

$$4000 = 1000e^{5t/100}$$

$$\Rightarrow \qquad e^{5t/100} = 4$$

$$\Rightarrow \qquad t = 20\ln 4 = 27.73 \text{ years.}$$

(c) To determine the amount of money in the account after $3\frac{1}{2}$ years we need to determine the value of each \$1000 deposit after $3\frac{1}{2}$ years has passed. This means that the initial \$1000 is in the account for the entire $3\frac{1}{2}$ years and grows to the amount given by $P_0 = 1000e^{\frac{5(3.5)}{100}}$. For the growth of the \$1000 deposited after 12 months, we take $t = 2.5$ in equation (3) because that is how long this \$1000 will be in the account. This gives $P_1 = 1000e^{\frac{5(2.5)}{100}}$. Using the above reasoning for the remaining deposits we arrive at $P_2 = 1000e^{\frac{5(1.5)}{100}}$ and $P_3 = 1000e^{\frac{5(0.5)}{100}}$. The total amount is determined by the sum of the P_i's.

$$P = 1000\left[e^{\frac{5(3.5)}{100}} + e^{\frac{5(2.5)}{100}} + e^{\frac{5(1.5)}{100}} + e^{\frac{5(0.5)}{100}}\right] = \$4{,}427.59\,.$$

39. Let $s(t)$, $t > 0$, denote the distance traveled by driver A from the time $t = 0$ when he ran out of gas to time t. Then driver A's velocity $v_A(t) = ds/dt$ is a solution to the initial value problem

$$\frac{dv_A}{dt} = -kv_A^2, \qquad v_A(0) = v_B,$$

where v_B is driver B's constant velocity, and $k > 0$ is a positive constant. Separating variables we get

$$\frac{dv_A}{v_A^2} = -k\,dt \qquad \Rightarrow \qquad \int \frac{dv_A}{v_A^2} = -\int k\,dt \qquad \Rightarrow \qquad \frac{1}{v_A(t)} = kt + C.$$

From the initial condition we find

$$\frac{1}{v_B} = \frac{1}{v_A(0)} = k \cdot 0 + C = C \qquad \Rightarrow \qquad C = \frac{1}{v_B}.$$

Thus

$$v_A(t) = \frac{1}{kt + 1/v_B} = \frac{v_B}{v_B kt + 1}.$$

The function $s(t)$ therefore satisfies

$$\frac{ds}{dt} = \frac{v_B}{v_B kt + 1}, \qquad s(0) = 0.$$

Integrating we obtain

$$s(t) = \int \frac{v_B}{v_B kt + 1}\,dt = \frac{1}{k}\ln\left(v_B kt + 1\right) + C_1\,.$$

To find C_1 we use the initial condition:

$$0 = s(0) = \frac{1}{k}\ln\left(v_B k \cdot 0 + 1\right) + C_1 = C_1 \qquad \Rightarrow \qquad C_1 = 0.$$

So,

$$s(t)=\frac{1}{k}\ln\left(v_{\mathrm{B}}kt+1\right).$$

At the moment $t=t_1$ when driver A's speed was halved, i.e., $v_{\mathrm{A}}(t_1)=\frac{1}{2}v_{\mathrm{A}}(0)=\frac{1}{2}v_{\mathrm{B}}$, we have

$$\frac{1}{2}v_{\mathrm{B}}=v_{\mathrm{A}}(t_1)=\frac{v_{\mathrm{B}}}{v_{\mathrm{B}}kt_1+1} \quad \text{and} \quad 1=s(t_1)=\frac{1}{k}\ln\left(v_{\mathrm{B}}kt_1+1\right)$$

$$\Rightarrow \quad v_{\mathrm{B}}kt_1+1=2 \quad \text{and so} \quad k=\ln\left(v_{\mathrm{B}}kt_1+1\right)=\ln 2$$

$$\Rightarrow \quad s(t)=\frac{1}{\ln 2}\ln\left(v_{\mathrm{B}}t\ln 2+1\right).$$

Since driver B was 3 miles behind driver A at time $t=0$, and his speed remained constant, he finished the race at time $t_{\mathrm{B}}=(3+2)/v_{\mathrm{B}}=5/v_{\mathrm{B}}$. At this moment, driver A had already gone

$$s(t_{\mathrm{B}})=\frac{1}{\ln 2}\ln\left(v_{\mathrm{B}}t_{\mathrm{B}}\ln 2+1\right)=\frac{1}{\ln 2}\ln\left(\frac{5}{v_{\mathrm{B}}}v_{\mathrm{B}}\ln 2+1\right)=\frac{1}{\ln 2}\ln\left(5\ln 2+1\right)\approx 2.1589>2 \text{ miles},$$

i.e., A won the race.

EXERCISES 2.3: Linear Equations, page 59

5. This is a linear equation with independent variable t and dependent variable y. This is also a separable equation because

$$\frac{dy}{dt}=\frac{y(t-1)}{t^2+1}=\left(\frac{t-1}{t^2+1}\right)y=g(t)p(y).$$

9. This is a linear equation with dependent variable r and independent variable θ. The method we will use to solve this equation is exactly the same as the method we use to solve an equation in the variables x and y since these variables are just dummy variables. Thus we have $P(\theta)=\tan\theta$ and $Q(\theta)=\sec\theta$ which are continuous on any interval not containing odd multiples of $\pi/2$. We proceed as usual to find the integrating factor $\mu(\theta)$. We have

$$\mu(\theta)=\exp\left(\int\tan\theta\, d\theta\right)=e^{(-\ln|\cos\theta|+C)}=K\left(\frac{1}{|\cos\theta|}\right)=K\,|\sec\theta|, \text{ where } K=e^{C}.$$

Thus we have

$$\mu(\theta)=\sec\theta,$$

where we can drop the absolute value sign by making $K=1$ if θ is in an interval where $\sec\theta$ is positive or by making $K=-1$ if $\sec\theta$ is negative. Multiplying the equation by the integrating factor yields

$$\sec\theta\frac{dr}{d\theta}+(\sec\theta\tan\theta)r=\sec^2\theta$$

$$\Rightarrow \qquad D_\theta(r\sec\theta)=\sec^2\theta.$$

Integrating with respect to θ yields

$$r\sec\theta=\int\sec^2\theta\,d\theta=\tan\theta+C$$

$$\Rightarrow \qquad r=\cos\theta\tan\theta+C\cos\theta$$

$$\Rightarrow \qquad r=\sin\theta+C\cos\theta.$$

Because of the continuity of $P(\theta)$ and $Q(\theta)$, this solution is valid on any open interval that has end points that are consecutive odd multiples of $\frac{\pi}{2}$.

15. To put this linear equation in standard form, we divide by (x^2+1) to obtain

$$\frac{dy}{dx}+\frac{x}{x^2+1}y=\frac{x}{x^2+1}. \tag{4}$$

Here $P(x)=\frac{x}{x^2+1}$, so

$$\int P(x)dx=\int\frac{x}{x^2+1}dx=\frac{1}{2}\ln\left[x^2+1\right].$$

Thus the integrating factor is

$$\mu(x)=e^{\frac{1}{2}\ln\left[x^2+1\right]}=e^{\ln\left[\left(x^2+1\right)^{1/2}\right]}=\left(x^2+1\right)^{1/2}.$$

Multiplying equation (4) by $\mu(x)$ yields

$$\left(x^2+1\right)^{1/2}\frac{dy}{dx}+\frac{x}{\left(x^2+1\right)^{1/2}}y=\frac{x}{\left(x^2+1\right)^{1/2}},$$

which becomes

$$\frac{d}{dx}\left(\left(x^2+1\right)^{1/2}y\right)=\frac{x}{\left(x^2+1\right)^{1/2}}.$$

Now we integrate both sides and solve for y to find

$$\left(x^2+1\right)^{1/2}y=\left(x^2+1\right)^{1/2}+C$$

$$\Rightarrow \qquad y = 1 + C\left(x^2 + 1\right)^{-1/2}.$$

This solution is valid for all x, since $P(x)$ and $Q(x)$ are continuous for all x.

17. This is a linear equation with $P(x) = -\frac{1}{x}$ and $Q(x) = xe^x$ which is continuous on any interval not containing 0. Therefore, the integrating factor is given by

$$\mu(x) = \exp\left[\int -\frac{1}{x}\,dx\right] = e^{-\ln x} = \frac{1}{x}, \qquad \text{for } x > 0.$$

Multiplying the equation by this integrating factor yields

$$\frac{1}{x}\frac{dy}{dx} - \frac{y}{x^2} = e^x$$

$$\Rightarrow \qquad D_x\left(\frac{y}{x}\right) = e^x.$$

Integrating gives

$$\frac{y}{x} = e^x + C$$

$$\Rightarrow \qquad y = xe^x + Cx.$$

Now applying the initial condition, $y(1) = e - 1$, we have

$$e - 1 = e + C$$

$$\Rightarrow \qquad C = -1.$$

Thus, the solution is

$$y = xe^x - x, \text{ on the interval } (0, \infty).$$

Note: This interval is the largest interval containing the initial value $x = 1$ in which $P(x)$ and $Q(x)$ are continuous.

25. **(a)** This is a linear problem and so an integrating factor is

$$\mu(x) = \exp\left(\int 2x\,dx\right) = \exp\left(x^2\right).$$

Multiplying the equation by this integrating factor yields

$$e^{x^2}\frac{dy}{dx} + 2xe^{x^2}y = e^{x^2}$$

$$\Rightarrow \qquad D_x\left(ye^{x^2}\right) = e^{x^2}$$

$$\Rightarrow \qquad \int_2^x D_t\left(ye^{t^2}\right)dt = \int_2^x e^{t^2}\,dt,$$

where we have changed the dummy variable x to t and integrated with respect to t from 2 (since the initial value for x in the initial condition is 2) to x. Thus, since $y(2)=1$,

$$ye^{x^2} - e^4 = \int_2^x e^{t^2}\,dt$$

$$\Rightarrow \qquad y = e^{-x^2}\left(e^4 + \int_2^x e^{t^2}\,dt\right) = e^{4-x^2} + e^{-x^2}\int_2^x e^{t^2}\,dt.$$

(b) We will use Simpson's rule (page A.3 of the Appendix B) to approximate the definite integral found in part (a) with upper limit $x=3$. Simpson's rule requires an even number of intervals, but we don't know how many are required to obtain the desired 3 place accuracy. Rather than make an error analysis, we will compute the approximate value of $y(3)$ using 4, 6, 8, 10, 12, ... intervals for Simpson's rule until the approximate values for $y(3)$ change by less than 5 in the fourth place. For $n=2$, we divide [2,3] into 4 equal subintervals. Thus, each subinterval will be of length $\dfrac{(3-2)}{4}=\dfrac{1}{4}$. Therefore, the integral is approximated by

$$\int_2^3 e^{t^2}\,dt \approx \frac{1}{12}\left[e^4 + 4e^{(2.25)^2} + 2e^{(2.5)^2} + 4e^{(2.75)^2} + e^{(3)^2}\right] \approx 1460.354350.$$

Dividing this by $e^{(3)^2}$, and adding $e^{\left(4-3^2\right)} = e^{(-5)}$, gives

$$y(3) \approx 0.186960.$$

Doing calculations for 6, 8, 10, and 12 intervals yields

Table 2-B. Successive approximations for $y(3)$ using Simpson's rule.

Number of Intervals	$y(3)$
6	0.183905
8	0.183291
10	0.183110
12	0.183043

Since the last 3 approximate values do not change by more than 5 in the fourth place, it appears that their first three places are accurate and the approximate solution is $y(3) \approx 0.183$.

31. **(a)** On the interval $0 \le x \le 2$, we have $P(x)=1$. Thus we are solving the equation

$$\frac{dy}{dx} + y = x, \qquad y(0)=1.$$

The integrating factor is given by

$$\mu(x) = \exp\left(\int dx\right) = e^{x}.$$

Multiplying the equation by the integrating factor, we obtain

$$e^{x}\frac{dy}{dx} + e^{x}y = xe^{x}$$

$$\Rightarrow \qquad D_x\left[e^{x}y\right] = xe^{x}$$

$$\Rightarrow \qquad e^{x}y = \int xe^{x}\,dx + C.$$

Calculating this integral by parts and dividing by e^{x} yields

$$y = e^{-x}\left(xe^{x} - e^{x} + C\right) = x - 1 + Ce^{-x}.$$

(b) Using the initial condition, $y(0) = 1$, we see that

$$1 = y(0) = 0 - 1 + C = -1 + C \qquad \Rightarrow \qquad C = 2.$$

Thus the equation becomes

$$y = x - 1 + 2e^{-x}.$$

(c) In the interval $x > 2$, we have $P(x) = 3$. Therefore, the integrating factor is given by

$$\mu(x) = \exp\left(\int 3\,dx\right) = e^{3x}.$$

Multiplying the equation by this factor and solving yields

$$e^{3x}\frac{dy}{dx} + 3e^{3x}y = xe^{3x}$$

$$\Rightarrow \qquad D_x\left(e^{3x}y\right) = xe^{3x}$$

$$\Rightarrow \qquad e^{3x}y = \int xe^{3x}\,dx.$$

Integrating by parts and dividing by e^{3x} gives

$$y = e^{-3x}\left[\frac{1}{3}xe^{3x} - \frac{1}{9}e^{3x} + C\right] = \frac{x}{3} - \frac{1}{9} + Ce^{-3x}.$$

(d) We want the value of the initial point for the solution in part (c) to be the value of the solution found in part (b) at the point $x = 2$. This value is given by

$$y(2) = 2 - 1 + 2e^{-2} = 1 + 2e^{-2}.$$

Thus the initial point we seek is

$$y(2) = 1 + 2e^{-2}.$$

Using this initial point to find the constant C given in part (c) yields

$$1+2e^{-2}=y(2)=\frac{2}{3}-\frac{1}{9}+Ce^{-6}$$

$$\Rightarrow \qquad C=\frac{4}{9}e^{6}+2e^{4}.$$

Thus, the solution of the equation on the interval $x>2$ is given by

$$y=\frac{x}{3}-\frac{1}{9}+\left[\frac{4}{9}e^{6}+2e^{4}\right]e^{-3x}.$$

Patching these two solutions together gives us a continuous solution to the original equation on the interval $x\geq 0$:

$$y=\begin{cases} x-1+2e^{-x}, & 0\leq x\leq 2 \\ \frac{x}{3}-\frac{1}{9}+\left[\frac{4}{9}e^{6}+2e^{4}\right]e^{-3x}, & 2<x. \end{cases}$$

(e) The graph of the solution is given in Figure B.18 of the answers in the text.

37. We are solving the equation

$$\frac{dx}{dt}+2x=1-\cos\left(\frac{\pi t}{12}\right), \qquad x(0)=10.$$

This is a linear problem with dependent variable x and independent variable t so that $P(t)=2$. Therefore, to solve this equation we first must find the integrating factor $\mu(t)$.

$$\mu(t)=\exp\left(\int 2\,dt\right)=e^{2t}.$$

Multiplying the equation by this factor yields

$$e^{2t}\frac{dx}{dt}+2xe^{2t}=e^{2t}\left[1-\cos\left(\frac{\pi t}{12}\right)\right]=e^{2t}-e^{2t}\cos\left(\frac{\pi t}{12}\right)$$

$$\Rightarrow \qquad xe^{2t}=\int e^{2t}dt-\int e^{2t}\cos\left(\frac{\pi t}{12}\right)dt=\frac{1}{2}e^{2t}-\int e^{2t}\cos\left(\frac{\pi t}{12}\right)dt.$$

The last integral can be found by integrating by parts twice which leads back to an integral similar to the original. Combining these two similar integrals and simplifying, we obtain

$$\int e^{2t}\cos\left(\frac{\pi t}{12}\right)dt=\frac{e^{2t}\left[2\cos\left(\frac{\pi t}{12}\right)+\frac{\pi}{12}\sin\left(\frac{\pi t}{12}\right)\right]}{4+\left(\frac{\pi}{12}\right)^{2}}+C.$$

Thus we see that

$$x(t)=\frac{1}{2}-\frac{2\cos\left(\frac{\pi t}{12}\right)+\frac{\pi}{12}\sin\left(\frac{\pi t}{12}\right)}{4+\left(\frac{\pi}{12}\right)^2}+Ce^{-2t}.$$

Using the initial condition, $t=0$ and $x=10$, to solve for C, we obtain

$$C=\frac{19}{2}+\frac{2}{4+\left(\frac{\pi}{12}\right)^2}.$$

Therefore, the desired solution is

$$x(t)=\frac{1}{2}-\frac{2\cos\left(\frac{\pi t}{12}\right)+\frac{\pi}{12}\sin\left(\frac{\pi t}{12}\right)}{4+\left(\frac{\pi}{12}\right)^2}+\left(\frac{19}{2}+\frac{2}{4+\left(\frac{\pi}{12}\right)^2}\right)e^{-2t}.$$

39. Let $T_j(t)$, $j=0,1,2,\ldots$, denote the temperature in the classroom for $9+j\le t<10+j$, where $t=13$ denotes 1:00 P.M., $t=14$ denotes 2:00 P.M., etc. Then

$$T_0(9)=0, \tag{5}$$

and the continuity of the temperature implies that

$$\lim_{t\to 10+j} T_j(t)=T_{j+1}(10+j), \qquad j=0,1,2,\ldots . \tag{6}$$

According to the work of the heating unit, the temperature satisfies the equation

$$\frac{dT_j}{dt}=\begin{cases}1-T_j, & \text{if } j=2k\\ -T_j, & \text{if } j=2k+1\end{cases}, \qquad 9+j<t<10+j,\ \ k=0,1,\ldots .$$

The general solutions of these equations are:
for j even

$$\frac{dT_j}{dt}=1-T_j \qquad \Rightarrow \qquad \frac{dT_j}{1-T_j}=dt$$

$$\Rightarrow \qquad \ln|1-T_j|=-t+c_j \qquad \Rightarrow \qquad T_j(t)=1-C_je^{-t};$$

for j odd

$$\frac{dT_j}{dt}=-T_j \qquad \Rightarrow \qquad \frac{dT_j}{-T_j}=dt$$

$$\Rightarrow \qquad \ln|T_j| = -t + c_j \qquad \Rightarrow \qquad T_j(t) = C_j e^{-t};$$

where $C_j \neq 0$ are constants. From (5) we have:

$$0 = T_0(9) = \left(1 - C_0 e^{-t}\right)\Big|_{t=9} = 1 - C_0 e^{-9} \qquad \Rightarrow \qquad C_0 = e^9.$$

Also from (6), for even values of j (say, $j = 2k$) we get

$$\left(1 - C_{2k} e^{-t}\right)\Big|_{t=9+(2k+1)} = C_{2k+1} e^{-t}\Big|_{t=9+(2k+1)}$$

$$\Rightarrow \quad 1 - C_{2k} e^{-(10+2k)} = C_{2k+1} e^{-(10+2k)}$$

$$\Rightarrow \quad C_{2k+1} = e^{10+2k} - C_{2k}.$$

Similarly from (6) for odd values of j (say, $j = 2k+1$) we get

$$C_{2k+1} e^{-t}\Big|_{t=9+(2k+2)} = \left(1 - C_{2k+2} e^{-t}\right)\Big|_{t=9+(2k+2)}$$

$$\Rightarrow \quad C_{2k+1} e^{-(11+2k)} = 1 - C_{2k+2} e^{-(11+2k)}$$

$$\Rightarrow \quad C_{2k+2} = e^{11+2k} - C_{2k+1}.$$

In general we see that for any integer j (even or odd) the following formula holds:

$$C_j = e^{9+j} - C_{j-1}.$$

Using this recurrence formula we successively compute

$$C_1 = e^{10} - C_0 = e^{10} - e^9 = e^9(e - 1)$$

$$C_2 = e^{11} - C_1 = e^{11} - e^{10} + e^9 = e^9\left(e^2 - e + 1\right)$$

$$\vdots$$

$$C_j = e^9 \sum_{k=0}^{j} (-1)^{j-k} e^k.$$

Therefore, the temperature at noon (when $t = 12$ and $j = 3$) is

$$T_3(12) = C_3 e^{-12} = e^{-12} e^9 \sum_{k=0}^{3} (-1)^{3-k} e^k = 1 - e^{-1} + e^{-2} - e^{-3} \approx 0.718 = 78.1^\circ\,\text{F}.$$

At 5 P.M. (when $t = 17$ and $j = 8$), we find

$$T_8(17) = 1 - C_8 e^{-17} = 1 - e^{-17} e^9 \sum_{k=0}^{8} (-1)^{8-k} e^k = \sum_{k=1}^{8} (-1)^{k+1} e^{-k}$$

$$= e^{-1} \cdot \frac{1 - \left(-e^{-1}\right)^8}{1 + e^{-1}} \approx 0.269 = 26.9^\circ\,\text{F}.$$

EXERCISES 2.4: Exact Equations, page 70

5. The differential equation is not separable because $(2xy + \cos y)$ cannot be factored. This equation can be put in standard form by defining x as the dependent variable and y as the independent variable. This gives

$$\frac{dx}{dy} + \frac{2}{y}x = \frac{-\cos y}{y^2},$$

so we see that the differential equation is linear. If we set $M(x,y) = y^2$ and $N(x,y) = 2xy + \cos y$ we are able to see that the differential equation is also exact because $M_y(x,y) = 2y = N_x(x,y)$.

9. We have that $M(x,y) = 2xy + 3$ and $N(x,y) = x^2 - 1$. Therefore, $M_y(x,y) = 2x = N_x(x,y)$ and so the equation is exact. We will solve this equation by first integrating $M(x,y)$ with respect to x, although integration of $N(x,y)$ with respect to y is equally easy. Thus

$$F(x,y) = \int(2xy + 3)\,dx = x^2y + 3x + g(y).$$

Differentiating $F(x,y)$ with respect to y gives $F_y(x,y) = x^2 + g'(y) = N(x,y) = x^2 - 1$. From this we see that $g'(y) = -1$. (As a partial check we note that $g'(y)$ does not involve x.) Integrating gives

$$g(y) = \int(-1)\,dy = -y.$$

Since the constant of integration will be incorporated into the parameter of the solution, it is not written here. Substituting this expression for $g(y)$ into the expression that we found for $F(x,y)$ yields

$$F(x,y) = x^2y + 3x - y.$$

Therefore, the solution of the differential equation is

$$x^2y + 3x - y = C \qquad \Rightarrow \qquad y = \frac{C - 3x}{x^2 - 1}.$$

The given equation could be solved by the method of grouping. To see this, express the differential equation in the form

$$\left(2xy\,dx + x^2dy\right) + \left(3dx - dy\right) = 0.$$

The first term of the left-hand side we recognize as the total differential of x^2y. The second term is the total differential of $(3x - y)$. Thus we again find that

$$F(x,y) = x^2y + 3x - y$$

and again the solution is $x^2y + 3x - y = C$.

15. This differential equation is expressed in the variables r and θ. Since the variables x and y are dummy variables, this equation is solved in exactly the same way as an equation in x and y. We will look for a solution with independent variable θ and dependent variable r. We see that the differential equation is expressed in the differential form

$$M(r,\theta)dr + N(r,\theta)d\theta = 0, \qquad \text{where } M(r,\theta) = \cos\theta \text{ and } N(r,\theta) = \left(-r\sin\theta + e^{\theta}\right).$$

This implies that

$$M_{\theta}(r,\theta) = -\sin\theta = N_r(r,\theta),$$

and so the equation is exact. Therefore, to solve the equation we need to find a function $F(r,\theta)$ that has $\cos\theta\, dr + \left(-r\sin\theta + e^{\theta}\right)d\theta$ as its total differential. Integrating $M(r,\theta)$ with respect to r we see that

$$F(r,\theta) = \int \cos\theta\, dr = r\cos\theta + g(\theta)$$

$$\Rightarrow \quad F_{\theta}(r,\theta) = -r\sin\theta + g'(\theta) = N(r,\theta) = -r\sin\theta + e^{\theta}.$$

Thus we have that

$$g'(\theta) = e^{\theta} \qquad \Rightarrow \qquad g(\theta) = e^{\theta},$$

where the constant of integration will be incorporated into the parameter of the solution. Substituting this expression for $g(\theta)$ into the expression we found for $F(r,\theta)$ yields

$$F(r,\theta) = r\cos\theta + e^{\theta}.$$

From this we see that the solution is given by the one parameter family $r\cos\theta + e^{\theta} = C$. Or solving for r we have

$$r = \frac{C - e^{\theta}}{\cos\theta} = \left(C - e^{\theta}\right)\sec\theta.$$

23. Here $M(t,y) = e^t y + te^t y$ and $N(t,y) = te^t + 2$. Thus $M_y(t,y) = e^t + te^t = N_t(t,y)$ and so the equation is exact. To find $F(t,y)$, we first integrate $N(t,y)$ with respect to y to obtain

$$F(t,y) = \int (te^t + 2)\, dy = \left(te^t + 2\right)y + h(t),$$

where we have chosen to integrate $N(t,y)$ because this integration is more easily accomplished. Thus

$$F_t(t,y) = e^t y + te^t y + h'(t) = M(t,y) = e^t y + te^t y$$

$$\Rightarrow \qquad h'(t) = 0$$

$$\Rightarrow \qquad h(t) = C.$$

We will incorporate this constant into the parameter of the solution. Combining these results gives

$F(t, y) = te^{t} y + 2y$. Therefore, the solution is given by $te^{t} y + 2y = C$. Solving for y yields $y = \dfrac{C}{te^{t} + 2}$. Now we use the initial condition $y(0) = -1$ to find the solution that passes through the point $(0, -1)$. Thus

$$y(0) = \frac{C}{0+2} = -1$$

$$\Rightarrow \quad \frac{C}{2} = -1$$

$$\Rightarrow \quad C = -2.$$

This gives us the solution $y = \dfrac{-2}{te^{t} + 2}$.

27. **(a)** We want to find $M(x, y)$ so that for

$$N(x, y) = \sec^2 y - \frac{x}{y}$$

we have

$$M_y(x, y) = N_x(x, y) = -\frac{1}{y}.$$

Therefore, we must integrate this last expression with respect to y. That is,

$$M(x, y) = \int -\frac{1}{y}\, dy = -\ln|y| + f(x),$$

where $f(x)$, the "constant" of integration, is a function only of x.

(b) We want to find $M(x, y)$ so that for

$$N(x, y) = \sin x \cos y - xy - e^{-y}$$

we have

$$M_y(x, y) = N_x(x, y) = \cos x \cos y - y.$$

Therefore, we must integrate this last expression with respect to y. That is

$$M(x, y) = \int [\cos x \cos y - y]\, dy = \cos x \int \cos y\, dy - \int y\, dy$$

$$= \cos x \sin y - \frac{y^2}{2} + f(x),$$

where $f(x)$, a function only of x, is the "constant" of integration.

29. **(a)** We have $M(x,y)=y^2+2xy$ and $N(x,y)=-x^2$. Therefore $M_y(x,y)=2y+2x$ and $N_x(x,y)=-2x$. Thus $M_y(x,y)\neq N_x(x,y)$, so the differential equation is not exact.

(b) If we multiply $(y^2+2xy)dx-x^2dy=0$ by y^{-2} we obtain $\left(1+2\frac{x}{y}\right)dx-\frac{x^2}{y^2}dy=0$. In this equation we have $M(x,y)=1+2\frac{x}{y}$ and $N(x,y)=-\frac{x^2}{y^2}$. Therefore $M_y(x,y)=-2\frac{x}{y^2}=N_x(x,y)$, so the new differential equation is exact.

(c) Following the method for solving exact equations we integrate $M(x, y)$ in part (b) with respect to x to obtain

$$F(x,y)=\int\left(1+2\frac{x}{y}\right)dx=x+\frac{x^2}{y}+g(y).$$

To determine $g(y)$, take the partial derivative of both sides of the above equation with respect to y to obtain

$$\frac{\partial F}{\partial y}=-\frac{x^2}{y^2}+g'(y).$$

Substituting N (given in part (b)) for $\frac{\partial F}{\partial y}$, we can now solve for $g'(y)$ to obtain

$$N=-\frac{x^2}{y^2}=-\frac{x^2}{y^2}+g'(y)$$

$$\Rightarrow \quad g'(y)=0.$$

The integral of $g'(y)$ will yield a constant and the choice of the constant of integration is not important so we can take $g(y)=0$. Hence we have $F(x,y)=x+\frac{x^2}{y}$ and the solution to the equation is given implicitly by

$$x+\frac{x^2}{y}=C.$$

Solving the above equation for y, we obtain

$$y=\frac{x^2}{C-x}.$$

(d) By dividing both sides by y^2 we lost the solution $y\equiv 0$.

32. **(a)** The slope of the orthogonal curves, say $m_{\perp}$, must be $-\frac{1}{m}$, where m is the slope of the original curves. Therefore, we have

$$m_{\perp} = \frac{F_y(x,y)}{F_x(x,y)}$$

$$\Rightarrow \quad \frac{dy}{dx} = \frac{F_y(x,y)}{F_x(x,y)}$$

$$\Rightarrow \quad F_y(x,y)\,dx - F_x(x,y)\,dy = 0.$$

(b) Let $F(x,y) = x^2 + y^2$. Then we have $F_x(x,y) = 2x$ and $F_y(x,y) = 2y$. Plugging these expressions into the final result of part (a) gives

$$2y\,dx - 2x\,dy = 0$$

$$\Rightarrow \quad y\,dx - x\,dy = 0.$$

To find the orthogonal trajectories, we must solve this differential equation. To this end, note that this equation is separable and thus

$$\int \frac{1}{x}\,dx = \int \frac{1}{y}\,dy \quad \Rightarrow \quad \ln|x| = \ln|y| + C$$

$$\Rightarrow \quad e^{\ln|x| - C} = e^{\ln|y|}$$

$$\Rightarrow \quad y = kx \quad \text{where } k = \pm e^{-C}.$$

Therefore, the orthogonal trajectories are lines through the origin.

(c) Let $F(x,y) = xy$. Then we have $F_x(x,y) = y$ and $F_y(x,y) = x$. Plugging these expressions into the final result of part (a) gives

$$x\,dx - y\,dy = 0.$$

To find the orthogonal trajectories, we must solve this differential equation. To this end, note that this equation is separable and thus

$$\int x\,dx = \int y\,dy \quad \Rightarrow \quad \frac{x^2}{2} = \frac{y^2}{2} + C$$

$$\Rightarrow \quad x^2 - y^2 = k,$$

where $k = 2C$. Therefore, the orthogonal trajectories are hyperbolas.

EXERCISES 2.5: Special Integrating Factors, page 76

3. This equation is not separable because of the factor $\left(y^2+2xy\right)$. It is not linear because of the factor y^2. To see if it is exact, we compute $M_y(x,y)$ and $N_x(x,y)$ and see that

$$M_y(x,y)=2y+2x \neq -2x=N_x(x,y).$$

Therefore, the equation is not exact. To see if we can find an integrating factor of the form $\mu(x)$, we compute

$$\frac{\dfrac{\partial M}{\partial y}-\dfrac{\partial N}{\partial x}}{N}=\frac{2y+4x}{-x^2},$$

which is not a function of x alone. To see if we can find an integrating factor of the form $\mu(y)$, we compute

$$\frac{\dfrac{\partial N}{\partial x}-\dfrac{\partial M}{\partial y}}{M}=\frac{-4x-2y}{y^2+2xy}=\frac{-2(y+2x)}{y(y+2x)}=\frac{-2}{y}.$$

Thus the equation has an integrating factor that is a function of y alone.

7. The equation $\left(3x^2+y\right)dx+\left(x^2y-x\right)dy=0$ is not separable or linear. To see if it is exact, we compute

$$\frac{\partial M}{\partial y}=1\neq 2xy-1=\frac{\partial N}{\partial x}.$$

Thus, the equation is not exact. To see if we can find an integrating factor, we compute

$$\frac{\dfrac{\partial M}{\partial y}-\dfrac{\partial N}{\partial x}}{N}=\frac{2-2xy}{x^2y-x}=\frac{-2(xy-1)}{x(xy-1)}=\frac{-2}{x}.$$

From this we see that the integrating factor will be

$$\mu(x)=\exp\left[\int\frac{-2}{x}\,dx\right]=\exp\left(-2\ln|x|\right)=x^{-2}.$$

To solve the equation, we multiply it by the integrating factor x^{-2} to obtain

$$\left(3+yx^{-2}\right)dx+\left(y-x^{-1}\right)dy=0.$$

This is now exact. Thus, we want to find $F(x,y)$. To do this, we integrate $M(x,y)=3+yx^{-2}$ with respect to x to get

$$F(x,y)=\int(3+yx^{-2})\,dx=3x-yx^{-1}+g(y)$$

$$\Rightarrow \qquad F_y(x,y) = -x^{-1} + g'(y) = N(x,y) = y - x^{-1}$$

$$\Rightarrow \qquad g'(y) = y \qquad \Rightarrow \qquad g(y) = \frac{y^2}{2}.$$

Therefore,

$$F(x,y) = 3x - x^{-1}y + \frac{y^2}{2}.$$

And so we see that an implicit solution is

$$\frac{y^2}{2} - \frac{y}{x} + 3x = C.$$

Since $\mu(x) = x^{-2}$, we must check to see if the solution $x \equiv 0$ was either gained or lost. The function $x \equiv 0$ is a solution to the original equation, but is not given by the above implicit solution for any choice of C. Hence, $\dfrac{y^2}{2} - \dfrac{y}{x} + 3x = C$ and $x \equiv 0$ are solutions.

13. We will multiply the equation by the factor $x^n y^m$ and try to make it exact. Thus, we have

$$\left(2x^n y^{m+2} - 6x^{n+1} y^{m+1}\right)dx + \left(3x^{n+1} y^{m+1} - 4x^{n+2} y^m\right)dy = 0.$$

We want $M_y(x,y) = N_x(x,y)$. Since

$$M_y(x,y) = 2(m+2)x^n y^{m+1} - 6(m+1)x^{n+1} y^m ,$$

$$N_x(x,y) = 3(n+1)x^n y^{m+1} - 4(n+2)x^{n+1} y^m,$$

we need

$$2(m+2) = 3(n+1) \qquad \text{and} \qquad 6(m+1) = 4(n+2).$$

Solving these equations simultaneously, we obtain $n = 1$ and $m = 1$. So,

$$\mu(x) = xy.$$

With these choices for n and m we obtain the exact equation

$$\left(2xy^3 - 6x^2y^2\right)dx + \left(3x^2y^2 - 4x^3y\right)dy = 0.$$

Solving this equation, we have

$$F(x,y) = \int \left(2xy^3 - 6x^2y^2\right)dx = x^2y^3 - 2x^3y^2 + g(y)$$

$$\Rightarrow \qquad F_y(x,y) = 3x^2y^2 - 4x^3y + g'(y) = N(x,y) = 3x^2y^2 - 4x^3y.$$

Therefore, $g'(y) = 0$. Since the constant of integration can be incorporated into the constant C of the solution, we can pick $g(y) \equiv 0$. Thus, we have

$$F(x,y) = x^2y^3 - 2x^3y^2$$

and the solution becomes

$$x^2y^3 - 2x^3y^2 = C.$$

Since we have multiplied the original equation by xy, we could have added the extraneous solutions $y \equiv 0$ or $x \equiv 0$. But, since $y \equiv 0$ implies that $\frac{dy}{dx} \equiv 0$ or $x \equiv 0$ implies $\frac{dx}{dy} \equiv 0$, $y \equiv 0$ and $x \equiv 0$ are solutions of the original equation as well as the transformed equation.

EXERCISES 2.6: Substitutions and Transformations, page 83

5. The given differential equation is not homogeneous due to the e^{-2x} terms. The equation $\left(ye^{-2x} + y^3\right)dx - e^{-2x}dy = 0$ is a Bernoulli equation because it can be written in the form $\frac{dy}{dx} + P(x)y = Q(x)y^n$ as follows:

$$\frac{dy}{dx} - y = e^{2x}y^3 .$$

The differential equation does not have linear coefficients nor is it of the form $y' = G(ax + by)$.

13. Since we can express $f(t,x)$ in the form $G(x/t)$, that is, (dividing numerator and denominator by t^2)

$$\frac{x^2 + t\sqrt{t^2 + x^2}}{tx} = \frac{\left(\frac{x}{t}\right)^2 + \sqrt{1 + \left(\frac{x}{t}\right)^2}}{\left(\frac{x}{t}\right)} = \left(\frac{x}{t}\right) + \frac{\sqrt{1 + \left(\frac{x}{t}\right)^2}}{\left(\frac{x}{t}\right)},$$

the equation is homogeneous. Substituting $v = \frac{x}{t}$ and $\frac{dx}{dt} = v + t\frac{dv}{dt}$ into the equation yields

$$v + t\frac{dv}{dt} = v + \frac{\sqrt{1+v^2}}{v}$$

$$\Rightarrow \qquad t\frac{dv}{dt} = \frac{\sqrt{1+v^2}}{v} .$$

This transformed equation is separable. Thus we have

$$\frac{v}{\sqrt{1+v^2}}dv = \frac{1}{t}dt$$

$$\Rightarrow \qquad \sqrt{1+v^2} = \ln|t| + C,$$

where we have integrated with the integration on the left hand side being accomplished by the substitution $u = 1 + v^2$. Substituting $\frac{x}{t}$ for v in this equation gives the solution to the original equation which is

$$\sqrt{1 + \frac{x^2}{t^2}} = \ln|t| + C.$$

17. With the substitutions $z = x + y$ and $\frac{dz}{dx} = 1 + \frac{dy}{dx}$ or $\frac{dy}{dx} = \frac{dz}{dx} - 1$, this equation becomes the separable equation

$$\frac{dz}{dx} - 1 = \sqrt{z} - 1$$

$$\Rightarrow \quad \frac{dz}{dx} = \sqrt{z}$$

$$\Rightarrow \quad z^{-1/2}\,dz = dx$$

$$\Rightarrow \quad 2z^{1/2} = x + C.$$

Substituting $x + y$ for z in this solution gives the solution of the original equation

$$2\sqrt{x + y} = x + C$$

which, on solving for y, yields

$$y = \left(\frac{x}{2} + \frac{C}{2}\right)^2 - x.$$

Thus, we have

$$y = \frac{(x + C)^2}{4} - x.$$

23. This is a Bernoulli equation with $n = 2$. Dividing it by y^2 and rewriting gives

$$y^{-2}\frac{dy}{dx} - 2x^{-1}y^{-1} = -x^2.$$

Making the substitution $v = y^{-1}$ and hence $\frac{dv}{dx} = -y^{-2}\frac{dy}{dx}$, the above equation becomes

$$\frac{dv}{dx} + 2\frac{v}{x} = x^2.$$

This is a linear equation in v and x. The integrating factor $\mu(x)$ is given by

$$\mu(x) = \exp\left(\int \frac{2}{x}\,dx\right) = \exp(2\ln|x|) = x^2.$$

Multiplying the linear equation by this integrating factor and solving, we have

$$x^2\frac{dv}{dx}+2vx=x^4$$

$$\Rightarrow \quad D_x\left(x^2v\right)=x^4$$

$$\Rightarrow \quad x^2v=\int x^4\,dx=\frac{x^5}{5}+C'$$

$$\Rightarrow \quad v=\frac{x^3}{5}+\frac{C'}{x^2}.$$

Substituting y^{-1} for v in this solution gives a solution to the original equation. Therefore, we find

$$y^{-1}=\frac{x^3}{5}+\frac{C'}{x^2}$$

$$\Rightarrow \quad y=\left[\frac{x^5+5C'}{5x^2}\right]^{-1}.$$

Letting $C=5C'$ and simplifying yields

$$y=\frac{5x^2}{x^5+C}.$$

Note: $y\equiv 0$ is also a solution to the original equation. It was lost in the first step when we divided by y^2.

29. Solving for h and k in the linear system

$$-3h+k-1=0$$
$$h+k+3=0$$

gives $h=-1$ and $k=-2$. Thus, we make the substitutions $x=u-1$ and $y=v-2$, so that $dx=du$ and $dy=dv$, to obtain

$$(-3u+v)\,du+(u+v)\,dv=0.$$

This is the same transformed equation that we encountered in Example 4 on page 82 of the text. There we found that its solution is

$$v^2+2uv-3u^2=C.$$

Substituting $x+1$ for u and $y+2$ for v gives the solution to the original equation

$$(y+2)^2+2(x+1)(y+2)-3(x+1)^2=C.$$

REVIEW PROBLEMS: page 87

3. The differential equation is an exact equation with $M=\left(2xy-3x^2\right)$ and $N=\left(x^2-2y^{-3}\right)$ because $M_y=2x=N_x$. To solve this problem we will follow the procedure for solving exact equations given in Section 2.4. First we integrate $M(x,y)$ with respect to x to get

$$F(x,y)=\int\left(2xy-3x^2\right)dx+g(y),$$

$$F(x,y)=x^2y-x^3+g(y). \qquad (7)$$

To determine $g(y)$, take the partial derivative with respect to y of both sides and substitute N for $\dfrac{\partial F}{\partial y}$ to obtain

$$N=x^2-2y^{-3}=x^2+g'(y).$$

Solving for $g'(y)$ we obtain

$$g'(y)=-2y^{-3}.$$

Since the choice of the constant of integration is arbitrary we will take $g(y)=\dfrac{1}{y^2}$. Hence, from equation (7) we have $F(x,y)=x^2y-x^3+\dfrac{1}{y^2}$ and the solution to the differential equation is given implicitly by $x^2y-x^3+y^{-2}=C$.

9. The given differential equation can be written in the form

$$\frac{dy}{dx}+\frac{1}{3x}y=\frac{-x}{3y}.$$

This is a Bernoulli equation with $n=-1$, $P(x)=\dfrac{1}{3x}$ and $Q(x)=\dfrac{-x}{3}$. To transform this equation into a linear equation, we first multiply by y to obtain

$$y\frac{dy}{dx}+\frac{1}{3x}y^2=-\frac{1}{3}x.$$

Next we make the substitution $v=y^2$. Since $\dfrac{dv}{dx}=2y\dfrac{dy}{dx}$, the transformed equation is

$$\frac{1}{2}\frac{dv}{dx}+\frac{1}{3x}v=-\frac{1}{3}x,$$

$$\frac{dv}{dx}+\frac{2}{3x}v=-\frac{2}{3}x. \qquad (8)$$

The above equation is linear, so we can solve it for v using the method for solving linear equations discussed in Section 2.3. Following this procedure, the integrating factor $\mu(x)$ is found to be

$$\mu(x) = \exp\left(\int \frac{2}{3x}\,dx\right) = \exp\left(\frac{2}{3}\ln|x|\right) = x^{2/3}.$$

Multiplying equation (8) by $x^{\frac{2}{3}}$ gives

$$x^{2/3}\frac{dv}{dx} + \frac{2}{3x^{1/3}}v = -\frac{2}{3}x^{5/3};$$

$$\frac{d}{dx}\left(x^{2/3}v\right) = -\frac{2}{3}x^{5/3}.$$

We now integrate both sides and solve for v to find

$$x^{2/3}v = \int \frac{-2}{3}x^{5/3}\,dx = \frac{-1}{4}x^{8/3} + C;$$

$$v = \frac{-1}{4}x^2 + Cx^{-2/3}.$$

Substituting $v = y^2$ gives the solution

$$y^2 = -\frac{1}{4}x^2 + Cx^{-2/3}$$

or $\left(x^2 + 4y^2\right)x^{2/3} = C_1$ or, cubing both sides, $(x^2 + 4y^2)^3 x^2 = C_2$.

15. The right-hand side of the differential equation $\frac{dy}{dx} = 2 - \sqrt{2x - y + 3}$ is a function of $2x - y$ and so can be solved using the method for equations of the form $\frac{dy}{dx} = G(ax + by)$ on page 79 of the text. By letting $z = 2x - y$ we can transform the equation into a separable one. To solve, we differentiate $z = 2x - y$ with respect to x to obtain

$$\frac{dz}{dx} = 2 - \frac{dy}{dx} \quad \Rightarrow \quad \frac{dy}{dx} = 2 - \frac{dz}{dx}.$$

Substituting $z = 2x - y$ and $\frac{dy}{dx} = 2 - \frac{dz}{dx}$ into the differential equation yields

$$2 - \frac{dz}{dx} = 2 - \sqrt{z + 3}$$

or

$$\frac{dz}{dx} = \sqrt{z + 3}.$$

To solve this equation we divide by $\sqrt{z+3}$, multiply by dx and integrate to obtain

$$\int (z+3)^{-1/2}\,dz = \int dx$$

$$2(z+3)^{1/2} = x + C\,.$$

Thus we get

$$z+3 = \frac{(x+C)^2}{4}\,.$$

Finally, replacing z by $2x-y$ yields

$$2x - y + 3 = \frac{(x+C)^2}{4}\,.$$

Solving for y, we obtain

$$y = 2x + 3 - \frac{(x+C)^2}{4}\,.$$

19. In the differential equation $M(x,y) = (x^2 - 3y^2)$ and $N(x,y) = 2xy$. The differential equation is not exact because

$$\frac{\partial M}{\partial y} = -6y \neq 2y = \frac{\partial N}{\partial x}\,.$$

However, because $\left(\frac{\partial M}{\partial y} - \frac{\partial N}{\partial x}\right)\Big/ N = \frac{-8y}{(2xy)} = \frac{-4}{x}$ depends only on x, we can determine $\mu(x)$ from equation (8) on page 75 of the text. This gives

$$\mu(x) = \exp\left(\int \frac{-4}{x}\,dx\right) = x^{-4}\,.$$

When we multiply the differential equation by $\mu = x^{-4}$ we get the exact equation

$$\left(x^{-2} - 3x^{-4}y^2\right)dx + 2x^{-3}y\,dy = 0\,.$$

To find $F(x,y)$, we integrate $\left(x^{-2} - 3x^{-4}y^2\right)$ with respect to x:

$$F(x,y) = \int \left(x^{-2} - 3x^{-4}y^2\right)dx$$

$$= -x^{-1} + x^{-3}y^2 + g(y)\,.$$

Next we take the partial derivative of $F(x,y)$ with respect to y and substitute $2x^{-3}y$ for $\frac{\partial F}{\partial y}$:

$$2x^{-3}y = 2x^{-3}y + g'(y)\,.$$

Thus $g'(y) = 0$ and since the choice of the constant of integration is not important, we will take $g(y) = 0$. Hence, we have $F(x, y) = -x^{-1} + x^{-3}y^2$ and the implicit solution to the differential equation is $-x^{-1} + x^{-3}y^2 = C$. Solving for y^2 we obtain $y^2 = x^2 + Cx^3$.

Finally we check to see if any solutions were lost in the process. We multiplied by the integrating factor $\mu = x^{-4}$, so we check $x \equiv 0$. This is also a solution to the original equation.

CHAPTER 3: Mathematical Models and Numerical Methods Involving First Order Equations

EXERCISES 3.2: Compartmental Analysis, page 104

3. Let $x(t)$ be the volume of nitric acid in the tank at time t. The tank initially held 200 L of a 0.5% nitric acid solution; therefore, $x(0)=200\times 0.005=1$. Since 6 L of 20% nitric acid solution are flowing into the tank per minute, the rate at which nitric acid is entering is $6\times 0.2=1.2$ L/min. Because the rate of flow out of the tank is 8 L/min and the rate of flow in is only 6 L/min, there is a net loss in the tank of 2 L of solution every minute. Thus, at any time t, the tank will be holding $200-2t$ liters of solution. Combining this with the fact that the volume of nitric acid in the tank at time t is $x(t)$, we see that the concentration of nitric acid in the tank at time t is $\frac{x(t)}{200-2t}$. Here we are assuming that the tank is kept well stirred. The rate at which nitric acid flows out of the tank is, therefore, $8\times\left[\frac{x(t)}{200-2t}\right]$ L/min. From all of these facts, we see that

$$\textit{Input rate} = 1.2 \text{ L/min},$$

$$\textit{Output Rate} = \frac{8x(t)}{200-2t} = \frac{4x(t)}{100-t}.$$

We know that $\frac{dx}{dt} =$ *input rate* $-$ *output rate,* thus we must solve the differential equation

$$\frac{dx}{dt}=1.2-\frac{4x}{100-t}, \qquad x(0)=1.$$

This is the linear equation

$$\frac{dx}{dt}+\left[\frac{4}{100-t}\right]x=1.2, \qquad x(0)=1.$$

An integrating factor for this equation has the form

$$\mu(t)=\exp\left[\int\frac{4}{100-t}\,dt\right]=e^{-4\ln(100-t)}=(100-t)^{-4}.$$

Multiplying the linear equation by the integrating factor yields

$$(100-t)^{-4}\frac{dx}{dt}+4x(100-t)^{-5}=(1.2)(100-t)^{-4}$$

$$\Rightarrow \qquad D_t\left[(100-t)^{-4}x\right]=(1.2)(100-t)^{-4}$$

$$\Rightarrow \quad (100-t)^{-4}x = 1.2\int(100-t)^{-4}\,dt = \frac{1.2}{3}(100-t)^{-3} + C$$

$$\Rightarrow \quad x(t) = (0.4)(100-t) + C(100-t)^4 .$$

To find the value of *C*, we use the initial condition $x(0)=1$. Therefore,

$$x(0) = (0.4)(100) + C(100)^4 = 1$$

$$\Rightarrow \quad C = \frac{-39}{100^4} = -3.9\times 10^{-7}.$$

This means that at time *t* there is

$$x(t) = (0.4)(100-t) - (3.9\times 10^{-7})(100-t)^4$$

liters of nitric acid in the tank. When the percentage of nitric acid in the tank is 10%, the concentration of nitric acid is 0.1. Thus we want to solve the equation

$$\frac{x(t)}{200-2t} = 0.1 .$$

Therefore, we divide the solution $x(t)$ that we found above by $2(100-t)$ and solve for *t*. That is, we solve

$$0.2 - [1.95\times 10^{-7}](100-t)^3 = 0.1$$

$$\Rightarrow \quad t = -\left[0.1\frac{10^7}{1.95}\right]^{1/3} + 100 \approx 19.96 \text{ min.}$$

7. Let $x(t)$ denote the mass of salt in the first tank at time t. Assuming that the initial mass is $x(0) = x_0$ we use the mathematical model described by equation (1) on page 96 of the text to find $x(t)$. We can determine the concentration of salt in the first tank by dividing $x(t)$ by the its volume, i.e., $x(t)/60$ kg/gal. Note that the volume of brine in this tank remains constant because the flow rate in is the same as the flow rate out. Then

$$\textit{output rate}_1 = (3 \text{ gal/min})\cdot\big(x(t)/60 \text{ kg/gal}\big) = x(t)/20 \text{ kg/min}.$$

Since the incoming liquid is pure water, we conclude that

$$\textit{input rate}_1 = 0 .$$

Therefore, $x(t)$ satisfies the initial value problem

$$\frac{dx}{dt} = \textit{input rate}_1 - \textit{output rate}_1 = -\frac{x}{20}, \qquad x(0) = x_0 .$$

This equation is linear and separable. Solving and using the initial condition to evaluate the arbitrary constant, we find

$$x(t) = x_0 e^{-t/20} .$$

Now, let $y(t)$ denote the mass of salt in the second tank at time t. Since initially this tank contained only pure water, we have $y(0)=0$. The function $y(t)$ can be described by the same mathematical model. We get

$$\textit{input rate}_2 = \textit{output rate}_1 = \frac{x(t)}{20} = \frac{x_0}{20}e^{-t/20} \text{ kg/min.}$$

Further since the volume of the second tank also remains constant, we have

$$\textit{output rate}_2 = (3 \text{ gal/min})\cdot\left(y(t)/60 \text{ kg/gal}\right) = y(t)/20 \text{ kg/min.}$$

Therefore, $y(t)$ satisfies the initial value problem

$$\frac{dy}{dt} = \textit{input rate}_2 - \textit{output rate}_2 = \frac{x_0}{20}e^{-t/20} - \frac{y(t)}{20}, \qquad y(0)=0,$$

or

$$\frac{dy}{dt} + \frac{y(t)}{20} = \frac{x_0}{20}e^{-t/20}, \qquad y(0)=0.$$

This is a linear equation in the standard form. Using the method given on page 56 of the text we find the general solution to be

$$y(t) = \frac{x_0}{20}te^{-t/20} + Ce^{-t/20}.$$

The constant C can be found from the initial condition:

$$0 = y(0) = \frac{x_0}{20}\cdot 0\cdot e^{-0/20} + Ce^{-0/20} \qquad \Rightarrow \qquad C=0.$$

Therefore, $y(t) = \frac{x_0}{20}te^{-t/20}$. To investigate $y(t)$ for maximum value we calculate

$$\frac{dy}{dt} = \frac{x_0}{20}e^{-t/20} - \frac{y(t)}{20} = \frac{x_0}{20}e^{-t/20}\left(1-\frac{t}{20}\right).$$

Thus

$$\frac{dy}{dt}=0 \qquad \Leftrightarrow \qquad 1-\frac{t}{20}=0 \qquad \Leftrightarrow \qquad t=20,$$

which is the point of global maximum (notice that $dy/dt>0$ for $t<20$ and $dy/dt<0$ for $t>20$). In other words, at this moment the water in the second tank will taste saltiest, and comparing concentrations, it will be

$$\frac{y(20)/60}{x_0/60} = \frac{y(20)}{x_0} = \frac{1}{20}\cdot 20\cdot e^{-20/20} = e^{-1}$$

as salty as the original brine.

9. Let $p(t)$ be the population of splake in the lake at time t. We start counting the population in 1980. Thus, we let $t = 0$ correspond to the year 1980. By the Malthusian law stated on page 100 of the text, we have

$$p(t) = p_0 e^{kt}.$$

Since $p_0 = p(0) = 1000$, we see that

$$p(t) = 1000e^{kt}.$$

To find k, we use the fact that the population of splake was 3000 in 1987. Therefore,

$$p(7) = 3000 = 1000e^{k7}$$

$$\Rightarrow \quad 3 = e^{k \cdot 7} \quad \Rightarrow \quad k = \frac{\ln 3}{7}.$$

Putting this value for k into the equation for $p(t)$ gives

$$p(t) = 1000e^{(t \ln 3)/7} = (1000)3^{t/7}.$$

To estimate the population in 2004 we plug $t = 24$ into this formula to get

$$p(24) = (1000)3^{24/7} \approx 43{,}236 \text{ splakes.}$$

13. Using the information given in Problem 9 and letting $t_0 = 0$ and $t_a = 7$, we have $p(0) = p_0 = 1000$ and $p(7) = p_a = 3000$. Here, we have $p(14) = p_b = 5000$. By the logistic model stated in equation (15) on page 102 of the text, we know that

$$p(t) = \frac{p_0 p_1}{p_0 + (p_1 - p_0)e^{-Ap_1 t}}.$$

By using the equations found in Problem 12, we can find p_1 and A. Thus

$$p_1 = \left[\frac{(3000)(5000) - 2(1000)(5000) + (1000)(3000)}{(3000)^2 - (1000)(5000)}\right](3000)$$

$$\Rightarrow \quad p_1 = 6000$$

and

$$A = \frac{1}{(6000)7}\ln\left[\frac{5000(3000 - 1000)}{1000(5000 - 3000)}\right] \quad \Rightarrow \quad A = \frac{1}{42{,}000}\ln 5.$$

Thus we have

$$p(t) = \frac{(1000)(6000)}{1000 + (6000 - 1000)e^{-(\ln 5/42{,}000)6000t}}$$

$$\Rightarrow \qquad p(t)=\frac{6000}{1+5e^{-(t\ln 5)/7}}=\frac{6000}{1+5^{(1-t/7)}}.$$

To estimate the population of splake in 2004, we plug 24 into this equation to obtain

$$p(24)=\frac{6000}{1+5^{(1-24/7)}}=\frac{6000}{1+5^{-17/7}}\approx 5882.$$

To find the limiting population, we look again at the equation we found for $p(t)$. Thus we see that

$$\lim_{t\to+\infty} p(t)=\lim_{t\to+\infty}\left[\frac{6000}{1+5^{(1-t/7)}}\right]=6000=p_1.$$

16. By definition

$$p'(t)=\lim_{h\to 0}\frac{p(t+h)-p(t)}{h}.$$

Replacing h by $-h$ in the above equation we obtain

$$p'(t)=\lim_{h\to 0}\frac{p(t-h)-p(t)}{-h}=\lim_{h\to 0}\frac{p(t)-p(t-h)}{h}.$$

Adding the previous two equations together we get

$$2p'(t)=\lim_{h\to 0}\left[\frac{p(t+h)-p(t)}{h}+\frac{p(t)-p(t-h)}{h}\right]$$

$$=\lim_{h\to 0}\left[\frac{p(t+h)-p(t-h)}{h}\right].$$

Thus

$$p'(t)=\lim_{h\to 0}\left[\frac{p(t+h)-p(t-h)}{2h}\right].$$

19. This problem can be regarded as a compartmental analysis problem for the population of fish. If we let $m(t)$ denote the mass in million tons of a certain species of fish, then the mathematical model for this process is given by

$$\frac{dm}{dt}=\text{increase rate} - \text{decrease rate}.$$

The increase rate of fish is given by $2m$ million tons/yr. The decrease rate of fish is given as 15 million tons/yr. Substituting these rates into the above equation we obtain

$$\frac{dm}{dt}=2m-15, \qquad m(0)=7 \text{ million tons/year}.$$

This equation is linear and separable. Using the initial condition, $m(0)=7$ to evaluate the arbitrary constant we obtain

$$m(t) = -\frac{1}{2}e^{2t} + \frac{15}{2}.$$

Knowing this equation we can now find when all the fish will be gone. To determine when all the fish will be gone we set $m(t) = 0$ and solve for t. This gives

$$0 = -\frac{1}{2}e^{2t} + \frac{15}{2}$$

and hence

$$t = \frac{1}{2}\ln(15) \approx 1.354 \text{ years.}$$

To determine the fishing rate required to keep the fish mass constant we solve the general problem

$$\frac{dm}{dt} = 2m - r, \qquad m(0) = 7$$

with r as the fishing rate. Thus we obtain

$$m(t) = Ke^{2t} + \frac{r}{2}.$$

The initial mass was given to be 7 million tons/year. Substituting this into the above equation we can find the arbitrary constant K:

$$m(0) = 7 = K + \frac{r}{2}$$

$$\Rightarrow \quad K = 7 - \frac{r}{2}.$$

Thus $m(t)$ is

$$m(t) = \left(7 - \frac{r}{2}\right)e^{2t} + \frac{r}{2}.$$

A fishing rate of $r = 14$ million tons/year will give a constant mass of fish by canceling out the coefficient of the e^{2t} term.

25. Let $M(t)$ denote the mass of carbon-14 present in the burnt wood of the campfire. Then since carbon-14 decays at a rate proportional to its mass, we have

$$\frac{dM}{dt} = -\alpha M,$$

where α is the proportionality constant. This equation is linear and separable. Using the initial condition, $M(0) = M_0$, we obtain

$$M(t) = M_0 e^{-\alpha t}.$$

Given the half-life of carbon-14 to be 5600 years we solve for α since we have

$$\frac{1}{2}M_0 = M_0 e^{-\alpha(5600)}$$

$$\Rightarrow \qquad \frac{1}{2} = e^{-\alpha(5600)}$$

which yields

$$\alpha = \frac{\ln(0.5)}{-5600} \approx 0.000123776 .$$

Thus,

$$M(t) = M_0 e^{-0.000123776t} .$$

Now we are told that after t years 2% of the original amount of carbon-14 remains in the campfire and we are asked to determine t. Thus

$$0.02M_0 = M_0 e^{-0.000123776t}$$

$$\Rightarrow \qquad 0.02 = e^{-0.000123776t}$$

which yields

$$t = \frac{\ln 0.02}{-0.000123776} \approx 31{,}606 \text{ years.}$$

EXERCISES 3.3: Heating and Cooling of Buildings, page 113

3. This problem is similar to one of cooling a building. In this problem we have no additional heating or cooling so we can say that the rate of change of the wine's temperature, $T(t)$, is given by Newton's law of cooling

$$\frac{dT}{dt} = K[M(t) - T(t)],$$

where $M(t) = 32$ is the temperature of ice. This equation is linear and is rewritten in the standard form as

$$\frac{dT}{dt} + KT(t) = 32K .$$

We find that the integrating factor is e^{Kt}. Multiplying both sides by e^{Kt} and integrating gives

$$e^{Kt}\frac{dT}{dt} + e^{Kt}KT(t) = e^{Kt}32K$$

$$\Rightarrow \qquad e^{Kt}T(t) = \int e^{Kt}32K\,dt + C$$

$$\Rightarrow \quad e^{Kt}T(t) = 32e^{Kt} + C$$

$$\Rightarrow \quad T(t) = 32 + Ce^{-Kt}.$$

By setting $t = 0$ and using the initial temperature 70°F, we find the constant, *C*, to be

$$70 = 32 + C$$

$$C = 38°\text{F}.$$

Knowing that it takes 15 minutes for the wine to chill to 60°F, we can find the constant, *K*:

$$60 = 32 + 38e^{-K(15)}.$$

Solving for *K* we obtain

$$K = \frac{-1}{15}\ln\left(\frac{60-32}{38}\right) = 0.02035.$$

Knowing this we can now determine how long it will take for the wine to reach 56°F. Using our equation for temperature, we set

$$56 = 32 + 38e^{-0.02035t}$$

and solving for *t* we get

$$t = \frac{-1}{0.02035}\ln\left(\frac{56-32}{38}\right) \approx 22.6\,\text{min}.$$

5. This problem can be treated as one similar to that of a cooling building. If we assume the air surrounding the body has not changed since the death, we can say that the rate of change of the body's temperature, $T(t)$, is given by Newton's law of cooling:

$$\frac{dT}{dt} = K[M(t) - T(t)],$$

where $M(t)$ represents the surrounding temperature which we've assumed to be a constant 16°C. This differential equation is linear and is solved using an integrating factor of e^{Kt}. Rewriting the above equation in standard form, multiplying both sides by e^{Kt} and integrating gives

$$\frac{dT}{dt} + KT(t) = K(16)$$

$$e^{Kt}\cdot\frac{dT}{dt} + e^{Kt}KT(t) = e^{Kt}K(16)$$

$$e^{Kt}T(t) = 16e^{Kt} + C$$

$$T(t) = 16 + Ce^{-Kt}.$$

Let's take $t = 0$ as the time at which the person died. Then $T(0) = 37°$ (normal body temperature) and we get

$$37 = 16 + C$$

$$C = 21.$$

Now we know that at sometime, say X hours after death, the body temperature was measured to be 34.5°C and that at $X + 1$ hours after death the body temperature was measured to be 33.7°C. Therefore we have

$$34.5 = 16 + 21e^{-KX}$$

and

$$33.7 = 16 + 21e^{-K(X+1)}.$$

Solving the first equation for KX we arrive at

$$KX = -\ln\left(\frac{34.5 - 16}{21}\right) = 0.12675.$$

Substituting this value into the second equation we can solve for K as follows

$$33.7 = 16 + 21e^{-0.12675 - K}$$

$$K = -\left(0.12675 + \ln\left(\frac{33.7 - 16}{21}\right)\right) = 0.04421.$$

This results in an equation for the body temperature of

$$T(t) = 16 + 21e^{-0.04421t}$$

To determine how much before 12 Noon the person died let's solve for X with $T = 34.5$°C,

$$34.5 = 16 + 21e^{-0.04421X}$$

$$X = \frac{-1}{0.04421}\ln\left(\frac{34.5 - 16}{21}\right) \approx 2.867 \text{ hours.}$$

Therefore the time of death is 2.867 hours (2 hours, 52 min) before Noon or 9:08 am.

9. Since we are evaluating the temperature in a warehouse, we can assume that any heat generated by people or equipment in the warehouse will be negligible. Therefore, we have $H(t) = 0$. Also, we are assuming that there is no heating or air conditioning in the warehouse. Therefore, we have that $U(t) = 0$. We are also given that the outside temperature has a sinusoidal fluctuation. Thus, as in Example 2, page 109, we see that

$$M(t) = M_0 - B\cos\omega t,$$

where M_0 is the average outside temperature, B is a positive constant for the magnitude of the temperature shift from this average, and $\omega = \pi/12$ radians per hour. To find M_0 and B, we are given that at 2:00 AM, $M(t)$ reaches a low of 16° C and at 2:00 PM it reaches a high of 32° C. This gives

$$M_0 = \frac{16+32}{2} = 24°\text{C}.$$

By letting $t = 0$ at 2:00 AM (so that low for the outside temperature corresponds to the low for the negative cosine function), we can calculate the constant B. That is

$$16 = 24 - B\cos 0 = 24 - B \qquad \Rightarrow \qquad B = 8.$$

Therefore, we see that

$$M(t) = 24 - 8\cos \omega t,$$

where $\omega = \pi/12$. As in Example 2, using the fact that $B_0 = M_0 + \frac{H_0}{K} = M_0 + \frac{0}{K} = M_0$, we see that

$$T(t) = 24 - 8F(t) + Ce^{-Kt},$$

where

$$F(t) = \frac{\cos \omega t + (\omega/K)\sin \omega t}{1 + (\omega/K)^2} = \left[1 + (\omega/K)^2\right]^{-1/2} \cos(\omega t - \alpha).$$

In the last expression, α is chosen such that $\tan\alpha = \omega/K$. By assuming that the exponential term dies off, we obtain

$$T(t) = 24 - 8\left[1 + (\omega/K)^2\right]^{-1/2} \cos(\omega t - \alpha).$$

This function will reach a minimum when $\cos(\omega t - \alpha) = 1$ and it will reach a maximum when $\cos(\omega t - \alpha) = -1$.

For the case when the time constant for the building is 1, we see that $\frac{1}{K} = 1$ which implies that $K = 1$. Therefore, the temperature will reach a maximum of K

$$T = 24 + 8\left[1 + (\pi/12)^2\right]^{-1/2} \approx 31.7° \text{ C}.$$

It will reach a minimum of

$$T = 24 - 8\left[1 + (\pi/12)^2\right]^{-1/2} \approx 16.3° \text{ C}.$$

For the case when the time constant of the building is 5, we have

$$\frac{1}{K} = 5 \qquad \Rightarrow \qquad K = \frac{1}{5}.$$

Then, the temperature will reach a maximum of

$$T = 24 + 8\left[1 + \left(\frac{5\pi}{12}\right)^2\right]^{-1/2} \approx 28.9° \text{ C},$$

and a minimum of

$$T = 24 - 8\left[1 + (5\pi/12)^2\right]^{-1/2} \approx 19.1° \text{ C}.$$

11. As in Example 3, page 111 of the text, this problem involves a thermostat to regulate the temperature in the van. Hence, we have

$$U(t) = K_U\left[T_D - T(t)\right],$$

where T_D is the desired temperature 16° C and K_U is a proportionality constant. We will assume that $H(t) = 0$ and that the outside temperature $M(t)$ is a constant 35° C. The time constant for the van is $\frac{1}{K} = 2$ hr, hence $K = 0.5$. Since the time constant for the van with its air conditioning system is $\frac{1}{K_1} = \frac{1}{3}$ hr, then $K_1 = K + K_U = 3$. Therefore, $K_U = 3 - 0.5 = 2.5$. The temperature in the van is governed by the equation

$$\frac{dT}{dt} = (0.5)(35 - T) + (2.5)(16 - T) = 57.5 - 3T.$$

Solving this separable equation yields

$$T(t) = 19.17 + Ce^{-3t}.$$

When $t = 0$, we are given $T(0) = 55$. Using this information to solve for C gives $C = 35.83$. Hence, the van temperature is given by

$$T(t) = 19.17 + 64.17e^{-3t}.$$

To find out when the temperature in the van will reach 27° C, we let $T(t) = 27$ and solve for t. Thus, we see that

$$27 = 19.17 + 35.83e^{-3t} \quad \Rightarrow \quad e^{-3t} = \frac{7.83}{35.83} \approx 0.2185$$

$$\Rightarrow \quad t \approx \frac{-\ln(0.2185)}{3} \approx 0.5070 \text{ hr} \quad \text{or} \quad 30.4 \text{ min}.$$

13. Since the time constant is 64, we have $K = \frac{1}{64}$. The temperature in the tank increases at the rate of 2° F for every 1000 Btu. Furthermore, every hour of sunlight provides an input of 2000 Btu to the tank. Thus,

$$H(t) = 2 \times 2 = 4° \text{ F per hr}.$$

We are given that $T(0) = 110$, and that the temperature $M(t)$ outside the tank is a constant 80° F. Hence the temperature in the tank is governed by

$$\frac{dT}{dt}=\frac{1}{64}[80-T(t)]+4=-\frac{1}{64}T(t)+5.25, \qquad T(0)=110.$$

Solving this separable equation gives

$$T(t)=336+Ce^{-t/64}.$$

To find C, we use the initial condition to see that

$$T(0)=110=336+C \qquad \Rightarrow \qquad C=-226.$$

This yields the equation

$$T(t)=336-226e^{-t/64}.$$

After 12 hours of sunlight, the temperature will be

$$T(12)=336-226e^{-12/64}\approx 148.6° \text{ F}.$$

EXERCISES 3.4: Newtonian Mechanics, page 121

3. For this problem, m = 500 kg, $v_0=0$, g = 9.81 m/sec^2, and b = 50 kg/sec. We also see that the object has 1000 m to fall before it hits the ground. Plugging these variables into equation (6) on page 117 of the text gives the equation

$$x(t)=\frac{(500)(9.81)}{50}t+\frac{500}{50}\left(0-\frac{(500)(9.81)}{50}\right)\left(1-e^{-50t/500}\right)$$

$$\Rightarrow \qquad x(t)=98.1t+981e^{-t/10}-981.$$

To find out when the object will hit the ground, we solve $x(t)=1000$ for t. Therefore, we have

$$1000=98.1t+981e^{-t/10}-981$$

$$\Rightarrow \qquad 98.1t+981e^{-t/10}=1981.$$

In this equation, if we ignore the term $981e^{-t/10}$, we will find that $t\approx 20.2$. But this means that we have ignored the term similar to $981e^{-2}\approx 132.8$, which we see is to large to ignore. Therefore, we must try to approximate t. We will use Newton's method on the equation

$$f(t)=98.1t+981e^{-t/10}-1981=0.$$

(If we can find a root to this equation, we will have found the t we want.) Newton's method generates a sequence of approximations given by the formula

$$t_{n+1}=t_n-\frac{f(t_n)}{f'(t_n)}.$$

Since $f'(t)==98.1-98.1e^{-t/10}=98.1\left(1-e^{-t/10}\right)$, the recursive equation above becomes

$$t_{n+1} = t_n - \frac{t_n + 10e^{-t_n/10} - \dfrac{1981}{98.1}}{1 - e^{-t_n/10}} . \qquad (1)$$

To start the process, let $t_0 = \dfrac{1981}{98.1} \approx 20.19368$, which was the approximation we obtained when we neglected the exponential term. Then, by equation (1) above we have

$$t_1 = 20.19368 - \frac{20.19368 + 10\,e^{-2.019368} - 20.19368}{1 - e^{-2.019368}}$$

$$\Rightarrow \qquad t_1 \approx 18.663121.$$

To find t_2, we plug this value for t_1 into equation (1). This gives $t_2 \approx 18.643753$. Continuing this process, we find that $t_3 \approx 18.643749$. Since t_2 and t_3 agree to four decimal places, an approximation for the time it takes the object to strike the ground will be given by $t \approx 18.6437$ sec.

9. This problem is similar to Example 1 on page 116 of the text with the addition of a buoyancy force of magnitude $\frac{1}{40}mg$. If we let $x(t)$ be the distance below the water at time t and $v(t)$ the velocity, then the total force acting on the object is

$$F = mg - bv - \frac{1}{40}mg.$$

We are given $m = 100$ kg, $g = 9.81$ m/sec^2, and $b = 10$ kg/sec. Applying Newton's Second Law gives

$$100\frac{dv}{dt} = (100)(9.81) - 10v - \frac{10}{4}(9.81)$$

$$\Rightarrow \quad \frac{dv}{dt} = 9.56 - (0.1)v.$$

Solving this equation by separation of variables, we have

$$v(t) = 95.65 + Ce^{-t/10}.$$

Since $v(0) = 0$, we find $C = -95.65$. Hence

$$v(t) = 95.65 - 95.65e^{-t/10}.$$

Integrating, we obtain

$$x(t) = 95.65t + 956.5e^{-t/10} + C_1.$$

Using the fact that $x(0) = 0$, we find $C_1 = -956.5$. Therefore, the equation of motion of the object is

$$x(t) = 95.65t + 956.5e^{-t/10} - 956.5\,\text{m}.$$

To determine when the object is traveling at the velocity of 70 m/sec, we solve $v(t) = 70$. That is,

$$70 = 95.65 - 95.65e^{-t/10} = (95.65)\left(1 - e^{-t/10}\right)$$

$$\Rightarrow \quad t = -10\ln\left(1 - \frac{70}{95.65}\right) \approx 13.2 \text{ sec.}$$

13. There are two forces acting on the shell: a constant force due to the downward pull of gravity and a force due to air resistance that acts in opposition to the motion of the shell. All of the motion occurs along a vertical axis. On this axis, we choose the origin to be the point where the shell was shot from and let $x(t)$ denote the position upward of the shell at time t. The forces acting on the object can be expressed in terms of this axis. The force due to gravity is

$$F_1 = -mg,$$

where g is the acceleration due to gravity near Earth. Note we have a minus force because our coordinate system was chosen with up as positive and gravity acts in a downward direction. The force due to air resistance is

$$F_2 = -(0.1)v^2.$$

The negative sign is present because air resistance acts in opposition to the motion of the object. Therefore the net force acting on the shell is

$$F = F_1 + F_2 = -mg - (0.1)v^2.$$

We now apply Newton's second law to obtain

$$m\frac{dv}{dt} = -\left(mg + 0.1v^2\right).$$

Because the initial velocity of the shell is 500 m/sec, a model for the velocity of the rising shell is expressed as the initial-value problem

$$m\frac{dv}{dt} = -\left(mg + 0.1v^2\right), \qquad v(t=0) = 500, \tag{2}$$

where $g = 9.81$. Using separation of variables, we get

$$\frac{dv}{10mg + v^2} = \frac{-dt}{10m}.$$

Integrating we obtain

$$\frac{1}{\sqrt{10mg}}\tan^{-1}\left(\frac{v}{\sqrt{10mg}}\right) = \frac{-t}{10m} + C.$$

Setting $m = 3$, $g = 9.81$ and $v = 500$ when $t = 0$, we find

$$C = \frac{1}{\sqrt{10(3)(9.81)}}\tan^{-1}\left(\frac{500}{\sqrt{10(3)(9.81)}}\right) = 0.08956.$$

Thus the equation of velocity as a function of time is

$$\frac{1}{\sqrt{10mg}}\tan^{-1}\left(\frac{v}{\sqrt{10mg}}\right)=\frac{-t}{10m}+0.08956\,.$$

From physics we know that when the shell reaches its maximum height the shell's velocity will be zero; therefore $t_{\max}$ will be

$$t_{\max}=-10(3)\left(\frac{1}{\sqrt{10(3)(9.81)}}\tan^{-1}\left(\frac{0}{\sqrt{10(3)(9.81)}}\right)-0.08956\right)=(30)(0.08956)$$

$$t_{\max}\approx 2.69 \text{ seconds.}$$

Using equation (2) and noting that $\frac{dv}{dt}=\frac{dv}{dx}\frac{dx}{dt}=v\frac{dv}{dx}$ we can determine the maximum height attained by the shell. With the above substitution, equation (2) becomes

$$mv\frac{dv}{dx}=-(mg+0.1v^2),\qquad v(x=0)=500.$$

Using separation of variables, we get

$$\frac{vdv}{\left(10mg+v^2\right)}=\frac{-dx}{10m}.$$

Integrating, we obtain

$$\frac{1}{2}\ln\left(10mg+v^2\right)=\frac{-x}{10m}+C$$

$$\Rightarrow\qquad 10mg+v^2=Ke^{-x/(5m)}.$$

Setting $v = 500$ when $x = 0$, we find

$$K=e^0\left(10(3)(9.81)+(500)^2\right)=250294.3\,.$$

Thus the equation of velocity as a function of distance is

$$v^2+10mg=(250294.3)e^{-x/(5m)}.$$

The maximum height will occur when the shell's velocity is zero, therefore $x_{\max}$ is

$$x_{\max}=-5(3)\ln\left(\frac{0+10(3)(9.81)}{250294.3}\right)=101.19 \text{ meters.}$$

15. The total torque exerted on the flywheel is the sum of the torque exerted by the motor and the retarding torque due to friction. Thus, by Newton's second law for rotation, we have

$$I\frac{d\omega}{dt} = T - k\omega \quad \text{with} \quad \omega(0) = \omega_0,$$

where I is the moment of inertia of the flywheel, $\omega(t)$ is the angular velocity, $\frac{d\omega}{dt}$ is the angular acceleration, T is the constant torque exerted by the motor, and k is a positive constant of proportionality for the torque due to friction. Solving this separable equation gives

$$\omega(t) = \frac{T}{k} + Ce^{-kt/I}.$$

Using the initial condition $\omega(0) = \omega_0$, we find $C = \left(\omega_0 - \frac{T}{k}\right)$. Hence,

$$\omega(t) = \frac{T}{k} + \left(\omega_0 - \frac{T}{k}\right)e^{-kt/I}.$$

17. Since the motor is turned off, its torque is $T = 0$, and the only torque acting on the flywheel is the retarding one, $-5\sqrt{\omega}$. Then Newton's second law for rotational motion becomes

$$I\frac{d\omega}{dt} = -5\sqrt{\omega} \qquad \text{with} \quad \omega(0) = \omega_0 = 225 \text{ rad/sec} \qquad \text{and} \qquad I = 50 \text{ kg - m}^2.$$

The general solution to this separable equation is

$$\sqrt{\omega(t)} = -\frac{5}{2I}t + C = -0.05t + C.$$

Using the initial condition we find

$$\sqrt{\omega(0)} = -0.05 \cdot 0 + C \qquad \Rightarrow \qquad C = \sqrt{\omega(0)} = \sqrt{225} = 15.$$

Thus

$$t = \frac{1}{0.05}\left(15 - \sqrt{\omega(t)}\right) = 20\left(15 - \sqrt{\omega(t)}\right).$$

At the moment $t = t_{\text{stop}}$ when the flywheel stops rotating we have $\omega\left(t_{\text{stop}}\right) = 0$ and so

$$t_{\text{stop}} = 20\left(15 - \sqrt{0}\right) = 300 \text{ sec.}$$

21. In this problem there are two forces acting on a sailboat: A constant horizontal force due to the wind and a force due to the water resistance that acts in opposition to the motion of the sailboat. All of the motion occurs along a horizontal axis. On this axis, we choose the origin to be the point where the hard blowing wind begins and $x(t)$ denotes the distance the sailboat travels in time t. The forces on the sailboat can be expressed in terms of this axis. The force due to the wind is

$$F_1 = 600 \text{ N}.$$

The force due to water resistance is

$$F_2 = -100v \text{ N}.$$

Applying Newton's second law we obtain

$$m\frac{dv}{dt} = 600 - 100v.$$

Since the initial velocity of the sailboat is 1 m/sec, a model for the velocity of the moving sailboat is expressed as the initial-value problem

$$m\frac{dv}{dt} = 600 - 100v, \qquad v(0) = 1\text{m/sec}.$$

Using separation of variables, we get with $m = 50$ kg,

$$\frac{dv}{6 - v} = 2\,dt$$

$$-\ln(6 - v) = 2t + C.$$

Therefore the velocity is given by

$$v = 6 - Ke^{-2t}.$$

Setting $v = 1$m/sec when $t = 0$, we find

$$1 = 6 - K \quad \Rightarrow \quad K = 5.$$

Thus the equation for velocity is

$$v = 6 - 5e^{-2t}.$$

The limiting velocity of the sailboat under these conditions is found by letting time approach infinity:

$$\lim_{t\to\infty} v(t) = \lim_{t\to\infty}\left(6 - 5e^{-2t}\right) = 6 \text{ m/sec}.$$

To determine the equation of motion we will use the equation of velocity obtained previously and substitute $\frac{dx}{dt}$ for v to obtain

$$\frac{dx}{dt} = 6 - 5e^{-2t}, \qquad x(0) = 0.$$

Integrating this equation we obtain

$$x = 6t + \frac{5}{2}e^{-2t} + C_1.$$

Setting $x = 0$ when $t = 0$, we find

$$0 = 0 + \frac{5}{2} + C_1 \qquad \Rightarrow \qquad C_1 = -\frac{5}{2}.$$

Thus the equation of motion for the sailboat is

$$x(t) = \frac{5}{2}e^{-2t} + 6t - \frac{5}{2}.$$

25. **(a)** From Newton's second law we have

$$m\frac{dv}{dt} = \frac{-GMm}{r^2}.$$

Dividing both sides by *m*, the mass of the rocket, and letting $g = \frac{GM}{R^2}$ we get

$$\frac{dv}{dt} = \frac{-gR^2}{r^2},$$

where *g* is the gravitational force of Earth, *R* is the radius of Earth and *r* is the distance between Earth and the projectile.

(b) Using the equation found in part (a), letting $\frac{dv}{dt} = \frac{dv}{dr}\frac{dr}{dt}$ and knowing that $\frac{dr}{dt} = v$ we get

$$v\frac{dv}{dr} = \frac{-gR^2}{r^2}.$$

(c) The differential equation found in part (b) is separable and can be written in the form

$$vdv = -\frac{gR^2}{r^2}dr.$$

If the projectile leaves Earth with a velocity of v_0 we have the initial value problem

$$vdv = -\frac{gR^2}{r^2}dr, \qquad v(r = R) = v_0.$$

Integrating we get

$$\frac{v^2}{2} = \frac{gR^2}{r} + K,$$

where *K* is an arbitrary constant. We can solve for the constant, *K*, using the initial value as follows:

$$K = \frac{v_0^2}{2} - \frac{gR^2}{R} = \frac{v_0^2}{2} - gR.$$

Substituting this formula for *K* and solving for the velocity we obtain

$$v^2 = \frac{2gR^2}{r} + v_0^2 - 2gR.$$

(d) In order for the velocity of the projectile to always remain positive, $\frac{2gR^2}{r}+v_0^2$ must be greater than $2gR$ as r approaches infinity. This means

$$\lim_{r\to\infty}\left(\frac{2gR}{r}+v_0^2\right)>2gR \qquad \Rightarrow \qquad v_0^2>2gR\,.$$

Therefore $v_0^2-2gR>0$.

(e) Using the equation from part (d) and converting meters to kilometers we have

$$v_e=\sqrt{2gR}=\sqrt{2\cdot 9.81\,\text{m/sec}^2\cdot(1\,\text{km}/1000\,\text{m})(6370\text{km})}=11.18\,\text{km/sec}\,.$$

(f) Using the equation from part (d) we have

$$v_e=\sqrt{2(g/6)R}=\sqrt{2(9.81/6)(1/1000)(1738)}=2.38\,\text{km/sec}\,.$$

EXERCISES 3.5: Improved Euler's Method, page 133

1. Given the step size h and considering equally spaced points we have

$$x_{n+1}=x_n+nh, \qquad n=0,1,2,\dots\,.$$

Euler's method is defined by equation (4) on page 126 of the text to be

$$y_{n+1}=y_n+hf(x_n,y_n), \qquad n=0,1,2,\dots\,,$$

where $f(x, y) = 5y$. Starting with the given value of $y_0 = 1$, we compute

$$y_1=y_0+h(5y_0)=1+5h.$$

We can then use this value to compute y_2 to be

$$y_2=y_1+h(5y_1)=(1+5h)y_1=(1+5h)^2.$$

Proceeding in this manner, we can generalize to y_n:

$$y_n=(1+5h)^n.$$

Referring back to our equation for x_n and using the given values of $x_0=0$ and $x_1=1$ we find

$$1=nh \qquad \Rightarrow \qquad n=\frac{1}{h}.$$

Substituting this back into the formula for y_n we find the approximation to the initial value problem

$$y'=5y, \qquad y(0)=1$$

at $x = 1$ to be $(1 + 5h)^{1/h}$.

5. In this problem, we have $f(x,y)=4y$. Thus, we have

$$f(x_n, y_n)=4y_n \qquad \text{and} \qquad f(x_n+h, y_n+hf(x_n,y_n))=4[\,y_n+h(4y_n)]=4y_n+16hy_n.$$

By equation (9) on page 129 of the text, we, have

$$y_{n+1}=y_n+\frac{h}{2}(4y_n+4y_n+16hy_n) = (1+4h+8h^2)y_n. \tag{3}$$

Since the initial condition $y(0)=\frac{1}{3}$ implies that $x_0=0$ and $y_0=\frac{1}{3}$, equation (3) above yields

$$y_1=(1+4h+8h^2)y_0 = \frac{1}{3}(1+4h+8h^2),$$

$$y_2=(1+4h+8h^2)y_1=(1+4h+8h^2)\left(\frac{1}{3}\right)(1+4h+8h^2),$$

$$\Rightarrow \qquad y_2=\frac{1}{3}(1+4h+8h^2)^2,$$

$$y_3=(1+4h+8h^2)y_2 = (1+4h+8h^2)\left(\frac{1}{3}\right)(1+4h+8h^2)^2,$$

$$\Rightarrow \qquad y_3=\frac{1}{3}(1+4h+8h^2)^3.$$

Continuing this way we see that

$$y_n=\frac{1}{3}(1+4h+8h^2)^n. \tag{4}$$

(This can be proved by induction using equation (3) above.) We are looking for an approximation to our solution at the point $x=\frac{1}{2}$. Therefore, we have

$$h=\frac{1/2-x_0}{n}=\frac{1/2-0}{n}=\frac{1}{2n} \qquad \Rightarrow \qquad n=\frac{1}{2h}.$$

Substituting this value for n into equation (4) yields

$$y_n=\frac{1}{3}(1+4h+8h^2)^{1/(2h)}.$$

7. For this problem, $f(x,y)=x-y^2$. We need to approximate the solution on the interval [1,1.5] using a step size of $h=0.1$. Thus the number of steps needed is $N=5$. The inputs to the subroutine on page 130 are $x_0=1$, $y_0=0$, $c=1.5$ and $N=5$. For step 3 of the subroutine we have

$$F=f(x,y)=x-y^2$$

and

$$G = f(x+h, y+hf(x,y)) = (x+h) - (y+hF)^2 = (x+h) - \left[y + h(x - y^2)\right]^2 .$$

Starting with $x = x_0 = 1$ and $y = y_0 = 0$ we get $h = 0.1$ (as specified) and

$$F = 1 - 0^2 = 1,$$

$$G = (1+0.1) - \left[0 + 0.1(1 - 0^2)\right]^2 = 1.1 - (0.1)^2 = 1.09 .$$

Hence in step 4 we compute

$$x = 1 + 0.1 = 1.1,$$

$$y = 0 + 0.05(1 + 1.09) = 0.1045 .$$

Thus the approximate value of the solution at 1.1 is 0.1045. Next we repeat step 3 with $x = 1.1$ and $y = 0.1045$ to obtain

$$F = 1.1 + (0.1045)^2 \approx 1.0891,$$

$$G = (1.1 + 0.1) - \left[0.1045 + (0.1)(1.1 - (0.145)^2)\right]^2 \approx 1.1545 .$$

Hence in step 4 we compute

$$x = 1.1 + 0.1 = 1.2,$$

$$y = 0.1045 + 0.05(1.0891 + 1.1545) \approx 0.21668 .$$

Thus the approximate value of the solution at 1.2 is 0.21668. By continuing in this way, we fill in Table 3-A below. (The reader can also use the software provided free with the text.)

Table 3-A. Improved Euler's method to approximate the solution of $y' = x - y^2$, $y(1) = 0$ with $h = 0.1$.

i	x	y
0	1	0
1	1.1	0.1045
2	1.2	0.21668
3	1.3	0.33382
4	1.4	0.45300
5	1.5	0.57135

---

13. We want to approximate the solution $\phi(x)$ to $y' = 1 - y + y^3$, $y(0) = 0$, at $x = 1$. (In other words, we want to find an approximate value for $\phi(1)$.) To do this, we will use the algorithm on page 131 of the text. (We assume that the reader has a programmable calculator or microcomputer available and can transform the step-by-step outline on page 131 into an executable program. Alternatively, the reader can use the software provided free with the text.)

The inputs to the program are $x_0 = 0$, $y_0 = 0$, $c = 1$, $\varepsilon = 0.003$, and $M = 100$ (say). Notice that by Step 6 of the Improved Euler's Method with Tolerance, the computations should terminate when two successive approximations differ by less that 0.003. The initial value for h in Step 1 of the Improved Euler's Method Subroutine is

$$h = (1-0)2^{-0} = 1.$$

For the given equation, we have $f(x, y) = 1 - y + y^3$, and so the numbers F and G in Step 3 of the Improved Euler's Method Subroutine are

$$F = f(x, y) = 1 - y + y^3,$$

$$G = f(x+h, y+hF) = 1 - (y+hF) + (y+hF)^3.$$

From Step 4 of the Improved Euler's Method Subroutine with $x_0 = 0$, $y = 0$, and $h = 1$, we get

$$x = x + h = 0 + 1 = 1,$$

$$y = y + \frac{h}{2}(F+G) = 0 + \frac{1}{2}\left[1 + \left(1 - 1 + 1^3\right)\right] = 1.$$

Thus,

$$\phi(1) \approx y(1;1) = 1.$$

The algorithm (see Step 1 of the Improved Euler's Method Subroutine) next sets $h = 2^{-1} = 0.5$. The inputs to the subroutine are $x = 0$, $y = 0$, $c = 1$ and $N = 2$. For step 3 of the subroutine we have

$$F = 1 - 0 + 0 = 1,$$

$$G = 1 - [0 + 0.5(1)] + [0 + 0.5(1)]^3 = 0.625.$$

Hence in step 4 we compute

$$x = 0 + 0.5 = 0.5,$$

$$y = 0 + 0.25(1 + 0.625) = 0.40625$$

Thus the approximate value of the solution at 0.5 is 0.40625. Next we repeat step 3 with $x = 0.5$ and $y = 0.40625$ to obtain

$$F = 1 - (0.40625) + (0.40625)^3 = 0.6607971,$$

$$G = 1 - [0.40625 + 0.5(0.6607971)] + [0.40625 + 0.5(0.6607971)]^3 \approx 0.6630946.$$

In step 4 we compute

$$x = 0.5 + 0.5 = 1,$$

$$y = 0.40625 + 0.25(0.6607971 + 0.6630946) \approx 0.7372229.$$

Thus the approximate value of the solution at $x = 1$ is 0.737229. Further outputs of the algorithm are given in Table 3-B.

Table 3-B. Improved Euler's method approximations to $\phi(1)$, where $\phi(x)$ is the solution to $y' = 1 - y + y^3$, $y(0) = 0$.

h	$y(1; h) \approx \phi(x)$
1	1.0
2^{-1}	0.7372229
2^{-2}	0.7194115
2^{-3}	0.7169839

Since

$$\left|y(1;2^{-3}) - y(1;2^{-2})\right| = |0.7169839 - 0.7194115| < 0.003,$$

the algorithm stops (see Step 6 of the Improved Euler's Method with Tolerance) and prints out that $\phi(1)$ is approximately 0.71698.

15. For this problem, $f(x, y) = (x + y + 2)^2$. We want to approximate the solution, satisfying $y(0) = -2$, on the interval [0, 1.4] to find the point, with two decimal places of accuracy, where it crosses the x-axis, that is $y = 0$. Our approach is to use a step size of 0.005 and look for a change in the sign of y. This requires 280 steps. For this procedure inputs to the improved Euler's method subroutine are $x_0 = 0$, $y_0 = -2$, $c = 1.4$ and $N = 280$. We will stop the subroutine when we see a sign change in the value of y. (The subroutine is implemented on the software package provided free with the text.)

For step 3 of the subroutine we have

$$F = f(x, y) = (x + y + 2)^2$$

and

$$G = f(x + h, y + hF) = (x + h + y + hF + 2)^2 = (x + y + 2 + h(1 + F))^2.$$

Starting with the inputs $x = x_0 = 0$, $y = y_0 = -2$, and $h = 0.005$ we obtain

$$F = (0 - 2 + 2)^2 = 0,$$

$$G = (0 - 2 + 2 + 0.005(1 + 0))^2 = 0.000025.$$

Thus in step 4 we compute

$$x = 0 + 0.005 = 0.005,$$

$$y = -2 + 0.005(0 + 0.000025)/2 \approx -2.$$

Thus the approximate value of the solution at $x = 0.005$ is -2. We continue with steps 3 and 4 of the improved Euler's method subroutine until we arrive at $x = 1.270$ and $y \approx -0.04658269$. The next iteration, with $x = 1.275$, yields $y \approx 0.006295411$. This tells us that $y = 0$ is occurs somewhere between $x = 1.270$ and $x = 1.275$. Therefore rounding off to two decimal places yields $x = 1.27$.

21. We will use the improved Euler's method with $h = \frac{2}{3}$ to approximate the solution of the problem

$$\left\{\left[75 - 20\cos\left(\frac{\pi t}{12}\right)\right] - T(t)\right\} + 0.1 + 1.5\left[70 - T(t)\right], \qquad T(0) = 65,$$

with $K = 0.2$. Since $h = \frac{2}{3}$, it will take 36 steps to go from $t = 0$ to $t = 24$. By simplifying the above expression, we obtain

$$\frac{dT}{dt} = (75K + 105.1) - 20K\cos\left(\frac{\pi t}{12}\right) - (K + 1.5)T(t), \qquad T(0) = 65.$$

(Note that here t takes the place of x and T takes the place of y.) Therefore, with $K = 0.2$ the inputs to the subroutine are $t_0 = 0$, $T_0 = 65$, $c = 24$ and $N = 36$. For step 3 of the subroutine we have

$$F = f(t, T) = (75K + 105.1) - 20K\cos\left(\frac{\pi t}{12}\right) - (K + 1.5)T, \qquad (5)$$

$$G = f(t + h, T + hF) = (75K + 105.1) - 20K\cos\left(\frac{\pi(t + h)}{12}\right) - (K + 1.5)\{T + hF\}. \qquad (6)$$

For step 4 in the subroutine we have

$$t = t + h,$$

$$T = T + \frac{h}{2}(F + G).$$

Now starting with $t = t_0 = 0$ and $T = T_0 = 65$ we get $h = \frac{2}{3}$ (as specified) we have step 3 of the subroutine to be

$$F = (75(0.2) + 105.1) - 20(0.2)\cos 0 - ((0.2) + 1.5)(65) = 5.6,$$

$$G = (75(0.2) + 105.1) - 20(0.2)\cos\left(\frac{\pi(0.6667)}{12}\right) - ((0.2) + 1.5)\{65 + (0.6667)(5.6)\},$$

$$G \approx -0.6862.$$

Hence in step 4 we compute

$$t = 0 + 0.6667 = 0.6667,$$

$$T = 65 + 0.3333(5.6 - 0.6862) \approx 66.638.$$

Recalling that t_0 is midnight, we see that these results imply that at 0.6667 hours after midnight (or 12:40 AM) the temperature is approximately 66.638. Continuing with this process for $n = 1,2,3,..., 35$ gives us the approximate temperatures in a building with $K = 0.2$ over a 24 hr period. These results are given in Table 3-C below. (This is just a partial table.)

Table 3-C: Improved Euler's method to approximate the temperature in a building over a 24-hour period (with $K = 0.2$).

Time	t_n	T_n
Midnight	0	65.00000
12:40 AM	0.6667	66.63803
1:20 AM	1.3333	67.52906
2:00 AM	2.0000	68.07270
2:40 AM	2.6667	68.46956
3:20 AM	3.3333	68.81808
4:00 AM	4.0000	69.16392
8:00 AM	8.0000	71.48357
Noon	12.000	72.90891
4:00 PM	16.000	72.07140
8:00 PM	20.000	69.80953
Midnight	24.000	68.38519

The next step is to redo the above work with $K = 0.4$. That is, we substitute $K = 0.4$ and $h = \frac{2}{3} \approx 0.6667$ into equations (5), and (6) above. This yields

$$F = 135.1 - 8\cos\left(\frac{\pi t}{12}\right) - 1.9T,$$

$$G = 135.1 - 8\cos\left(\frac{\pi(t + 0.6667)}{12}\right) - 1.9(T + 0.6667F),$$

and

$$T = T + (0.333)[F + G].$$

Then, using these equations, we go through the process of first finding F then using this result to find G and finally using both results to find T. (This process must be done for $n = 0,1,2,...,35$.)

Lastly, we redo this work with $K=0.6$ and $h=\frac{2}{3}$. By so doing, we obtain the results given in the table in the answers on page B-6 of the text. (Note that the values for T_0, T_6, T_{12}, T_{18}, T_{24}, T_{30}, T_{36} are given in the answers.)

EXERCISES 3.6: Higher Order Numerical Methods: Taylor and Runge-Kutta, page 143

5. For the Taylor method of *order* 2, we need to find (see equation (4) on page 136 of the text)

$$f_2(x,y)=\frac{\partial f(x,y)}{\partial x}+\frac{\partial f(x,y)}{\partial y}f(x,y)$$

for $f(x,y)=x+1-y$. Thus, we have

$$f_2(x,y)=1+(-1)(x+1-y)=y-x.$$

Therefore, by equations (5) and (6) on page 136 of the text, we see that the recursive formulas with $h=0.25$ become

$$x_{n+1}=x_n+0.25$$

and

$$y_{n+1}=y_n+0.25(x_n+1-y_n)+\frac{(0.25)^2}{2}(y_n-x_n).$$

By starting with $x_0=0$ and $y_0=1$ (the initial values for the problem), we find

$$y_1=1+\frac{0.0625}{2}\approx 1.03125.$$

Plugging this value into the recursive formulas yields

$$y_2=1.03125+0.25(0.25+1-1.03125)+\left(\frac{0.0625}{2}\right)(1.03125-0.25)\approx 1.11035.$$

By continuing in this way, we can fill in the first three columns in Table 3-D below.

For the Taylor method of *order* 4, we need to find f_3 and f_4. Thus, we have

$$f_3(x,y)=\frac{\partial f_2(x,y)}{\partial x}+\left[\frac{\partial f_2(x,y)}{\partial y}\right]f(x,y)=-1+1(x+1-y)=x-y,$$

$$f_4(x,y)=\frac{\partial f_3(x,y)}{\partial x}+\left[\frac{\partial f_3(x,y)}{\partial y}\right]f(x,y)=1+(-1)(x+1-y)=y-x.$$

Hence, by equation (6) on page 136 of the text, we see that the recursive formula for y_{n+1} for the Taylor method of order 4 with $h=0.25$ is given by

$$y_{n+1}=y_n+0.25(x_n+1-y_n)+\frac{(0.25)^2}{2}(y_n-x_n)+\frac{(0.25)^3}{6}(x_n-y_n)+\frac{(0.25)^4}{24}(y_n-x_n).$$

By starting with $x_0 = 0$ and $y_0 = 1$, we can fill in the fourth column of Table 3-D.

Table 3-D. Taylor approximations of order 2 and 4 for the equation $y'=x+1-y$.

n	x_n	y_n**(order 2)**	y_n**(order 4)**
0	0	1	1
1	0.25	1.03125	1.02881
2	0.50	1.11035	1.10654
3	0.75	1.22684	1.22238
4	1.00	1.37253	1.36789

Thus, the approximation (rounded to 4 decimal places) of the solution by the Taylor method at the point $x = 1$ is given by $\phi_2(1)=1.3725$ if we use order 2 and by $\phi_4(1)=1.3679$ if we use order 4. The actual solution is $y=x+e^{-x}$ and so has the value $y(1)=1+e^{-1}\approx 1.3678794$ at $x=1$. Comparing these results, we see that $|y(1)-\phi_2(1)|=0.00462$ and $|y(1)-\phi_4(1)|=0.00002$.

9. For this problem we will use the Fourth order Runge-Kutta subroutine with $f(x,y)=x+1-y$. Using the step size of $h=0.25$ the number of steps needed is $N = 4$ to approximate the solution at $x=1$. For step 3 we have

$$k_1=hf(x,y)=0.25(x+1-y),$$

$$k_2=hf\left(x+\frac{h}{2},y+\frac{k_1}{2}\right)=0.25(0.875x+1-0.875y),$$

$$k_3=hf\left(x+\frac{h}{2},y+\frac{k_2}{2}\right)=0.25(0.890625x+1-0.890625y),$$

$$k_4=hf(x+h,y+k_3)=0.25(0.77734375x+1-0.77734375y).$$

Hence in step 4 we have

$$x=x+0.25,$$

$$y=y+\frac{1}{6}(k_1+2k_2+2k_3+k_4).$$

Using the initial conditions $x_0 = 0$ and $y_0 = 1$, $c = 1$ and $N = 4$ for step 3 we have

$$k_1=0.25(0+1-1)=0,$$

$$k_2 = 0.25(0.875(0) + 1 - 0.875(1)) = 0.03125,$$

$$k_3 = 0.25(0.890625(0) + 1 - 0.890625(1)) \approx 0.0273438,$$

$$k_4 = 0.25\ (0.77734375(0) + 1 - 0.77734375(1)) \approx 0.0556641.$$

Thus, in step 4 we have

$$x = 0 + 0.25 = 0.25,$$

and

$$y \approx 1 + \frac{1}{6}[0 + 2(0.03125) + 2(0.0273438) + 0.0556641] \approx 1.02881.$$

Thus the approximate value of the solution at 0.25 is 1.02881. By repeating steps 3 and 4 of the algorithm we fill in the following Table 3–E.

Table 3-E. Fourth order Runge-Kutta subroutine approximations for $y' = x + 1 - y$ at $x = 1$ with $h = 0.25$.

x	0	0.25	0.50	0.75	1.0
y	1	1.02881	1.10654	1.22238	1.36789

Thus, our approximation at $x = 1$ is approximately 1.36789. Comparing this with Problem 5, we see we have obtained accuracy to four decimal places as we did with the Taylor method of order four, but without having to compute any partial derivatives.

13. For this problem $f(x, y) = y^2 - 2e^x y + e^{2x} + e^x$. We want to find the vertical asymptote located in the interval [0,2] within two decimal places of accuracy using the Forth Order Runge-Kutta subroutine. One approach is to use a step size of 0.005 and look for y to approach infinity. This would require 400 steps. We will stop the subroutine when the value of y ("blows up") becomes very large. For step 3 we have

$$k_1 = hf(x, y) = 0.005\left(y^2 - 2e^x y + e^{2x} + e^x\right),$$

$$k_2 = hf\left(x + \frac{h}{2}, y + \frac{k_1}{2}\right) = 0.005\left[\left(y + \frac{k_1}{2}\right)^2 - 2e^{(x+h/2)}\left(y + \frac{k_1}{2}\right) + e^{2(x+h/2)} + e^{(x+h/2)}\right],$$

$$k_3 = hf\left(x + \frac{h}{2}, y + \frac{k_2}{2}\right) = 0.005\left[\left(y + \frac{k_2}{2}\right)^2 - 2e^{(x+h/2)}\left(y + \frac{k_2}{2}\right) + e^{2(x+h/2)} + e^{(x+h/2)}\right],$$

$$k_4 = hf(x+h, y+k_3) = 0.005\left[(y+k_3)^2 - 2e^{(x+h)}(y+k_3) + e^{2(x+h)} + e^{(x+h)}\right].$$

Hence in step 4 we have

$$x = x + 0.5,$$

$$y = y + \frac{1}{6}(k_1 + 2k_2 + 2k_3 + k_4).$$

Using the initial conditions $x_0 = 0$, $y_0 = 3$, $c = 2$ and $N = 400$ for step 3 we have

$$k_1 = 0.005(3^2 - 2e^0(3) + e^{2(0)} + e^0) = 0.025,$$

$$k_2 = 0.005\left[(3+0.0125)^2 - 2e^{(0+0.0025)}(3+0.0125) + e^{2(0+0.0025)} + e^{(0+0.0025)}\right] \approx 0.025213,$$

$$k_3 = 0.005\left[(3+0.0126065)^2 - 2e^{(0+0.0025)}(3+0.0126065) + e^{2(0+0.0025)} + e^{(0+0.0025)}\right] \approx 0.0252151,$$

$$k_4 = 0.005\left[(3+0.0252151)^2 - 2e^{(0+0.0025)}(3+0.0252151) + e^{2(0+0.0025)} + e^{(0+0.0025)}\right] \approx 0.0254312.$$

Thus in step 4 we have

$$x = 0 + 0.005$$

and

$$y \approx 3 + \frac{1}{6}(0.025 + 2(0.0252213) + 2(0.025215) + 0.0254312) \approx 3.0252145.$$

Thus the approximate value at $x = 0.005$ is 3.0252145. By repeating steps 3 and 4 of the subroutine we find that at $x = 0.505$, $y = 2.0201 \cdot 10^{13}$. The next iteration gives a floating point overflow. This would lead one to think the asymptote occurs at $x = 0.51$.

As a check lets apply the Fourth Order Runge-Kutta subroutine with the initial conditions $x_0 = 0$, $y_0 = 3$, $c = 1$ and $N = 400$. This gives a finer step size of $h = 0.0025$. With these inputs we find $y(0.05025) = 4.0402 \cdot 10^{13}$.

Repeating the subroutine one more time with a step size of 0.00125 yields $y(0.50125) = 8.0804 \cdot 10^{13}$. Therefore we conclude that the vertical asymptote occurs at $x = 0.50$ and not at 0.51 as was earlier thought.

17. Taylor's method of order 2 has recursive formulas given by equations (5) and (6) on page 136 of the text: that is

$$x_{j+1} = x_j + h \qquad \text{and} \qquad y_{j+1} = y_j + hf(x_j, y_j) + \frac{h^2}{2!} f_2(x_j, y_j),$$

when $f(x, y) = y$, then

$$f_2(x, y) = y'' = \frac{\partial f(x,y)}{\partial x} + \left[\frac{\partial f(x,y)}{\partial y}\right] f(x, y) = 0 + 1 \cdot (y) = y.$$

Therefore, since $h = \frac{1}{n}$, the recursive formula for y_{j+1} is given by the equation

$$y_{j+1} = y_j + \frac{1}{n} y_j + \frac{1}{2n^2} y_j = \left(1 + \frac{1}{n} + \frac{1}{2n^2}\right) y_j .$$

We are starting the process at $x_0 = 0$ and $y_0 = 1$, and we are taking steps of size $\frac{1}{n}$ until we reach $x = 1$. This means that we will take n steps. Thus, y_n will be an approximation for the solution to the differential equation at $x = 1$. Since the actual solution is given by the equation $y = e^x$, this means that $y_n \approx e$. To find the equation we are looking for, we see that

$$y_1 = \left(1 + \frac{1}{n} + \frac{1}{2n^2}\right) y_0 = \left(1 + \frac{1}{n} + \frac{1}{2n^2}\right),$$

$$y_2 = \left(1 + \frac{1}{n} + \frac{1}{2n^2}\right) y_1 = \left(1 + \frac{1}{n} + \frac{1}{2n^2}\right)^2,$$

$$y_3 = \left(1 + \frac{1}{n} + \frac{1}{2n^2}\right) y_2 = \left(1 + \frac{1}{n} + \frac{1}{2n^2}\right)^3,$$

$$\vdots$$

and, in general,

$$y_n = \left(1 + \frac{1}{n} + \frac{1}{2n^2}\right) y_{n-1} = \left(1 + \frac{1}{n} + \frac{1}{2n^2}\right)^n .$$

(This can be proved rigorously by mathematical induction.) As we observed above, $y_n \approx e$ and so we have $e \approx \left(1 + \frac{1}{n} + \frac{1}{2n^2}\right)^n$.

CHAPTER 4: Linear Second Order Equations

EXERCISES 4.2: Linear Differential Operators, page 166

3. This equation is nonlinear because the coefficient of y' is y, not a function of x alone.

9. **(a)** Here we want L to operate on the function $y = \cos x$. Therefore, we need to find $y' = -\sin x$ and $y'' = -\cos x$. Plugging these results into the appropriate places in the operator L, we have

$$L[\cos x] = x^2(-\cos x) - 3x(-\sin x) - 5(\cos x) = 3x\sin x - \left(x^2 + 5\right)\cos x .$$

(b) This time $y = x^{-1}$, so $y' = -x^{-2}$ and $y'' = 2x^{-3}$. Plugging these values into the operator L yields

$$L\left[x^{-1}\right] = x^2\left(2x^{-3}\right) - 3x\left(-x^{-2}\right) - 5\left(x^{-1}\right) = 2x^{-1} + 3x^{-1} - 5x^{-1} = 0.$$

(c) Here $y = x^r$ and so $y' = rx^{r-1}$ and $y'' = r(r-1)x^{r-2}$. Plugging these values into the operator L yields

$$L\left[x^r\right] = x^2\left(r(r-1)x^{r-2}\right) - 3x\left(rx^{r-1}\right) - 5\left(x^r\right) = \left(r^2 - 4r - 5\right)x^r .$$

11. To show that T is a nonlinear operator, we will show that property (7) of Lemma 1 does not hold. Let's try $y(x) \equiv 1$. Then $T[cy] = T[c] = 0 - 0 + c^2 = c^2$, but $cT[y] = cT[1] = c(0 - 0 + 1) = c$. Since $c^2 \neq c$ for all constants c, property (7) does not hold.

13. From Theorem 1 on page 163 any linear combination

$$y = c_1 e^{2x}\cos x + c_2 e^{2x}\sin x$$

with c_1 and c_2 as arbitrary constants, is a solution to the differential equation. We want to select c_1 and c_2 to satisfy the initial conditions. Differentiating the above equation we find

$$y' = c_1(2e^{2x}\cos x - e^{2x}\sin x) + c_2(2e^{2x}\sin x + e^{2x}\cos x) .$$

(a) Substituting the initial conditions, $y(0) = 2$ and $y'(0) = 1$, into the above equations we obtain the following system of equations

$$y(0) = 2 = c_1$$

$$y'(0) = 1 = 2c_1 + c_2 .$$

The solution is given by $c_1 = 2$ and $c_2 = -3$. Hence the solution to the differential equation that satisfies the initial conditions is

$$y(x) = 2e^{2x}\cos x - 3e^{2x}\sin x .$$

(b) Substituting the initial conditions, $y(\pi)=4e^{2\pi}$ and $y'(\pi)=5e^{2\pi}$, into the above equations we obtain the following system of equations

$$y(\pi)=4e^{2\pi}=c_1 e^{2\pi}(-1)$$

$$y'(\pi)=e^{2\pi}=c_1 2e^{2\pi}(-1)+c_2 e^{2\pi}(-1).$$

The solution to this system of equations is given by $c_1=-4$ and $c_2=3$. Thus the solution to the differential equations that satisfies the initial conditions is

$$y(x)=-4e^{2x}\cos x+3e^{2x}\sin x.$$

17. In standard form this equation becomes

$$y''-\left[e^x(x-3)\right]^{-1}y'+e^{-x}y=e^{-x}\ln x.$$

The term $\left[e^x(x-3)\right]^{-1}$ is continuous at every real number except 3. The term e^{-x} is continuous everywhere and the term $e^{-x}\ln x$ is continuous for all $x>0$. Therefore, Theorem 2 on page 165 of the text applies on the interval (0, 3) and on the interval (3, ∞). Since we must have the initial point $x_0=1$ in the interval, the interval we want is (0, 3); that is, there will be a unique solution on (0, 3).

25. Since $L[y]=\left(D^2+x\right)[y]$, we get $L=D^2+x$.

27. Here $L[y]=\left(D^2-xD+2\right)[y]=y''-xy'+2y.$

(a) If $y=x^2$, then we have $y'=2x$ and $y''=2$. Therefore,

$$L[y]=2-x(2x)+2x^2=2.$$

(b) If $y=\cos x$, we have $y'=-\sin x$ and $y''=-\cos x$. This means that

$$L[y]=-\cos x-x(-\sin x)+2(\cos x)=\cos x+x\sin x.$$

29. To see that L is linear, we need to show that properties (6) and (7) of Lemma 1 are satisfied. For property (6) we have

$$L[y_1+y_2]=\int_0^1(x-t)^2[y_1(t)+y_2(t)]\,dt+x^2\left(y_1(x)+y_2(x)\right)'$$

$$=\int_0^1(x-t)^2y_1(t)\,dt+x^2y_1'(x)+\int_0^1(x-t)^2y_2(t)\,dt+x^2y_2'(x)$$

$$=L[y_1]+L[y_2].$$

Thus property (6) holds. Next we consider property (7).

$$L[cy]=\int_0^1 (x-t)^2 cy\,dt + x^2(cy)' = c\left\{\int_0^1 (x-t)^2 y\,dt + x^2 y'\right\} = cL[y].$$

Property (7) also holds, therefore the integro-differential operator L is linear.

EXERCISES 4.3: Fundamental Solutions of Homogeneous Equations, page 176

3. To see whether $y_1(x)=xe^{2x}$ and $y_2(x)=e^{2x}$ are linearly dependent, we must examine the equation

$$c_1 xe^{2x} + c_2 e^{2x} = 0$$

$$\Rightarrow \qquad e^{2x}(c_1 x + c_2) = 0.$$

Since $e^{2x} \neq 0$ for all x, we must have $c_1 x + c_2 = 0$ for all x in (0, 1). But this can be true only if $c_1 = c_2 = 0$. Thus y_1 and y_2 are linearly independent. (Alternatively, it is clear that neither function is a constant multiple of the other.)

To find the Wronskian for y_1 and y_2, we see that $y_1'(x) = 2xe^{2x} + e^{2x}$ and $y_2'(x) = 2e^{2x}$. Thus, the Wronskian becomes

$$W[y_1, y_2](x) = y_1(x)y_2'(x) - y_1'(x)y_2(x) = xe^{2x}\left(2e^{2x}\right) - \left(2xe^{2x} + e^{2x}\right)e^{2x} = e^{4x}(2x - 2x - 1) = -e^{4x}.$$

7. **(a)** To verify that $y_1 = x^2$ is a solution to the ODE, we must find y_1' and y_1''. Thus, we compute

$$y_1'(x) = 2x\,,$$

$$y_1''(x) = 2.$$

By plugging these functions into the ODE, we obtain

$$x^2 y_1'' - 2y_1 = x^2(2) - 2\left(x^2\right) = 0.$$

Thus, y_1 is a solution to the ODE. In the same way, we can verify that y_2 is also a solution to the ODE.

It is intuitively clear that neither solution is a constant multiple of the other, and hence they are linearly independent. To give a rigorous argument we can proceed as follows. Since y_1 and y_2 are solutions to our differential equation, by Theorem 4 on page 173 of the text, we can show that they are linearly independent by showing that their initial vectors are linearly independent. This can be done, as explained on page 173 of the text, by showing that the determinant of the initial vectors is not zero. Since $y_2' = -x^{-2}$, the initial vectors of y_1 and y_2 at the point $x = 1$ are

$$\left(y_1(1), y_1'(1)\right) = \left((1), 2(1)\right) = (1, 2) \qquad \text{and} \qquad \left(y_2(1), y_2'(1)\right) = \left((1)^1, -(1)^{-2}\right) = (1, -1).$$

Thus their determinant is given by

$$(1\times(-1))-(1\times 2)=-3\neq 0.$$

Therefore, y_1 and y_2 are linearly independent and so form a fundamental set of solutions for the differential equation.

(b) As we saw above, y_1 and y_2 form a fundamental set for the differential equation and so a general solution is given by the equation

$$y(x)=c_1x^2+c_2x^{-1}.$$

(c) To find the solution that satisfies the initial conditions, we must first find the derivative of the general solution. This is given by

$$y'(x)=2c_1x-c_2x^{-2}.$$

Plugging the initial conditions $y(1)=-2$ and $y'(1)=-7$ into the equations for y and y' yields the system

$$c_1+c_2=-2 \qquad \text{and} \qquad 2c_1-c_2=-7.$$

Adding yields

$$3c_1=-9 \qquad \Rightarrow \qquad c_1=-3 \qquad \text{and} \qquad c_2=1.$$

Thus, the solution we seek is $y=-3x^2+x^{-1}$.

13. **(a)** We must first show that $y_1(x)=e^x$ and $y_2(x)=e^x-e^{-6x}$ are solutions to the ODE. Plugging y_1, y_1', and y_1'' into the given equation gives

$$y_1''+5y_1'-6y_1=e^x+5e^x-6e^x=0.$$

Thus, y_1 is a solution. Similarly, plugging in $y_2(x)$, $y_2'(x)=e^x+6e^{-6x}$, and $y_2''(x)=e^x-36e^{-6x}$ into the given equation yields

$$y_2''+5y_2'-6y_2=e^x-36e^{-6x}+5e^x+30e^{-6x}-6e^x+6e^{-6x}=0.$$

Thus y_2 is also a solution to the ODE. Next we must show that y_1 and y_2 are linearly independent or, equivalently that the Wronskian of y_1 and y_2 is not equal to zero. To this end, we have

$$W\left[e^x,e^x-e^{-6x}\right]=e^x\left(e^x+6e^{-6x}\right)-e^x\left(e^x-e^{-6x}\right)=7e^{-5x}\neq 0.$$

Thus $S_1:=\{e^x,e^x-e^{-6x}\}$ is a fundamental solution set for the ODE.

(b) As above, we can show that $y_3(x)=3e^x+e^{-6x}$ is a solution to the ODE. To see if y_1 and y_3 form another fundamental set of solutions to our equation, we must check that their Wronskian is not equal to zero. Thus, we have

$$W\left[e^x,3e^x+e^{-6x}\right]=e^x\left(3e^x-6e^{-6x}\right)-e^x\left(3e^x+e^{-6x}\right)=-7e^{-5x}\neq 0.$$

Hence, $S_2 := \{e^x, 3e^x + e^{-6x}$ is another fundamental solution set to the equation. Note that fundamental solution sets are not unique.

(c) To see that $\phi(x) = e^{-6x}$ is a solution, we substitute $\phi(x)$, $\phi'(x) = -6e^{-6x}$, and $\phi''(x) = 36e^{-6x}$ into the equation to obtain

$$\phi''(x) + 5\phi'(x) - 6\phi(x) = 36e^{-6x} - 30e^{-6x} - 6e^{-6x} = 0.$$

Thus, $\phi(x)$ is also a solution to our ODE. Therefore, by Theorem 3 on page 169 of the text, we can express $\phi(x)$ as a linear combination of the solutions in S_1. Hence, we need to find c_1 and c_2 so that

$$e^{-6x} = c_1 e^x + c_2\left(e^x - e^{-6x}\right) = \left(c_1 + c_2\right)e^x - c_2 e^{-6x}.$$

By equating coefficients we see that

$$c_1 + c_2 = 0 \qquad \text{and} \qquad -c_2 = 1$$

$$\Rightarrow \qquad c_1 = 1 \qquad \text{and} \qquad c_2 = -1.$$

This yields $\phi(x) = e^{-6x} = (1)e^x + (-1)\left(e^x - e^{-6x}\right)$.

In the same way, we can express $\phi(x)$ as a linear combination of the solutions given in S_2. That is, we have

$$e^{-6x} = c_3 e^x + c_4\left(3e^x + e^{-6x}\right) = \left(c_3 + 3c_4\right)e^x + c_4 e^{-6x}.$$

Equating coefficients yields

$$c_3 + 3c_4 = 0 \qquad \text{and} \qquad c_4 = 1$$

$$\Rightarrow \qquad c_3 = -3 \qquad \text{and} \qquad c_4 = 1.$$

Thus, we have

$$\phi(x) = e^{-6x} = (-3)e^x + (1)\left(3e^x + e^{-6x}\right).$$

15. **(a)** The function $w(x) = 6e^{4x}$ can be the Wronskian of solutions because $w(x) \neq 0$ on the interval $-1 < x < 1$.

(b) The function $w(x) = x^3$ cannot be a Wronskian of solutions because $w(0) = 0$ but w is not identically zero.

(c) The function $w(x) = (x+1)^{-1}$ can be the Wronskian of solutions because $w(x) \neq 0$ on the interval $-1 < x < 1$.

(d) The function $w(x) \equiv 0$ can be the Wronskian of solutions (in fact linearly dependent ones).

17. **(a)** True. If y_1 and y_2 are linearly dependent on $[a, b]$, there exist constants c_1 and c_2 not both zero such that

$$c_1y_1(x) + c_2y_2(x) = 0,$$

for all x in $[a,b]$ and thus for all x in $[c,d] \subset [a,b]$. Therefore, y_1 and y_2 are also linearly dependent on $[c,d]$.

(b) False. Let

$$y_1(x) = \begin{cases} x, & x \le 2 \\ 2, & x > 2 \end{cases} \qquad \text{and} \qquad y_2(x) = x .$$

These two functions are linearly dependent on [0, 2] since they are equal there. But they are linearly independent on [0, 5] because $y_2(x)$ is not a constant multiple of $y_1(x)$ for *all* x in [0, 5].

18. **(a)** Here the two functions $y_1(x) = x^3$ and $y_2(x) = |x^3|$ are identical on the interval $[0, \infty)$. Therefore y_1 and y_2 are linearly dependent on $[0, \infty)$.

(b) Here the two functions are not identical but we see that $y_1(x) = -y_2(x)$ on the interval $(-\infty, 0]$. Hence y_1 and y_2 are linearly dependent on $(-\infty, 0]$.

(c) Although y_1 and y_2 are linearly dependent on the two separate intervals $[0, \infty)$ and $(-\infty, 0]$, the functions are linearly independent on the interval $(-\infty, \infty)$. This is because neither function is a fixed constant multiple of the other for all x.

(d) From Definition 1 the Wronskian is defined to be

$$W[y_1, y_2](x) := y_1(x)y_2'(x) - y_1'(x)y_2(x).$$

For $x \ge 0$ we have $y_2(x) = x^3$. Thus

$$W[y_1, y_2](x) = x^3\left(3x^2\right) - \left(3x^2\right)x^3 = 0 \qquad \text{for all } x \ge 0.$$

For $x < 0$ we have $y_2(x) = -x^3$. Thus

$$W[y_1, y_2](x) = x^3\left(-3x^2\right) - 3x^2\left(-x^3\right) = 0 \qquad \text{for } x < 0.$$

Hence $W[y_1, y_2](x) = 0$ for all $x \in (-\infty, \infty)$.

21. We will assume that p and q are continuous on (a,b). Since y_1 and y_2 are linearly independent solutions of $y'' + py' + qy = 0$ on (a,b), we know, by Corollary 1 on page 174 of the text, that the Wronskian is never zero on (a,b). We will prove by contradiction that y_1 and y_2 cannot both be zero at the same point in (a,b). That is we will assume that there exists a point x_0 in (a,b) such that $y_1(x_0) = 0 = y_2(x_0)$. Then we have for this x_0 in (a,b) that

$$W[y_1, y_2](x_0) = y_1(x_0)y_2'(x_0) - y_1'(x_0)y_2(x_0) = 0 \cdot y_2'(x_0) - y_1'(x_0) \cdot 0 = 0.$$

But this is a contradiction to the fact, as stated above, that $W[y_1, y_2](x) \ne 0$ for all x in (a,b). Thus, there can be no such point x_0 in (a,b).

25. From Problem 24 we have that three functions are said to be linearly dependent on an interval I if there exists constants, C_1, C_2 and C_3, not all zero such that

$$C_1 y_1(x) + C_2 y_2(x) + C_3 y_3(x) = 0$$

for all x in I. Notice that this is equivalent to saying that one of the three functions is a linear combination of the other two.

(a) For $y_1 = 1$, $y_2 = x$ and $y_3 = x^2$ it is clear that none of the functions is a linear combination of the other two on $(-\infty, \infty)$. To be more rigorous suppose that

$$C_1 + C_2 x + C_3 x^2 = 0 \qquad \text{for all } x \in (-\infty, \infty).$$

There are several ways to show that this equation implies that $C_1 = C_2 = C_3 = 0$. One way is to choose three values for x, say $x = 0$, $x = 1$, and $x = 2$, and substitute them into the equation to obtain a system of three linear homogeneous equations in the three unknowns C_1, C_2 and C_3. One then solves for C_1, C_2 and C_3 and obtains $C_1 = C_2 = C_3 = 0$. A quicker way is to differentiate twice the equation

$$C_1 + C_2 x + C_3 x^2 = 0$$

to obtain

$$C_2 + 2C_3 x = 0,$$

$$2C_3 = 0.$$

We immediately get $C_3 = C_2 = C_1 = 0$, so the functions are linearly independent.

(c) For $y_1 = e^x$, $y_2 = xe^x$ and $y_3 = x^2 e^x$ we suppose that

$$C_1 e^x + C_2 x e^x + C_3 x^2 e^x = 0 \qquad \text{for all } x \in (-\infty, \infty).$$

Dividing by e^x (which is never zero) we get

$$C_1 + C_2 x + C_3 x^2 = 0 \qquad \text{for all } x \in (-\infty, \infty).$$

But from part (a) we know that the last equation implies that $C_1 = C_2 = C_3 = 0$. Hence $y_1(x)$, $y_2(x)$, and $y_3(x)$ are linearly independent on $(-\infty, \infty)$.

EXERCISES 4.4: Reduction of Order, page 181

4. This equation is of the form

$$ax^2 y'' + bxy' + cy = g(x),$$

where a, b, and c are real constants. An equation of this type is called a Cauchy-Euler equation. Cauchy-Euler equations will be studied in Section 4.5 and again in Section 8.5 of the text. We will first solve this equation by assuming the solution we seek is of the form

$$y = v(x)f(x) = v(x)x^{-2}.$$

Thus, we have

$$y' = v'x^{-2} - 2vx^{-3} \qquad \text{and} \qquad y'' = v''x^{-2} - 4v'x^{-3} + 6vx^{-4}.$$

Substituting these expressions into the differential equation yields

$$x^2y'' + 6xy' + 6y = v'' - 4v'x^{-1} + 6vx^{-2} + 6v'x^{-1} - 12vx^{-2} + 6vx^{-2} = 0$$

$$\Rightarrow \qquad v'' + 2v'x^{-1} = 0.$$

(Note that the coefficient of the v term is zero. This must always be the case at this stage when we use the reduction of order method. Therefore, this serves as a partial check.) We make the substitution $w = v'$ in the above equation and use the fact that $x > 0$ to obtain

$$w' + 2wx^{-1} = 0$$

$$\Rightarrow \qquad \frac{dw}{w} = -2\frac{dx}{x} \qquad \Rightarrow \qquad \ln w = \ln x^{-2} + C \qquad \Rightarrow \qquad w = C_1x^{-2},$$

where we have solved the differential equation by the method of separation of variables. Next, we make the substitution $v' = w$ and see that

$$v' = C_1x^{-2} \qquad \Rightarrow \qquad v = C_2x^{-1} + C_3.$$

We let $C_2 = 1$ and $C_3 = 0$ to obtain $v = x^{-1}$. This implies that a linearly independent solution is given by

$$y = v(x)x^{-2} = x^{-1}x^{-2} = x^{-3}.$$

Alternatively, we can solve the original differential equation by using the reduction of order formula on page 179 of the text. By first writing the equation in standard form, we see that $p(x) = 6x^{-1}$. Thus, we have

$$\exp\left(-\int 6x^{-1}dx\right) = \exp\left(-6\ln|x| + C\right) = C_1x^{-6}.$$

Therefore, using the fact that $x > 0$, we see that

$$y(x) = x^{-2}\int C_1\frac{x^{-6}}{x^{-4}}dx = C_1x^{-2}\int x^{-2}dx = C_2x^{-2}x^{-1} + C_3x^{-2}$$

$$\Rightarrow \qquad y(x) = C_2x^{-3} + C_3x^{-2}.$$

Again we let $C_2 = 1$ and $C_3 = 0$ to obtain the linearly independent solution $y(x) = x^{-3}$.

5. We will first find a second solution by letting $x = v(t)f(t) = v(t)e^t$. Then

$$x' = ve^t + v'e^t$$

$$x'' = ve^t + 2v'e^t + v''e^t.$$

Substituting these expressions into the differential equation gives

$$t(ve^t + 2v'e^t + v''e^t) - (t+1)(ve^t + v'e^t) + ve^t = 0,$$

which simplifies to $tv'' + (t-1)v' = 0$. Next we make the substitution $w = v'$ to obtain

$$tw' + (t-1)w = 0.$$

Solving this equation by separation of variables yields

$$\frac{dw}{w} = \left(\frac{1}{t} - 1\right)dt$$

$$\Rightarrow \qquad \ln w = \ln t - t + C \qquad \Rightarrow \qquad w = C_1 te^{-t}.$$

Thus

$$v' = C_1 te^{-t} \qquad \Rightarrow \qquad v = C_1 \int te^{-t}\, dt = C_1 e^{-t}(-t-1) + C_2.$$

We set $C_1 = -1$ and $C_2 = 0$ to obtain $v = e^{-t}(t+1)$. This implies that a linearly independent is given by

$$x = v(t)e^t = t+1.$$

(It's a good idea to check this answer by substituting into the original differential equation.)

As an alternative approach we can use the reduction of order formula (6) on page 179. By first writing the equation in standard form, we have $p(t) = -\left(\frac{t+1}{t}\right)$. Thus we have

$$e^{\int \frac{t+1}{t} dt} = e^{(t + \ln t + C)} = C_1 te^t.$$

Therefore, using the fact that $t > 0$, we see that

$$x(t) = e^t \int C_1 \frac{te^t}{e^{2t}}\, dt = C_1 e^t \int te^{-t}\, dt = -C_1(t+1) + C_2 e^t.$$

Choosing $C_1 = -1$ and $C_2 = 0$, we again obtain the linearly independent solution $x(t) = t+1$.

9. **(a)** In this problem, $p(x) = -x$. Therefore, we see that

$$\exp\left(\int x\, dx\right) = \exp\left(\frac{x^2}{2} + C\right).$$

Taking $C = 0$, and plugging this expression into the reduction of order formula yields

$$y(x) = x\int x^{-2}e^{x^2/2}dx\,.$$

(b) As stated on page 181 of the text, the Maclaurin expansion for e^x is

$$e^x = 1 + x + \frac{x^2}{2!} + \frac{x^3}{3!} + \cdots + \frac{x^n}{n!} + \cdots = \sum_{n=0}^{\infty}\frac{x^n}{n!}.$$

Thus, we have

$$e^{x^2/2} = 1 + \frac{x^2}{2} + \frac{x^4}{2^2(2!)} + \frac{x^6}{2^3(3!)} + \cdots + \frac{2n}{2^n(n!)} + \cdots = \sum_{n=0}^{\infty}\frac{x^{2n}}{2^n(n!)}$$

$$\Rightarrow \qquad x^{-2}e^{x^2/2} = x^{-2} + \frac{1}{2} + \frac{x^2}{2^2(2!)} + \frac{x^4}{2^3(3!)} + \cdots + \frac{x^{2n-2}}{2^n(n!)} + \cdots = \sum_{n=0}^{\infty}\frac{x^{2n-2}}{2^n(n!)}.$$

Therefore, $y(x)$ is given by the series

$$y(x) = x\left\{\sum_{n=0}^{\infty}\int\frac{x^{2n-2}}{2^n(n!)}dx\right\} = x\left\{\sum_{n=0}^{\infty}\frac{x^{2n-1}}{2^n(2n-1)n!}\right\},$$

where we have assumed that we can integrate the series term by term. Thus, we have

$$y(x) = \sum_{n=0}^{\infty}\frac{x^{2n}}{2^n(2n-1)n!} = -1 + \frac{x^2}{2} + \frac{x^4}{24} + \frac{x^6}{240} + \cdots .$$

11. Assume that both f and y are solutions to the equation $y'' + p(x)y' + q(x)y = 0$. Using the quotient rule, $(y/f)'$ can be written as

$$\left(\frac{y}{f}\right)' = \frac{y'f - yf'}{f^2} = \frac{W[f,y]}{f^2}.$$

From Abel's identity on page 175 we know that

$$W[y_1, y_2](x) = C\exp\left(-\int p(x)dx\right),$$

so

$$\left(\frac{y}{f}\right)' = \frac{C\exp\left(-\int p(x)dx\right)}{f^2}.$$

Integrating with respect to x we obtain

$$\frac{y}{f}=\int \frac{C\exp\left(-\int p(x)dx\right)}{f^2}dx .$$

Multiplying by f yields

$$y(x)=f(x)\int \frac{C\exp\left(-\int p(x)dx\right)}{f^2}dx$$

and taking $C=1$ we obtain the reduction of order formula.

15. To find $p(x)$ we first need to write Laguerre's equation in standard form by multiplying by x^{-1} to get

$$y''+x^{-1}(1-x)y'+x^{-1}\lambda y=0 .$$

This gives $p(x)=x^{-1}(1-x)$.

(a) Using the reduction of order formula, equation (6) on page 179 of the text, with $p(x)$ defined above and $f(x)=x-1$ we get

$$y(x)=(x-1)\int \frac{e^{-\int\frac{1-x}{x}dx}}{(x-1)^2}dx=(x-1)\int \frac{e^{-\int\left(\frac{1}{x}-1\right)dx}}{(x-1)^2}dx = (x-1)\int\frac{e^{-\ln x+x}}{(x-1)^2}dx ;$$

$$y(x)=(x-1)\int e^{x}x^{-1}(x-1)^{-2}dx .$$

(b) Repeating the same procedure used in part (a) with $f(x)=x^2-4x+2$ we get

$$y(x)=(x^2-4x+2)\int \frac{e^{-\int\frac{1-x}{x}dx}}{(x^2-4x+2)^2}dx =(x^2-4x+2)\int e^{x}x^{-1}(x^2-4x+2)^{-2}dx .$$

EXERCISES 4.5: Homogeneous Linear Equations with Constant Coefficients, page 189

1. The auxiliary equation for this problem is $r^2+5r+6=(r+2)(r+3)=0$, which has the roots $r=-2$ and $r=-3$. Thus $\{e^{-2x},e^{-3x}\}$ is a fundamental solution set for this differential equation. Therefore a general solution is

$$y(x)=c_1e^{-2x}+c_2e^{-3x},$$

where c_1 and c_2 are arbitrary constants.

5. The auxiliary equation for this problem is $r^2+r-1=0$. By the quadratic formula, we have

$$r=\frac{-1\pm\sqrt{1+4}}{2}=\frac{-1\pm\sqrt{5}}{2}.$$

Therefore, a general solution is given by

$$z(x)=c_1e^{\left(-1+\sqrt{5}\right)x/2}+c_2e^{\left(-1-\sqrt{5}\right)x/2}.$$

15. The auxiliary equation for this equation is $r^2+2r+1=(r+1)^2=0$. We see that $r=-1$ is a repeated root. Thus, two linearly independent solutions are $y_1(x)=e^{-x}$ and $y_2(x)=xe^{-x}$. This means that a general solution is given by

$$y(x)=c_1e^{-x}+c_2xe^{-x}.$$

To find the constants c_1 and c_2, we substitute the initial conditions into the general solution and its derivative $y'(x)=-c_1e^{-x}+c_2\left(e^{-x}-xe^{-x}\right)$ to obtain

$$y(0)=1=c_1e^0+c_2\cdot 0=c_1$$

and

$$y'(0)=-3=-c_1e^0+c_2\left(e^0-0\right)=-c_1+c_2.$$

We see that $c_1 = 1$ and $c_2 = -2$. Therefore, the solution that satisfies the initial conditions is given by

$$y(x)=e^{-x}-2xe^{-x}.$$

19. Here the auxiliary equation is $r^2-4r-5=(r-5)(r+1)=0$, which has roots $r = 5$, and $r = -1$. Consequently, a general solution to the differential equation is

$$y(x)=c_1e^{5x}+c_2e^{-x},$$

where c_1 and c_2 are arbitrary constants. To find the solution that satisfies the initial conditions $y(-1)=3,\ y'(-1)=9$, we first differentiate the solution found above, then plug in y and y' into the initial conditions. This gives

$$y(-1)=3=c_1e^{-5}+c_2e^{1}$$

$$y'(-1)=9=5c_1e^{-5}-c_2e^{1}.$$

Solving this system yields $c_1=2e^5$ and $c_2=e^{-1}$. Thus

$$y(x)=2e^{5(x+1)}+e^{-(x+1)}$$

is the desired solution.

27. The auxiliary equation associated with this differential equation is given by $r^3-6r^2-r+6=0$. We see, by inspection, that $r = 1$ is a root. Dividing the cubic polynomial by $r - 1$ we find

$$r^3-6r^2-r+6=(r-1)\left(r^2-5r-6\right)=(r-1)(r+1)(r-6).$$

Hence $r = -1, 1, 6$ are the roots to the auxiliary equation. Thus, a general solution is given by

$$y(x) = c_1 e^{-x} + c_2 e^{x} + c_3 e^{6x}.$$

35. First multiply this equation through by x^2 (which we can do since $x > 0$) to transform it into the Cauchy-Euler equation given by

$$x^2 w''(x) + 6xw'(x) + 4w(x) = 0.$$

Then by making the substitution $x = e^t$ (and using equation (17) on page 188 of the text), we transform the Cauchy-Euler equation into an equation with constant coefficients given by

$$\left(\frac{d^2 w}{dt^2} - \frac{dw}{dt}\right) + 6\frac{dw}{dt} + 4w = 0,$$

or

$$\frac{d^2 w}{dt^2} + 5\frac{dw}{dt} + 4w = 0. \qquad (1)$$

This is a linear equation with constant coefficients and has the associated auxiliary equation $r^2 + 5r + 4 = 0$, which has roots $r = -1, -4$. Therefore, a general solution to equation (1) is

$$w(t) = c_1 e^{-t} + c_2 e^{-4t} = c_1 \left(e^t\right)^{-1} + c_2 \left(e^t\right)^{-4}.$$

To change this equation back into one with the independent variable x, we again use the substitution $x = e^t$. Therefore, the solution becomes

$$w(x) = c_1 x^{-1} + c_2 x^{-4}.$$

39. The differential equation $x^2 y'' - 4xy' + 4y = 0$ is a Cauchy-Euler equation. To solve this equation we substitute $x = e^t$ to transform the differential equation into an equation with a new independent variable t. With $x = e^t$, it follows by the chain rule (see equations (12) and (13) on pages 187 and 188 of the text) that

$$x\frac{dy}{dx} = \frac{dy}{dt} \qquad \text{and} \qquad x^2 \frac{d^2 y}{dx^2} = \frac{d^2 y}{dt^2} - \frac{dy}{dt}.$$

Substituting into the differential equation for $x\dfrac{dy}{dx}$ and $x^2 \dfrac{d^2 y}{dx^2}$ yields

$$\left(\frac{d^2 y}{dt^2} - \frac{dy}{dt}\right) - 4\left(\frac{dy}{dt}\right) + 4y = 0$$

$$\Rightarrow \quad \frac{d^2 y}{dt^2} - 5\frac{dy}{dt} + 4y = 0 \qquad \text{(compare with equation (17), page 188, of the text).} \qquad (2)$$

The auxiliary equation for the above is

$$r^2 - 5r + 4 = (r-1)(r-4) = 0 .$$

The roots of this equation are $r = 1$ and $r = 4$. Therefore the general solution to equation (2) is

$$y(t) = c_1 e^t + c_2 e^{4t} .$$

Expressing y in terms of the original variable x results in

$$y(x) = c_1 x + c_2 x^4 , \qquad \text{for } x > 0,$$

where c_1 and c_2 are arbitrary constants. To find the solution that satisfies the initial conditions $y(1) = -2$ and $y'(1) = -11$, we first differentiate the solution found above, then plug in our initial conditions. This gives

$$y(1) = -2 = c_1 + c_2 ,$$

$$y'(1) = -11 = c_1 + 4c_2 .$$

Solving this system yields $c_1 = 1$ and $c_2 = -3$. Thus

$$y(x) = x - 3x^4$$

is the desired solution.

45. Using the method of Problem 43, we first want to solve this differential equation in the case when $x > 0$. To do this we will make the substitution $x = e^t$. Using equation (17) on page 188, we obtain

$$2\frac{d^2y}{dt^2} + (7-2)\frac{dy}{dt} + 2y = 0 ,$$

or

$$2\frac{d^2y}{dt^2} + 5\frac{dy}{dt} + 2y = 0.$$

This equation has the associated auxiliary equation $2r^2 + 5r + 2 = 0$, which has the roots $r = -\frac{1}{2}, -2$. Therefore, a general solution of this constant coefficient equation is given by

$$y(t) = c_1 e^{-2t} + c_2 e^{-t/2} .$$

To transform this into a solution of the original, variable coefficient equation with the independent variable $x\ (>0)$, we again use the substitution $x = e^t$ to obtain

$$y(x) = c_1 x^{-2} + c_2 x^{-1/2} , \qquad \text{for } x > 0 .$$

From Problem 21, we see that $\phi(x) := y(-x)$ is a solution to the original equation in the case when $x < 0$. Thus, the solution we are looking for is given by

$$\phi(x) = c_1 (-x)^{-2} + c_2 (-x)^{-1/2} , \qquad \text{for } x < 0 .$$

Note that $(-x)^{-1/2}$ makes sense since $-x>0$ when $x<0$. Also note that $(-x)^{-2}=x^{-2}$. Therefore, we can write this solution as

$$\phi(x)=c_1x^{-2}+c_2(-x)^{-1/2}, \qquad \text{for } x<0.$$

47. From Problem 46 we have that a differential equation of the form

$$ax^3y'''(x)+bx^2y''(x)+cxy'(x)+dy(x)=0 \qquad \text{for } x>0$$

translates into

$$ay'''(t)+(b-3a)y''(t)+(2a-b+c)y'(t)+dy(t)=0$$

using the substitution $x=e^t$. In our differential equation we have $a=1$, $b=-2$, $c=3$ and $d=-3$, which gives

$$y'''(t)-5y''(t)+7y'(t)-3y(t)=0.$$

The auxiliary equation for this is

$$r^3-5r^2+7r-3=0$$

with roots $r_1=r_2=1$ and $r_3=3$. Therefore, the general solution to this constant coefficient equation is given by

$$y(t)=c_1e^t+c_2te^t+c_3e^{3t}.$$

Expressing y in terms of the original variable x, that is, replacing e^t by x and t by $\ln x$, gives

$$y(t)=c_1x+c_2x\ln x+c_3x^3, \qquad \text{for } x>0$$

where c_1, c_2 and c_3 are arbitrary constants.

53. Damped vibration will be studied in more detail in Chapter 5. For a smooth, vibration-free ride we do not want the vertical displacement, $x(t)$, to oscillate. Therefore, we do not want the auxiliary equation of our differential equation to have complex roots. The auxiliary equation is given by

$$mr^2+br+k=0,$$

with roots $r=\dfrac{-b\pm\sqrt{b^2-4mk}}{2m}$, where we require $b^2-4mk\geq 0$ to avoid complex roots. For the first part of this problem, $m=1000$ and $k=3000$. Therefore, we want

$$b\geq\sqrt{4mk}=2000\sqrt{3}.$$

Thus, the minimum value of b which will eliminate oscillation in the vertical displacement is $b=2000\sqrt{3}$. Doubling the spring constant k means that we want

$$b\geq\sqrt{2\cdot 4mk}=\sqrt{2}\sqrt{4mk}.$$

From this we see that doubling the spring constant means that we must multiply b by a factor of $\sqrt{2}$.

EXERCISES 4.6: Auxiliary Equations with Complex Roots, page 198

5. This equation has auxiliary equation $r^2 + 4r + 6 = 0$. The roots to this auxiliary equation are

$$r = \frac{-4 \pm \sqrt{16 - 24}}{2} = -2 \pm \sqrt{2}\, i.$$

From this we see that $\alpha = -2$ and $\beta = \sqrt{2}$. Thus, a general solution to the differential equation is given by

$$w(x) = c_1 e^{-2x} \cos \sqrt{2}x + c_2 e^{-2x} \sin \sqrt{2}x.$$

7. The auxiliary equation for this problem is given by

$$4r^2 - 4r + 26 = 0 \qquad \Rightarrow \qquad 2r^2 - 2r + 13 = 0.$$

This equation has roots $r = \dfrac{2 \pm \sqrt{4 - 104}}{4} = \dfrac{1}{2} \pm \left(\dfrac{5}{2}\right) i$. Therefore, $\alpha = \dfrac{1}{2}$ and $\beta = \dfrac{5}{2}$. Thus, a general solution is given by

$$y(x) = c_1 e^{x/2} \cos\left(\frac{5x}{2}\right) + c_2 e^{x/2} \sin\left(\frac{5x}{2}\right).$$

11. The auxiliary equation for this problem is $r^2 + 10r + 25 = (r+5)^2 = 0$. We see that $r = -5$ is a repeated root. Thus two linearly independent solutions are $z_1(x) = e^{-5x}$ and $z_2(x) = xe^{-5x}$. This means a general solution is given by

$$z(x) = c_1 e^{-5x} + c_2 x e^{-5x},$$

where c_1 and c_2 are arbitrary constants.

23. The auxiliary equation for this problem is $r^2 - 4r + 2 = 0$. The roots to this equation are

$$r = \frac{4 \pm \sqrt{16 - 8}}{2} = 2 \pm \sqrt{2},$$

which are real numbers. A general solution is given by

$$w(x) = c_1 e^{(2+\sqrt{2})x} + c_2 e^{(2-\sqrt{2})x},$$

where c_1 and c_2 are arbitrary constants. To find the solution that satisfies the initial conditions $w(0) = 0$ and $w'(0) = 1$, we first differentiate the solution found above, then plug in our initial conditions. This gives

$$w(0) = 0 = c_1 + c_2$$

$$w'(0) = 1 = \left(2 + \sqrt{2}\right)c_1 + \left(2 - \sqrt{2}\right)c_2 .$$

Solving this system of equations yields $c_1 = \dfrac{1}{2\sqrt{2}}$ and $c_2 = \dfrac{-1}{2\sqrt{2}}$. Thus

$$w(x) = \frac{1}{2\sqrt{2}} e^{(2+\sqrt{2})x} - \frac{1}{2\sqrt{2}} e^{(2-\sqrt{2})x} = \frac{\sqrt{2}}{4}\left(e^{(2+\sqrt{2})x} - e^{(2-\sqrt{2})x}\right)$$

is the desired solution.

29. **(a)** As was stated in Section 4.5, third order linear homogeneous differential equations with constant coefficients can be handled in the same way as second order equations. Therefore, we look for the roots of the auxiliary equation $r^3 - r^2 + r + 3 = 0$. By the Rational Root Theorem, the only possible rational roots are $r = \pm 1$ and ± 3. By checking these, we find that one of the roots is $r = -1$. Therefore, factors of the auxiliary equation are given by

$$r^3 - r^2 + r + 3 = (r+1)\left(r^2 - 2r + 3\right).$$

Using the quadratic formula, we find that the other roots are

$$r = \frac{2 \pm \sqrt{4 - 12}}{2} = 1 \pm \sqrt{2}\, i .$$

A general solution is, therefore,

$$y(x) = c_1 e^{-x} + c_2 e^{x} \cos\sqrt{2}x + c_3 e^{x} \sin\sqrt{2}x .$$

33. Making the substitution $x = e^t$ (and using equation (17) on page 188 of the text), we transform the Cauchy-Euler equation into an equation with constant coefficients given by

$$\left(\frac{d^2y}{dt^2} - \frac{dy}{dt}\right) - 3\frac{dy}{dt} + 6y = 0,$$

or

$$\frac{d^2y}{dt^2} - 4\frac{dy}{dt} + 6y = 0 . \qquad (3)$$

This is a linear equation with constant coefficients and has the associated auxiliary equation $r^2 - 4r + 6 = 0$, which has roots $r = 2 \pm \sqrt{2}\, i$. Therefore, a general solution to equation (3) is

$$y(t) = c_1 e^{2t} \cos\left(\sqrt{2}\, t\right) + c_2 e^{2t} \sin\left(\sqrt{2}\, t\right) = c_1 \left(e^t\right)^2 \cos\left(\sqrt{2}\, t\right) + c_2 \left(e^t\right)^2 \sin\left(\sqrt{2}\, t\right).$$

To change this equation back into one with the independent variable x, we again use the substitution $x = e^t$ or, solving this expression for t, $t = \ln x$. Therefore, the solution becomes

$$y(x) = c_1 x^2 \cos\left(\sqrt{2}\ln x\right) + c_2 x^2 \sin\left(\sqrt{2}\ln x\right).$$

37. From Example 2 on page 194 we see that in the study of a vibrating spring with damping, we have the initial value problem

$$my''(t)+by'(t)+ky(t)=0; \qquad y(0)=y_0, \qquad y'(0)=v_0,$$

where m is the mass of the spring system, b is the damping constant, k is the spring constant, $y(0)$ is the initial displacement, $y'(0)$ is the initial velocity and $y(t)$ is the displacement from the equilibrium of the mass at time t.

(a) We want to determine the equation of motion for a spring system with $m=10\,\text{kg}$, $b=60\,\text{kg/sec}$, $k=250\,\text{kg/sec}^2$, $y(0)=0.3\,\text{m}$, $y'(0)=-0.1\,\text{m/sec}$. That is, we seek the solution to

$$10y''(t)+60y'(t)+250y(t)=0; \qquad y(0)=0.3, \qquad y'(0)=-0.1.$$

The auxiliary equation for the above equation is

$$10r^2+60r+250=0$$

which has roots

$$r=\frac{-60\pm\sqrt{3600-10000}}{20}=\frac{-60\pm i\sqrt{6400}}{20}=-3\pm 4i.$$

Hence $\alpha=-3$ and $\beta=4$. The displacement $y(t)$ is expressed in the form

$$y(t)=c_1e^{-3t}\cos 4t+c_2e^{-3t}\sin 4t.$$

We find c_1 and c_2 by using the initial conditions. We first differentiate $y(t)$ to get

$$y'(t)=(-3c_1+4c_2)e^{-3t}\cos 4t+(-4c_1-3c_2)e^{-3t}\sin 4t.$$

Substituting in the initial conditions we obtain the system

$$y'(0)=-0.1=-3c_1+4c_2$$

$$y(0)=0.3=c_1.$$

Upon solving, we find $c_1=0.3$ and $c_2=0.2$. Therefore the equation of motion is given by

$$y(t)=0.3e^{-3t}\cos 4t+0.2e^{-3t}\sin 4t.$$

(b) From Problem 36 we see the frequency of oscillation is given by $\beta/2\pi$. From part (a) we found $\beta=4$. Therefore the frequency of oscillation is $2/\pi$.

(c) We see a decrease in the frequency of oscillation. We also have the introduction of the factor e^{-3t} which causes the solution to decay to zero. This is a result of energy loss due to the damping.

41. **(a)** The auxiliary equation for this problem is $r^4+2r^2+1=(r^2+1)^2=0$. This equation has the roots $r_1=r_2=-i$ and $r_3=r_4=i$. Thus, $\cos x$ and $\sin x$ are solutions and since the complex root is repeated we get two more solutions by multiplying by x, that is $x\cos x$ and $x\sin x$ are also solutions. This gives a general solution of

$$y(x) = (c_1 + c_3 x)\cos x + (c_2 + c_4 x)\sin x .$$

(b) The auxiliary equation for this problem is

$$r^4 + 4r^3 + 12r^2 + 16r + 16 = (r^2 + 2r + 4)^2 = 0 .$$

The roots of the quadratic $r^2 + 2r + 4 = 0$ are

$$r = \frac{-2 \pm \sqrt{4-16}}{2} = -1 \pm \sqrt{3}\, i .$$

Hence the roots of the auxiliary equation are $r_1 = r_2 = -1 - \sqrt{3}\, i$ and $r_3 = r_4 = -1 + \sqrt{3}\, i$. Thus two linearly independent solutions are $e^{-x}\cos\sqrt{3}\, x$ and $e^{-x}\sin\sqrt{3}\, x$ and we get two more by multiplying them by x. This gives a general solution of

$$y(x) = (c_1 + c_3 x)e^{-x}\cos\sqrt{3}\, x + (c_2 + c_4 x)e^{-x}\sin\sqrt{3}\, x .$$

EXERCISES 4.7: Superposition and Nonhomogeneous Equations, page 202

1. The corresponding homogeneous equation, $y'' - y = 0$, has the associated auxiliary equation $r^2 - 1 = (r-1)(r+1) = 0$. This gives $r = \pm 1$ as roots of this equation. A general solution to the homogeneous equation is $c_1 e^x + c_2 e^{-x}$. Combining this with the particular solution, $y_p(x) = -x$, we find that a general solution is

$$y(x) = c_1 e^x + c_2 e^{-x} - x .$$

5. Since this equation can be written as

$$y'' + 2y' + 4y = 4\cos 2x ,$$

the corresponding homogeneous equation is $y'' + 2y' + 4y = 0$. This homogeneous equation has associated auxiliary equation $r^2 + 2r + 4 = 0$, which has the complex roots

$$r = \frac{-2 \pm \sqrt{4-16}}{2} = -1 \pm \sqrt{3}\, i .$$

Therefore, $y_1(x) = e^{-x}\cos\sqrt{3}\, x$ and $y_2(x) = e^{-x}\sin\sqrt{3}\, x$ linearly independent solutions of the homogeneous equation. Thus, by Theorem 6 on page 201 of the text, a general solution of the nonhomogeneous equation is given by

$$y(x) = c_1 e^{-x}\cos\sqrt{3}\, x + c_2 e^{-x}\sin\sqrt{3}\, x + \sin 2x .$$

11. Let $L[y] := y'' - y' + y$. We are given that $L[y_1](x) = g_1(x) = \sin x$ and that

$$L[y_2](x) = g_2(x) = e^{2x}.$$

(a) By the superposition principle stated on page 200 of the text, we have

$$L[5y_1](x) = 5g_1(x) = 5\sin x.$$

Thus, $y(x) = 5y_1(x) = 5\cos x$ is a solution of the equation

$$L[y](x) = y'' - y' + y = 5\sin x.$$

(b) Since $\sin x - 3e^{2x} = g_1(x) - 3g_2(x)$, the desired solution is $y_1 - 3y_2$, that is by the superposition principle

$$L[y_1 - 3y_2](x) = g_1(x) - 3g_2(x) = \sin x - 3e^{2x}.$$

Therefore, $y(x) = y_1(x) - 3y_2(x) = \cos x - e^{2x}$ is a solution to the equation

$$L[y](x) = y'' - y' + y = \sin x - 3e^{2x}.$$

13. (a) Let $L[y] := y'' + py' + qy$. According to the Superposition Principle on page 200 of the text, the functions $f_1(x) = x^2 - x$ and $f_2(x) = x^3 - x$ are solutions to the equation

$$L[y](x) = g(x) - g(x) = 0,$$

i.e., to the corresponding homogeneous differential equation. These two functions are linearly independent (one is not a constant multiple of the other). More rigorously, notice that

$$c_1 f_1(x) + c_2 f_2(x) \equiv 0 \quad \Leftrightarrow \quad c_2 x^3 + c_1 x^2 - (c_1 + c_2)x \equiv 0 \quad \Leftrightarrow \quad c_1 = c_2 = 0.$$

(b) By Theorem 6 on page 201 of the text, a general solution to the mysterious differential equation can be written, for instance, in the form

$$y(x) = x + c_1 f_1(x) + c_2 f_2(x).$$

Then

$$y'(x) = 1 + c_1 f_1'(x) + c_2 f_2'(x) = 1 + c_1(2x - 1) + c_2(3x^2 - 1).$$

From given initial conditions we obtain the system

$$\begin{aligned} 2 = y(2) &= 2 + c_1(2^2 - 2) + c_2(2^3 - 2) \\ 5 = y'(2) &= 1 + c_1(2\cdot 2 - 1) + c_2(3\cdot 2^2 - 1) \end{aligned} \quad \Leftrightarrow \quad \begin{aligned} 2c_1 + 6c_2 &= 0 \\ 3c_1 + 11c_2 &= 4. \end{aligned}$$

Upon solving we get $c_1 = -6$, $c_2 = 2$. So, the required solution satisfying the given initial conditions is

$$y(x) = x - 6(x^2 - x) + 2(x^3 - x) = 2x^3 - 6x^2 + 5x.$$

(c) From the formula (17), page 175 of the text, we conclude (after taking the log of both sides and differentiating) that

$$p(x) = -\left(\ln\{W[f_1, f_2](x)\}\right)'.$$

Now we compute

$$W[f_1, f_2](x) = W\left[x^2 - x, x^3 - x\right] = \begin{vmatrix} x^2 - x & x^3 - x \\ 2x - 1 & 3x^2 - 1 \end{vmatrix} = x^2(x-1)^2$$

$$\Rightarrow \qquad p(x) = -\left[\ln\left(x^2(x-1)^2\right)\right]' = -2\left(\ln|x| + \ln|x-1|\right)' = -2\left(\frac{1}{x} + \frac{1}{x-1}\right) = \frac{2-4x}{x(x-1)}.$$

19. **(a)** We first solve $(D + 1)v = e^x$. That is, we solve $v' + v = e^x$. This equation is linear with integration factor $\mu(x) = e^x$. Thus we have

$$D_x\left(e^x v\right) = e^{2x} \qquad \Rightarrow \qquad v = e^{-x}\int e^{2x}dx = \frac{e^x}{2} + ce^{-x}.$$

(b) We now want to solve $(D-2)y = \frac{e^x}{2} + ce^{-x}$. That is, we solve

$$y' - 2y = \frac{e^x}{2} + ce^{-x}.$$

This equation is also linear. It has an integrating factor of $\mu(x) = e^{-2x}$. Therefore,

$$D_x\left(e^{-2x}y\right) = \frac{e^{-x}}{2} + ce^{-3x}$$

$$\Rightarrow \qquad y = e^{2x}\int\left(\frac{e^{-x}}{2} + ce^{-3x}\right)dx$$

$$\Rightarrow \qquad y = -\frac{e^x}{2} + c_1e^{-x} + c_2e^{2x}, \qquad \text{where } c_1 = \frac{c}{-3}.$$

EXERCISES 4.8: Method of Undetermined Coefficients, page 211

5. This equation is a linear first order equation and can be solved by the methods of Chapter 2, but the methods of this section may also be applied. We will proceed with the methods of this section. To do this, note that the corresponding homogeneous equation is given by $2x' + x = 0$. This equation has a solution $x_h(t) = Ce^{-t/2}$. To find a particular solution of the nonhomogeneous equation, we see that the nonhomogeneous term of the original differential equation is a polynomial of degree two. Thus, according to Table 4.1 on page 208 of the text, the nonhomogeneous term is of Type I. Therefore, we look for a particular solution of the form

$$x_p(t) = t^s\left(A_2t^2 + A_1t + A_0\right).$$

Since no term of this particular solution x_p is a solution of the corresponding homogeneous equation, we take $s = 0$. Thus, we see that x_p and its first derivative have the respective forms

$$x_p(t) = A_2t^2 + A_1t + A_0, \quad \text{and} \quad x_p'(t) = 2A_2t + A_1.$$

Substituting these expressions into the original differential equation yields

$$2x_p'(t) + x_p(t) = 2(2A_2t + A_1) + A_2t^2 + A_1t + A_0$$

$$= A_2t^2 + (4A_2 + A_1)t + (2A_1 + A_0) = 3t^2 + 10t.$$

By equating coefficients we obtain

$A_2 = 3,$

$4A_2 + A_1 = 10 \quad \Rightarrow \quad A_1 = -2,$

$2A_1 + A_0 = 0 \quad \Rightarrow \quad A_0 = 4.$

Therefore, a particular solution is $x_p(t) = 3t^2 - 2t + 4$.

9. For this problem, the corresponding homogeneous equation is $\theta'' - 5\theta' + 6\theta = 0$, which has the associated auxiliary equation $\rho^2 - 5\rho + 6 = 0$. This auxiliary equation has roots $\rho = 3, 2$. Thus, a general solution of this homogeneous equation is given by

$$\theta_h(r) = C_1e^{2r} + C_2e^{3r}.$$

The nonhomogeneous term of the original differential equation is re^r. According to Table 4.1 on page 208 of the text, this nonhomogeneous term is of Type IV. Thus, we want a particular solution of this differential equation to have the form $\theta_p(r) = r^s\left(A_1r + A_0\right)e^r$. Since neither re^r nor e^r are solutions of the corresponding homogeneous equation, we let $s = 0$. Therefore, θ_p, θ_p', and θ_p'' are given by

$$\theta_p(r) = A_1re^r + A_0e^r,$$

$$\theta_p'(r) = A_1re^r + \left(A_1 + A_0\right)e^r, \quad \text{and}$$

$$\theta_p''(r) = A_1re^r + \left(2A_1 + A_0\right)e^r.$$

Substituting these expressions into the original differential equation yields

$$\theta_p'' - 5\theta_p' + 6\theta_p = A_1re^r + \left(2A_1 + A_0\right)e^r - 5\left[A_1re^r + \left(A_1 + A_0\right)e^r\right] + 6\left(A_1re^r + A_0e^r\right)$$

$$= \left(A_1 - 5A_1 + 6A_1\right)re^r + \left(2A_1 + A_0 - 5A_1 - 5A_0 + 6A_0\right)e^r = re^r.$$

By equating coefficients, we obtain

$$2A_1 = 1 \quad \Rightarrow \quad A_1 = \frac{1}{2},$$

$$-3A_1 + 2A_0 = 0 \quad \Rightarrow \quad A_0 = \frac{3}{4}.$$

Therefore, a particular solution of the nonhomogeneous differential equation $\theta'' - 5\theta' + 6\theta = re^r$ is given by $\theta_p(r) = \dfrac{re^r}{2} + \dfrac{3e^r}{4}$.

12. The corresponding homogeneous equation for this equation is $\theta'' - \theta = 0$. This homogeneous equation has auxiliary equation $r^2 - 1 = 0$, with roots $r = \pm 1$. Thus, a general solution of the homogeneous equation is

$$\theta_h(t) = C_1 e^t + C_2 e^{-t}.$$

The nonhomogeneous term, according to Table 4.1 on page 208 of the text, is of Type V with $p_n(t) = 0$ and $q_m(t) = t$. Therefore, the form of the particular solution will be

$$\theta_p(t) = t^s\{(A_1 t + A_0)\cos t + (B_1 t + B_0)\sin t\}.$$

We take $s = 0$ since none of the terms on the right-hand side of θ_p is a solution of the homogeneous equation. Thus, the particular solution will have the form

$$\theta_p(t) = (A_1 t + A_0)\cos t + (B_1 t + B_0)\sin t$$

$$\Rightarrow \quad \theta_p'(t) = A_1 \cos t - (A_1 t + A_0)\sin t + B_1 \sin t + (B_1 t + B_0)\cos t$$

$$= (B_1 t + A_1 + B_0)\cos t + (-A_1 t - A_0 + B_1)\sin t$$

$$\Rightarrow \quad \theta_p''(t) = B_1 \cos t - (B_1 t + B_0 + A_1)\sin t - A_1 \sin t + (-A_1 t - A_0 + B_1)\cos t$$

$$= (-A_1 t - A_0 + 2B_1)\cos t + (-B_1 t - B_0 - 2A_1)\sin t.$$

Substituting these expressions into the original differential equation yields

$$\theta_p'' - \theta_p = (-A_1 t - A_0 + 2B_1)\cos t + (-B_1 t - B_0 - 2A_1)\sin t - (A_1 t + A_0)\cos t - (B_1 t + B_0)\sin t$$

$$= -2A_1 t\cos t + (-2A_0 + 2B_1)\cos t - 2B_1 t\sin t + (-2A_1 - 2B_0)\sin t$$

$$= t\sin t.$$

By equating coefficients, we see that

$$-2A_1 = 0 \quad \Rightarrow \quad A_1 = 0,$$

$$-2A_0 + 2B_1 = 0 \quad \Rightarrow \quad B_1 = A_0,$$

$$-2B_1 = 1 \qquad \Rightarrow \qquad B_1 = -\frac{1}{2} \qquad \text{and so} \quad A_0 = -\frac{1}{2},$$

$$-2A_1 - 2B_0 = 0 \qquad \Rightarrow \qquad B_0 = 0.$$

Therefore, a particular solution of the nonhomogeneous equation $\theta'' - \theta = t\sin t$ is given by

$$\theta_p(t) = -\frac{t\sin t + \cos t}{2}.$$

17. The homogeneous differential equation corresponding to this equation is $y'' - 3y' + 2y = 0$. The associated auxiliary equation has roots $r = 1, 2$. Therefore, a general solution of the homogeneous equation is given by

$$y_h(t) = C_1 e^t + C_2 e^{2t}.$$

The nonhomogeneous term of the original equation is $e^t \sin t$. According to Table 4.1 on page 208 of the text, this term has the form of Type VI with $\alpha = 1$, $\beta = 1$, $a = 0$, and $b = 1$. Thus, a particular solution of the nonhomogeneous equation will have the form $y_p(t) = t^s e^t \{A\cos t + B\sin t\}$. Since neither $e^t \cos t$ nor $e^t \sin t$ is a solution of the homogeneous equation, we can let $s = 0$. Thus, a particular solution of the nonhomogeneous differential equation is given by

$$y_p(t) = e^t (A\cos t + B\sin t)$$

$$\Rightarrow \qquad y_p'(t) = e^t (A\cos t + B\sin t) + e^t(-A\sin t + B\cos t)$$

$$= e^t \left[(A + B)\cos t + (-A + B)\sin t\right]$$

$$\Rightarrow \qquad y_p''(t) = e^t \left[(A + B)\cos t + (-A + B)\sin t\right] + e^t\left[-(A + B)\sin t + (-A + B)\cos t\right]$$

$$= e^t \left[2B\cos t - 2A\sin t\right].$$

Substituting these expressions into the original differential equation yields

$$y_p'' - 3y_p' + 2y_p$$

$$= e^t\left[2B\cos t - 2A\sin t\right] - 3\left\{e^t\left[(A + B)\cos t + (-A + B)\sin t\right]\right\} + 2\left[e^t (A\cos t + B\sin t)\right]$$

$$= (-B - A)e^t \cos t + (A - B)e^t \sin t = e^t \sin t.$$

Equating coefficients yields

$$-B - A = 0 \qquad \text{and} \qquad A - B = 1$$

$$\Rightarrow \qquad B = -\frac{1}{2} \qquad \text{and} \qquad A = \frac{1}{2}.$$

Therefore, $y_p(t) = e^t \dfrac{\cos t - \sin t}{2}$ is a particular solution to the nonhomogeneous equation. Thus

$$y(t) = y_h(t) + y_p(t) = C_1 e^t + C_2 e^{2t} + e^t \frac{\cos t - \sin t}{2}$$

is a general solution of the original, nonhomogeneous differential equation.

21. The homogeneous differential equation corresponding to this equation is $x'' - 4x' + 4x = 0$. The associated auxiliary equation has a double root of $r = 2$. Therefore, a general solution of the homogeneous equation is given by

$$x_h(t) = C_1 e^{2t} + C_2 t e^{2t}.$$

The nonhomogeneous term of the original equation is te^{2t}. According to Table 4.1 on page 208 of the text, this term has the form of Type IV with $t^s P_n(t) e^{\alpha t} = t^s\{A_1 t + A_0\}e^{2t}$. Thus $x_p(t) = t^s\{A_1 t + A_0\}e^{2t}$. The nonnegative integer s is chosen to be the smallest integer so that no term in the particular solution, $x_p(t)$, is a solution to the corresponding homogeneous equation. Therefore, s is chosen to be 2. Thus a particular solution of the nonhomogeneous equation will have the form

$$x_p(t) = t^2\{A_1 t + A_0\}e^{2t} = \{A_1 t^3 + A_0 t^2\}e^{2t}.$$

We compute

$$x_p'(t) = \{3A_1 t^2 + 2A_0 t\}e^{2t} + 2\{A_1 t^3 + A_0 t^2\}e^{2t},$$

$$x_p''(t) = \{6A_1 t + 2A_0\}e^{2t} + 4\{3A_1 t^2 + 2A_0 t\}e^{2t} + 4\{A_1 t^3 + A_0 t^2\}e^{2t}.$$

Substituting these equations into the original differential equation yields

$$\begin{aligned} x_p''(t) - 4x_p'(t) + 4x_p(t) &= \{6A_1 t + 2A_0\}e^{2t} + 4\{3A_1 t^2 + 2A_0 t\}e^{2t} + 4\{A_1 t^3 + A_0 t^2\}e^{2t} \\ &\quad - 4\{3A_1 t^2 + 2A_0 t\}e^{2t} - 8\{A_1 t^3 + A_0 t^2\}e^{2t} + 4\{A_1 t^3 + A_0 t^2\}e^{2t} \\ &= \{6A_1 t + 2A_0\}e^{2t} = te^{2t}. \end{aligned}$$

Equating coefficients yields $A_0 = 0$ and $A_1 = \frac{1}{6}$. Therefore $x_p(t) = t^3 \frac{e^{2t}}{6}$ is a particular solution to the nonhomogeneous equation. Thus

$$x(t) = x_h(t) + x_p(t) = C_1 e^{2t} + C_2 t e^{2t} + t^3 \frac{e^{2t}}{6}$$

is a general solution of the original nonhomogeneous differential equation.

27. The homogeneous differential equation corresponding to this equation is $z'' + z = 0$. The associated auxiliary equation has roots $r = \pm i$. Therefore, a general solution of the homogeneous equation is given by

$$z_h(x) = A \sin x + B \cos x.$$

The nonhomogeneous term of the original equation is $2e^{-x}$. According to Table 4.1 on page 208 of the text, this term has the form of Type II with $s=0$ and $\alpha=-1$. Thus a particular solution of the nonhomogeneous equation has the form

$$z_p(x)=Ce^{-x}$$

$$\Rightarrow \quad z_p'(x)=-Ce^{-x}$$

$$\Rightarrow \quad z_p''(x)=Ce^{-x}.$$

Substituting these equations into the original differential equation yields

$$Ce^{-x}+Ce^{-x}=2e^{-x}$$

$$\Rightarrow \quad 2Ce^{-x}=2e^{-x}$$

$$\Rightarrow \quad C=1.$$

Therefore $z_p(x)=e^{-x}$, is a particular solution to the nonhomogeneous equation. Thus

$$z(x)=e^{-x}+A\sin x+B\cos x$$

is a general solution of the original nonhomogeneous differential equation. Applying the initial condition, $z(0)=0$, yields

$$z(0)=0=e^{-0}+A\sin 0+B\cos 0$$

$$\Rightarrow \quad 0=1+B$$

$$\Rightarrow \quad B=-1.$$

Applying the second initial condition, $z'(0)=0$, yields

$$z'(0)=0=-e^{-0}+A\cos 0+B\sin 0$$

$$\Rightarrow \quad 0=-1+A$$

$$\Rightarrow \quad A=1.$$

Therefore, the solution to the initial value problem is

$$z(x)=e^{-x}+\sin x-\cos x.$$

33. The corresponding homogeneous equation for this problem is $y''+y=0$. The associated auxiliary equation has the roots $r=\pm i$. Therefore, this homogeneous equation has a general solution given by

$$y_h(x)=C_1\cos x+C_2\sin x.$$

The nonhomogeneous term is the sum of the terms $\sin x+x\cos x$, which is of Type V with $n=1$ and $m=0$, and $10^x=\exp\left(\ln 10^x\right)=e^{x\ln 10}$, which is of Type II with $a=1$ and $\alpha=\ln 10$. Therefore, by the superposition principle (Theorem 5, page 200), a particular solution of the nonhomogeneous equation will be the sum of the terms $x^{s_1}\left[(A_1x+A_0)\cos x+(B_1x+B_0)\sin x\right]$ and $x^{s_2}Ce^{x\ln 10}$. Since $\sin x$ and $\cos x$ are solutions to the homogeneous equation, we will let $s_1=1$. Thus, the particular solution we are looking for will contain the terms $A_1x^2\cos x$, $A_0x\cos x$, $B_1x^2\sin x$, and $B_0x\sin x$. None of these terms is a solution of the homogeneous equation. Since $e^{x\ln 10}$ is not a solution of the homogeneous equation, we shall let $s_2=0$. This means that a particular solution will contain the term $Ce^{x\ln 10}=C10^x$. Therefore, a particular solution of the nonhomogeneous equation has the form

$$y_p(x)=\left(A_1x^2+A_0x\right)\cos x+\left(B_1x^2+B_0x\right)\sin x+C10^x.$$

37. The corresponding homogeneous equation for this problem is $y''-4y'+5y=0$. The associated auxiliary equation has the roots $r=2\pm i$. Therefore, this homogeneous equation has a general solution given by

$$y_h(x)=C_1e^{2x}\cos x+C_2e^{2x}\sin x,$$

where C_1 and C_2 are arbitrary constants. The nonhomogeneous term is the sum of the terms e^{5x}, which is of Type II with $\alpha=5$, and $x\sin 3x-\cos 3x$, which is of Type V with $n=0$ and $m=1$. Therefore, by the superposition principle (Theorem 5), a particular solution of the nonhomogeneous equation will be the sum of the terms

$$x^{s_1}Ae^{5x} \qquad \text{and} \qquad x^{s_2}\left[(Bx+C)\cos 3x+(Dx+E)\sin 3x\right].$$

Since none of these terms are solutions to the homogeneous equation we take $s_1=s_2=0$. Hence, a particular solution of the nonhomogeneous equation has the form

$$y_p(x)=Ae^{5x}+(Bx+C)\cos 3x+(Dx+E)\sin 3x.$$

39. We cannot use the method of undetermined coefficients to find a particular solution because of the x^{-1} term.

45. Since the nonhomogeneous term is $\sin x$, the form of a particular solution is $y_p(x)=x^s\{A\cos x+B\sin x\}$. To determine s we need to consider the corresponding homogeneous equation

$$y'''-y''+y=0$$

which has auxiliary equation $r^3-r^2+1=0$. Unfortunately the roots of this equation are not immediately found by factoring. (The roots consist of one real and two complex conjugate numbers.) However, to determine s it is sufficient to observe only that $r=i$ is <u>not</u> a root of the auxiliary equation; in other words, $\sin x$ and $\cos x$ are <u>not</u> solutions to the homogeneous equation. Hence we take $s=0$ and seek a particular solution of the form

$$y_p(x)=A\cos x+B\sin x.$$

This gives

$$y_p'(x) = -A\sin x + B\cos x\,,$$

$$y_p''(x) = -A\cos x - B\sin x\,,$$

$$y_p'''(x) = A\sin x - B\cos x\,.$$

Substituting these equations into the original differential equation yields

$$A\sin x - B\cos x - \left(-A\cos x - B\sin x\right) + A\cos x + B\sin x = \sin x\,.$$

$$(A+2B)\sin x + (2A-B)\cos x = \sin x$$

$$\Rightarrow \qquad A+2B = 1 \quad \text{and} \quad 2A - B = 0$$

$$\Rightarrow \qquad A = \frac{1}{5} \quad \text{and} \quad B = \frac{2}{5}\,.$$

Thus a particular solution to the nonhomogeneous differential equation is

$$y_p(x) = \left(\frac{1}{5}\right)\cos x + \left(\frac{2}{5}\right)\sin x\,.$$

49. The differential equation becomes a Cauchy-Euler equation if we multiply the equation by x, giving $x^2y''(x) + 3xy'(x) - 3y(x) = x^3$. To solve this equation we substitute $x = e^t$ to transform the differential equation into an equation with a new independent variable t. With $x = e^t$, it follows by the chain rule (see equations (12) and (13) on pages 187 and 188 of the text) that

$$x\frac{dy}{dx} = \frac{dy}{dt} \qquad \text{and} \qquad x^2\frac{d^2y}{dx^2} = \frac{d^2y}{dt^2} - \frac{dy}{dt}\,.$$

Substituting into the differential equation for $x\dfrac{dy}{dx}$ and $x^2\dfrac{d^2y}{dx^2}$ yields

$$\left(\frac{d^2y}{dt^2} - \frac{dy}{dt}\right) + 3\left(\frac{dy}{dt}\right) - 3y = (e^t)^3 = e^{3t}\,,$$

$$\frac{d^2y}{dt^2} + 2\frac{dy}{dt} - 3y = e^{3t}\,. \qquad (4)$$

The auxiliary equation for the corresponding homogeneous equation is

$$r^2 + 2r - 3 = (r-1)(r+3) = 0\,,$$

so the roots of this equation are $r = 1$ and $r = -3$. Therefore the general solution to the homogeneous equation is

$$y_h(t) = c_1 e^t + c_2 e^{-3t}\,.$$

To find a particular solution to (4) we take $y_p(t) = A\,e^{3t}$ ($s = 0$ because e^{3t} is not a solution to the corresponding homogeneous equation). Substituting $A\,e^{3t}$ into equation (4) gives

$$9A\,e^{3t} + 6A\,e^{3t} - 3A\,e^{3t} = e^{3t}$$

$$12A\,e^{3t} = e^{3t}$$

$$\Rightarrow \qquad A = 1/12\ .$$

Hence $y_p(t) = (1/12)e^{3t}$ and a general solution to (4) is

$$y(t) = y_h(t) + y_p(t) = c_1\,e^t + c_2\,e^{-3t} + (1/12)e^{3t}\,.$$

Finally replacing e^t by x we get

$$y(x) = c_1 x + c_2 x^{-3} + \frac{1}{12}x^3\ .$$

55. **(a)** The corresponding homogeneous equation for this problem is $y'' + 2y' + 5y = 0$. The associated auxiliary equation has roots $r = -1 \pm 2i$. Therefore, the general solution to this homogeneous equation is

$$y_h(x) = e^{-x}[A\sin 2x + B\cos 2x].$$

The nonhomogeneous term of the original equation is 10 between 0 and $\frac{3\pi}{2}$. This term is a Type I with $n = 0$ and $s = 0$. Thus a particular solution of the nonhomogeneous equation has the form

$$y_p(x) = A_0\,.$$

Substituting $y_p(x)$ into the original differential equation yields

$$0 + 0 + 5A_0 = 10$$

$$\Rightarrow \qquad A_0 = 2\,.$$

Therefore $y_p(x) = 2$ is a particular solution to the nonhomogeneous equation. Thus

$$y(x) = e^{-x}A\sin 2x + e^{-x}B\cos 2x + 2$$

is a general solution of the original differential equation on $0 \le x \le \frac{3\pi}{2}$. Applying the initial condition, $y(0) = 0$, yields

$$y(0) = 0 = e^{-0}A\sin 0 + e^{-0}B\cos 0 + 2$$

$$\Rightarrow \qquad 0 = B + 2$$

$$\Rightarrow \qquad B = -2\,.$$

Applying the second initial condition, $y'(0)=0$, yields

$$y'(0)=0=-e^{-0}A\sin 0+2e^{-0}A\cos 0-e^{-0}B\cos 0-2e^{-0}B\sin 0$$

$$\Rightarrow \qquad 0=2A-B$$

$$\Rightarrow \qquad A=\frac{B}{2}=-1.$$

Thus the solution to the initial value problem in the interval $0\le x\le\frac{3\pi}{2}$ is

$$y_a(x)=-e^{-x}\sin 2x-2e^{-x}\cos 2x+2.$$

(b) The general solution to the differential equation for $x>3\pi/2$ is the same as for the homogeneous equation, that is

$$y_b(x)=e^{-x}\left[c_1\sin 2x+c_2\cos 2x\right].$$

(c) To determine the values of c_1 and c_2 in part (b) we first find the value of the $y_a(x)$ found in part (a) at $x=\frac{3\pi}{2}$.

$$y_a\left(\frac{3\pi}{2}\right)=-e^{-\frac{3\pi}{2}}\sin 3\pi-2e^{-\frac{3\pi}{2}}\cos 3\pi+2=2e^{-\frac{3\pi}{2}}+2.$$

The value of $y(x)$ found in part (b) at $x=\frac{3\pi}{2}$ is given by

$$y_b\left(\frac{3\pi}{2}\right)=e^{-\frac{3\pi}{2}}\left[c_1\sin 3\pi+c_2\cos 3\pi\right]=e^{-\frac{3\pi}{2}}c_2.$$

By setting $y_a(3\pi/2)=y_b(3\pi/2)$, we can then solve for c_2:

$$c_2=\frac{2e^{-\frac{3\pi}{2}}+2}{-e^{-\frac{3\pi}{2}}}=-\left(2+2e^{\frac{3\pi}{2}}\right)=-2\left(1+e^{\frac{3\pi}{2}}\right).$$

To determine c_1 we set $y_a'(3\pi/2)=y_b'(3\pi/2)$, and solve for c_1:

$$y_a'\left(\frac{3\pi}{2}\right)=e^{-\frac{3\pi}{2}}\sin 3\pi-2e^{-\frac{3\pi}{2}}\cos 3\pi+2e^{-\frac{3\pi}{2}}\cos 3\pi+4e^{-\frac{3\pi}{2}}\sin 3\pi=0$$

$$y_b'\left(\frac{3\pi}{2}\right)=-e^{-\frac{3\pi}{2}}\left[c_1\sin 3\pi+c_2\cos 3\pi\right]+e^{-\frac{3\pi}{2}}\left[2c_1\cos 3\pi-2c_2\sin 3\pi\right]=e^{-\frac{3\pi}{2}}c_2-2e^{-\frac{3\pi}{2}}c_1.$$

Therefore

$$0 = e^{-\frac{3\pi}{2}} c_2 - 2e^{-\frac{3\pi}{2}} c_1$$

$$c_1 = \frac{1}{2} c_2 = -\left(1 + e^{\frac{3\pi}{2}}\right).$$

Thus the general solution for the differential equation so that the solutions from part (a) and part (b) agree is given by

$$y(x) = \begin{cases} -e^{-x}\sin 2x - 2e^{-x}\cos 2x + 2, & 0 \le x \le \dfrac{3\pi}{2} \\ -\left(1 + e^{\frac{3\pi}{2}}\right)e^{-x}\sin 2x - 2\left(1 + e^{\frac{3\pi}{2}}\right)e^{-x}\cos 2x, & x > \dfrac{3\pi}{2}. \end{cases}$$

57. **(a)** With $m = k = 1$ and $L = \pi$ given initial value problem becomes

$$y(t) = 0, \qquad t \le -\frac{\pi}{2V},$$

$$y'' + y = \begin{cases} \cos Vt, & -\dfrac{\pi}{2V} < t < \dfrac{\pi}{2V} \\ 0, & t \ge \dfrac{\pi}{2V}. \end{cases}$$

The corresponding homogeneous equation $y'' + y = 0$ is the simple harmonic equation whose general solution is

$$y_h(t) = C_1 \cos t + C_2 \sin t. \qquad (5)$$

First, we find the solution to the given problem for $-\frac{\pi}{2V} < t < \frac{\pi}{2V}$. The nonhomogeneous term, $\cos Vt$, is of the Type III in the Table 4.1 page 208 of the text. Therefore, the particular solution has the form

$$y_p(t) = A\cos Vt + B\sin Vt.$$

Substituting $y_p(t)$ into the equation yields

$$(A\cos Vt + B\sin Vt)'' + (A\cos Vt + B\sin Vt) = \cos Vt$$

$$\Rightarrow \qquad (-V^2 A\cos Vt + V^2 B\sin Vt) + (A\cos Vt + B\sin Vt) = \cos Vt$$

$$\Rightarrow \qquad (1 - V^2)A\cos Vt + (1 + V^2)B\sin Vt = \cos Vt.$$

Equating coefficients, we get

$$A = \frac{1}{1 - V^2}, \qquad B = 0,$$

Thus the general solution on $\left(-\frac{\pi}{2V}, \frac{\pi}{2V}\right)$ is

$$y_1(t) = y_h(t) + y_p(t) = C_1 \cos t + C_2 \sin t + \frac{1}{1-V^2} \cos Vt. \qquad (6)$$

Since $y(t) \equiv 0$ for $t \le -\frac{\pi}{2V}$, initial conditions for the solution obtained are

$$y_1\left(-\frac{\pi}{2V}\right) = y_1'\left(-\frac{\pi}{2V}\right) = 0.$$

From (6) we obtain

$$y_1\left(-\frac{\pi}{2V}\right) = C_1 \cos\left(-\frac{\pi}{2V}\right) + C_2 \sin\left(-\frac{\pi}{2V}\right) = 0$$

$$y_1'\left(-\frac{\pi}{2V}\right) = C_1 \cos\left(-\frac{\pi}{2V}\right) + C_2 \sin\left(-\frac{\pi}{2V}\right) + \frac{V}{1-V^2} = 0.$$

Solving the system yields

$$C_1 = \frac{V}{V^2-1} \sin\frac{\pi}{2V}, \qquad C_2 = \frac{V}{V^2-1} \cos\frac{\pi}{2V},$$

and

$$y_1(t) = \frac{V}{V^2-1} \sin\frac{\pi}{2V} \cos t + \frac{V}{V^2-1} \cos\frac{\pi}{2V} \sin t + \frac{1}{1-V^2} \cos Vt$$

$$= \frac{V}{V^2-1} \sin\left(t + \frac{\pi}{2V}\right) - \frac{1}{V^2-1} \cos Vt, \qquad -\frac{\pi}{2V} < t < \frac{\pi}{2V}.$$

For $t > \frac{\pi}{2V}$ given equation is homogeneous, and its general solution, $y_2(t)$, is given by (5). That is,

$$y_2(t) = C_3 \cos t + C_4 \sin t.$$

From the initial conditions

$$y_2\left(\frac{\pi}{2V}\right) = y_1\left(\frac{\pi}{2V}\right),$$

$$y_2'\left(\frac{\pi}{2V}\right) = y_1'\left(\frac{\pi}{2V}\right),$$

we conclude that

$$C_3 \cos\frac{\pi}{2V} + C_4 \sin\frac{\pi}{2V} = \frac{V}{V^2-1} \sin\frac{\pi}{V}$$

$$-C_3 \sin\frac{\pi}{2V} + C_4 \cos\frac{\pi}{2V} = \frac{V}{V^2-1}\cos\frac{\pi}{V} + \frac{V}{V^2-1} = \frac{2V}{V^2-1}\cos^2\frac{\pi}{2V}.$$

The solution of this system is

$$C_3 = 0, \qquad C_4 = \frac{2V}{V^2-1}\cos\frac{\pi}{2V}.$$

So,

$$y_2(t) = \frac{2V}{V^2-1}\cos\frac{\pi}{2V}\sin t.$$

(b) The graph of the function

$$|A(V)| = \left|\frac{2V}{V^2-1}\cos\frac{\pi}{2V}\right|$$

is given in Figure 4-A. From this graph, we find that the most violent shaking of the vehicle (the maximum of $|A|$) happens when the speed $V \approx 0.73$.

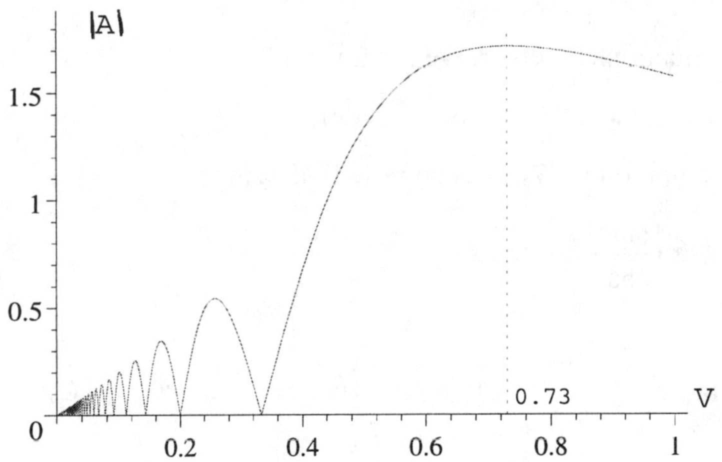

Figure 4-A. The graph of the function $|A(V)|$.

EXERCISES 4.9: Variation of Parameters, page 217

2. As in Example 1 on page 215 of the text we know that the fundamental solution set for the corresponding homogeneous equation is $y_1(x) = \cos x$, $y_2(x) = \sin x$ and its general solution is given by

$$y_h(x) = C_1 \cos x + C_2 \sin x .$$

Now we apply the method of variation of parameters to find a particular solution of the original equation. By the formula (3) on page 214 of the text, y_p has the form

$$y_p(x) = v_1(x)y_1(x) + v_2(x)y_2(x) .$$

Since

$$y_1'(x) = (\cos x)' = -\sin x , \quad y_2'(x) = (\sin x)' = \cos x ,$$

the system (9) on page 214 becomes

$$\begin{aligned} v_1'(x)\cos x + v_2'(x)\sin x &= 0 \\ -v_1'(x)\sin x + v_2'(x)\cos x &= \sec x . \end{aligned} \qquad (7)$$

Multiplying the first equation by $\sin x$ and the second one by $\cos x$ yields

$$\begin{aligned} v_1'(x)\sin x\cos x + v_2'(x)\sin^2 x &= 0 \\ -v_1'(x)\sin x\cos x + v_2'(x)\cos^2 x &= 1 . \end{aligned}$$

Adding these equations together, we obtain

$$v_2'(x)\left(\cos^2 x + \sin^2 x\right) = 1 \qquad \text{or} \qquad v_2'(x) = 1 .$$

From the first equation in (7), we can now find $v_1'(x)$:

$$v_1'(x) = -v_2'(x)\frac{\sin x}{\cos x} = -\tan x .$$

So,

$$\begin{aligned} v_1'(x) &= -\tan x \\ v_2'(x) &= 1 \end{aligned} \qquad \Rightarrow \qquad \begin{aligned} v_1(x) &= -\int \tan x\, dx = \ln|\cos x| + C_3 \\ v_2(x) &= \int dx = x + C_4 . \end{aligned}$$

Since we are looking for a particular solution, we can take $C_3 = C_4 = 0$ and get

$$y_p(x) = \cos x \ln|\cos x| + x\sin x .$$

Thus the general solution of the given equation is

$$y_p(x) + y_h(x) = \cos x \ln|\cos x| + x\sin x + C_1 \cos x + C_2 \sin x .$$

Alternatively, one can find a particular solution by using the formulas (10) on page 215.

5. This equation has associated homogeneous equation $y'' - 2y' + y = 0$. A general solution of this equation is

$$y_h(x) = C_1 e^x + C_2 x e^x .$$

For the variation of parameters method, we let

$$y_p = v_1 y_1 + v_2 y_2 , \qquad \text{where } y_1(x) = e^x \text{ and } y_2(x) = xe^x .$$

Thus, $y_1'(x) = e^x$ and $y_2'(x) = xe^x + e^x$. This means that we want to solve the system (see system (9) on page 214 of text)

$$e^x v_1' + xe^x v_2' = 0 ,$$

$$e^x v_1' + \left(xe^x + e^x\right) v_2' = x^{-1} e^x .$$

Subtracting the two equations gives

$$e^x v_2' = x^{-1} e^x$$

$$\Rightarrow \qquad v_2' = x^{-1} .$$

So

$$v_2(x) = \int x^{-1} dx = \ln|x| + K_1 .$$

Also we have from the first equation of the system

$$e^x v_1' = -xe^x v_2' = -xe^x x^{-1} = -e^x$$

$$\Rightarrow \qquad v_1' = -1 .$$

So

$$v_1(x) = -x + K_2 .$$

By letting K_1 and K_2 equal zero, and plugging the expressions found above for v_1 and v_2 into the equation defining y_p, we obtain a particular solution

$$y_p(x) = -xe^x + xe^x \ln|x| .$$

We obtain a general solution of the nonhomogeneous equation by adding this expression for y_p to the expression for y_h. Therefore, we obtain

$$y(x) = C_1 e^x + C_2 xe^x - xe^x + xe^x \ln|x| = C_1 e^x + (C_2 - 1) xe^x + xe^x \ln|x| .$$

If we let $c_1 = C_1$ and $c_2 = C_2 - 1$, we can express this general solution in the form

$$y(x) = c_1 e^x + c_2 xe^x + xe^x \ln|x| .$$

17. The associated homogeneous equation is $y'' + y = 0$. This equation has roots $r = \pm i$. Hence, a general solution to the corresponding homogeneous problem is given by

$$y_h(x) = c_1 \cos x + c_2 \sin x .$$

We will find a particular solution to the original problem by first finding a particular solution for each of two problems, one with the nonhomogeneous term $g_1(x) = 3\sec x$ and the other with the nonhomogeneous term $g_2(x) = -x^2 + 1$. Then we will use the superposition principle to obtain a particular solution for the original problem. The term $3\sec x$ is not in a form that allows us to use the method of undetermined coefficients. Therefore, we will use the method of variation of parameters. To do this let $y_1(x) = \cos x$ and $y_2(x) = \sin x$ (linearly independent solutions of the corresponding homogeneous problem). Then a particular solution y_{p1} to $y'' + y = 3\sec x$ has the form

$$y_{p1} = v_1 y_1 + v_2 y_2 = v_1 \cos x + v_2 \sin x ,$$

where v_1 and v_2 are determined by the system

$$(\cos x)v_1' + (\sin x)v_2' = 0$$

$$-(\sin x)v_1' + (\cos x)v_2' = 3\sec x .$$

Multiplying the first equation by $\cos x$ and the second equation by $\sin x$ and subtracting gives

$$\left(\cos^2 x + \sin^2 x\right)v_1' = -3\sec x \sin x = -3\tan x$$

$$\Rightarrow \qquad v_1' = -3\tan x .$$

Hence

$$v_1(x) = -3\int \tan x\, dx = 3\ln|\cos x| + C_1 .$$

To find v_2 we multiply the first equation of the above system by $\sin x$, the second by $\cos x$ and add to obtain

$$\left(\sin^2 x + \cos^2 x\right)v_2' = 3\sec x \cos x = 3$$

$$\Rightarrow \qquad v_2' = 3 .$$

Hence

$$v_2(x) = 3x + C_2 .$$

Therefore, for this first equation (with $g_1(x) = 3\sec x$), by letting $C_1 = C_2 = 0$, we have a particular solution given by

$$y_{p1}(x) = 3(\cos x)\ln|\cos x| + 3x\sin x .$$

The nonhomogeneous term $g_2(x) = -x^2 + 1$ is in a form that allows us to use the method of undetermined coefficients. Thus, a particular solution to this nonhomogeneous equation will have the form

$$y_{p2}(x) = A_2 x^2 + A_1 x + A_0$$

$$\Rightarrow \qquad y'_{p2}(x) = 2A_2x + A_1$$

$$\Rightarrow \qquad y''_{p2}(x) = 2A_2 .$$

Plugging these expressions into the equation $y'' + y = -x^2 + 1$, yields

$$y''_{p2} + y_{p2} = 2A_2 + A_2x^2 + A_1x + A_0 = A_2x^2 + A_1x + (2A_2 + A_0) = -x^2 + 1 .$$

By equating coefficients, we obtain

$$A_2 = -1, \qquad A_1 = 0, \qquad 2A_2 + A_0 = 1 \qquad \Rightarrow \qquad A_0 = 3 .$$

Therefore, we have

$$y_{p2}(x) = -x^2 + 3 .$$

By the superposition principle, we see that a particular solution to the original problem is given by

$$y_p(x) = y_{p1}(x) + y_{p2}(x) = 3(\cos x)\ln|\cos x| + 3x\sin x - x^2 + 3 .$$

Combining this with the general solution to the homogeneous equation found above, yields a general solution of the original differential equation given by

$$y(x) = c_1 \cos x + c_2 \sin x - x^2 + 3 + 3x\sin x + 3(\cos x)\ln|\cos x| .$$

21. The given differential equation is a Cauchy-Euler equation. We make the substitution $x = e^t$ to transform the nonhomogeneous equation to

$$z'' - 2z' + z = e^t\left(1 + \frac{3}{t}\right)$$

(see page 188 of the text). The auxiliary equation for the corresponding homogeneous equation is $r^2 - 2r + 1 = (r-1)^2 = 0$, which has the double root $r = 1$. Therefore the homogeneous solution is

$$z_h(t) = C_1e^t + C_2te^t ,$$

where C_1 and C_2 are arbitrary constants. For the variation of parameters method, we let

$$z_p = v_1z_1 + v_2z_2 , \qquad \text{where} \quad z_1 = e^t \quad \text{and} \quad z_2 = te^t .$$

Thus we have $z_1' = e^t$ and $z_2' = e^t(t+1)$. This means we want to solve the system (see system (9) on page 214 of the text)

$$e^tv_1' + te^tv_2' = 0$$

$$e^tv_1' + e^t(t+1)v_2' = e^t\left(1 + \frac{3}{t}\right).$$

Subtracting these two equations yields

$$e^tv_2' = e^t\left(1 + \frac{3}{t}\right)$$

$$\Rightarrow \qquad v_2' = 1 + \frac{3}{t}$$

$$\Rightarrow \qquad v_2 = \int \left(1 + \frac{3}{t}\right) dt = t + 3\ln t + K_1 .$$

From the first equation of the system we get

$$e^t v_1' = -te^t v_2' = -te^t\left(1 + \frac{3}{t}\right) = -e^t(t+3)$$

$$\Rightarrow \qquad v_1' = -(t+3)$$

$$\Rightarrow \qquad v_1 = -\int (t+3)\,dt = -\frac{t^2}{2} - 3t + K_2 .$$

By letting K_1 and K_2 equal zero, and plugging the expressions for v_1 and v_2 into the above equation defining z_p, we obtain the particular solution

$$z_p(t) = \left(-\frac{t^2}{2} - 3t\right)e^t + (t + 3\ln t)te^t = \frac{t^2}{2}e^t - 3te^t + 3te^t \ln t .$$

We obtain a general solution of the nonhomogeneous equation by adding the particular solution, z_p, and the homogeneous solution, z_h. Therefore, we get

$$z(t) = C_1 e^t + C_2 te^t + \frac{t^2}{2}e^t - 3te^t + 3te^t \ln t = C_1 e^t + (C_2 - 3)te^t + \frac{1}{2}t^2 e^t + 3te^t \ln t .$$

With $x = e^t$ (so that $t = \ln x$), $c_1 = C_1$ and $c_2 = C_2 - 3$, we express the general solution in the form

$$z(t) = c_1 x + c_2 x \ln x + \frac{1}{2}x(\ln x)^2 + 3x(\ln x)[\ln(\ln x)] \qquad \text{for } x > 0 .$$

27. A general solution of the corresponding homogeneous equation is given by

$$y_h(x) = C_1 e^{-x} + C_2 e^x .$$

We will try to find a particular solution of the nonhomogeneous equation of the form $y_p(x) = v_1 y_1 + v_2 y_2$, where $y_1 = e^{-x}$ and $y_2 = e^x$. We see that

$$W[y_1, y_2](x) = e^{-x}e^x - e^x\left(-e^{-x}\right) = 2 .$$

Next we use formulas (10) on page 215 of the text. Replacing the dummy variable x by t and integrating from 1 to x yields, with $g(t) = \frac{1}{t}$, the following:

$$v_1(x) = \int_1^x \frac{-g(t)y_2(t)}{W[y_1, y_2](t)}\,dt = -\frac{1}{2}\int_1^x \left(\frac{1}{t}\right)e^t\,dt ,$$

$$v_2(x) = \int_1^x \frac{g(t)y_1(t)}{W[y_1, y_2](t)}\, dt = \frac{1}{2}\int_1^x \left(\frac{1}{t}\right) e^{-t}\, dt .$$

(Notice that we have chosen the lower limit of integration to be equal to 1 because the initial conditions are given at 1. We could have chosen any other value for the lower limit, but the choice of 1 will make the determination of the constants C_1 and C_2 easier.)

Thus

$$y_p(x) = \frac{-e^{-x}}{2}\int_1^x \frac{e^t}{t}\, dt + \frac{e^x}{2}\int_1^x \frac{e^{-t}}{t}\, dt ,$$

and so a general solution of the original differential equation is

$$y(x) = C_1 e^{-x} + C_2 e^x - \frac{e^{-x}}{2}\int_1^x \frac{e^t}{t}\, dt + \frac{e^x}{2}\int_1^x \frac{e^{-t}}{t}\, dt .$$

By plugging in the first initial condition (and using the fact that the integral of a function from a to a is zero which is why we choose the lower limit of integration to be the initial point, 1), we find that

$$y(1) = C_1 e^{-1} + C_2 e^1 = 0 .$$

Differentiating $y(x)$ yields

$$y'(x) = -C_1 e^{-x} + C_2 e^x + \frac{e^{-x}}{2}\int_1^x \frac{e^t}{t}\, dt + \left(\frac{-e^{-x}}{2}\right)\left(\frac{e^x}{x}\right) + \frac{e^x}{2}\int_1^x \frac{e^{-t}}{t}\, dt + \left(\frac{e^x}{2}\right)\left(\frac{e^{-x}}{x}\right),$$

where we have used the product rule and the Fundamental Theorem of Calculus to differentiate the

last two terms of $y(x)$. We now plug the second initial condition into the equation we just found for y' to obtain

$$y'(1) = -C_1 e^{-1} + C_2 e^1 + \left(\frac{-e^{-1}}{2}\right)\left(\frac{e^1}{1}\right) + \left(\frac{e^1}{2}\right)\left(\frac{e^{-1}}{1}\right) = -C_1 e^{-1} + C_2 e^1 - \frac{1}{2} + \frac{1}{2} = -2 .$$

Solving the system

$$C_1 e^{-1} + C_2 e^1 = 0 ,$$

$$-C_1 e^{-1} + C_2 e^1 = -2 ,$$

simultaneously, yields $C_2 = -e^{-1}$ and $C_1 = e^1$. Therefore, the solution to our problem is given by

$$y(x) = e^{1-x} - e^{x-1} - \frac{e^{-x}}{2}\int_1^x \frac{e^t}{t}\, dt + \frac{e^x}{2}\int_1^x \frac{e^{-t}}{t}\, dt . \qquad (8)$$

Simpson's rule is implemented on the software package provided free with the text (see also the discussion of the solution to Problem 25 in Exercises 2.3). Simpson's rule requires an even number of intervals, but we don't know how many are required to obtain the 2 place accuracy desired. We will compute the approximate value of $y(x)$ at $x = 2$ using 2, 4, 6, ... intervals for Simpson's rule until the approximate value changes by less than five in the third place. For $n = 2$, we divide [1, 2] into 4 equal subintervals. Thus each interval will be of length $\frac{2-1}{4} = \frac{1}{4}$. Therefore the integrals are approximated by

$$\int_1^2 \frac{e^t}{t}\,dt = \frac{1}{12}\left[\frac{e^1}{1} + 4\frac{e^{1.25}}{1.25} + 2\frac{e^{1.50}}{1.50} + 4\frac{e^{1.75}}{1.75} + \frac{e^2}{2}\right] \approx 3.0592,$$

$$\int_1^2 \frac{e^{-t}}{t}\,dt = \frac{1}{12}\left[\frac{e^{-1}}{1} + 4\frac{e^{-1.25}}{1.25} + 2\frac{e^{-1.50}}{1.50} + 4\frac{e^{-1.75}}{1.75} + \frac{e^{-2}}{2}\right] \approx 0.1706.$$

Substituting these values into equation (8) we obtain

$$y(2) \approx e^{1-2} - e^{2-1} - \frac{e^{-2}}{2}[3.0592] + \frac{e^2}{2}[0.1706] = -1.9271.$$

Repeating these calculations for $n = 3$, 4, and 5 yields the approximations in Table 4-A.

Table 4-A. Successive approximations for $y(2)$ using Simpson's rule.

Intervals	$y(2) \approx$
6	−1.9275
8	−1.9275
10	−1.9275

Since these values do not change in the third place, we can expect that the first three places are accurate and we obtained an approximate solution of $y(2) = -1.93$.

EXERCISES 4.10: Qualitative Considerations for Variable-Coefficient and Nonlinear Equations, page 228

2. Comparing the given equation with (13) on page 221 of the text, we conclude that

inertia $m = 1$, damping $b = 0$, stiffness "k" $= -6y$.

For $y > 0$, the stiffness "k" is negative, and it tends to reinforce displacements. So, we should expect that the solutions $y(t)$ grow without bound.

6. Rewriting given equation in the equivalent form $y''=-\frac{k}{m}y$, we see that the function $f(y)$ in the energy integral lemma is $-\frac{k}{m}y$. So,

$$F(y)=\int\left(-\frac{k}{m}y\right)dy=-\frac{k}{2m}y^2+C.$$

With $C=0$, $F(y)=-\frac{k}{2m}y^2$, and the energy

$$E(t)=\frac{1}{2}(y'(t))^2-F(y(t))=\frac{1}{2}(y'(t))^2-\left(-\frac{k}{2m}y^2\right)=\frac{1}{2}(y'(t))^2+\frac{k}{2m}y^2.$$

By the energy integral lemma,

$$\frac{1}{2}(y'(t))^2+\frac{k}{2m}y^2=\text{const}.$$

Multiplying both sides by $2m$, we get the stated equation.

9. According to the Problem 8, with $\ell=g$, the function $\theta(t)$ satisfies the identity

$$\frac{(\theta')^2}{2}-\cos\theta=C=\text{constant}. \tag{9}$$

Our first purpose is to determine the constant C. Let t_a denote the moment when pendulum is in the apex point, i.e., $\theta(t_a)=\pi$. Since it doesn't cross the apex over, we also have $\theta'(t_a)=0$. Substituting these two values into (9), we obtain

$$\frac{(0)^2}{2}-\cos\pi=C \qquad \Rightarrow \qquad C=1.$$

Thus (8) becomes

$$\frac{(\theta')^2}{2}-\cos\theta=1.$$

In particular, at the initial moment, $t=0$,

$$\frac{(\theta'(0))^2}{2}-\cos\theta(0)=1.$$

Since $\theta(0)=0$, we get

$$\frac{(\theta'(0))^2}{2}-\cos 0=1$$

$$\Rightarrow \qquad (\theta'(0))^2=4$$

$$\Rightarrow \qquad \theta'(0)=2 \qquad \text{or} \qquad \theta'(0)=-2\,.$$

17. For the radius, $r(t)$, we have the initial value problem

$$r''(t)=-GMr^{-2}\,, \qquad r(0)=a\,, \qquad r'(0)=0\,.$$

Thus in the energy integral lemma, page 220 of the text, $f(r)=-GMr^{-2}$. Since

$$\int f(r)\,dr=\int\left(-GMr^{-2}\right)dr=GMr^{-1}+C,$$

we can take $F(r)=GMr^{-1}$, and the energy integral lemma yields

$$\frac{1}{2}\left(r'(t)\right)^2-\frac{GM}{r(t)}=C_1=\text{const.}$$

To find the constant C_1, we use the initial conditions.

$$C_1=\frac{1}{2}\left(r'(0)\right)^2-\frac{GM}{r(0)}=\frac{1}{2}\cdot 0^2-\frac{GM}{a}=-\frac{GM}{a}\,.$$

Therefore, $r(t)$ satisfies

$$\frac{1}{2}\left(r'(t)\right)^2-\frac{GM}{r(t)}=-\frac{GM}{a}$$

$$\Rightarrow \qquad \frac{1}{2}\left(r'\right)^2=\frac{GM}{r}-\frac{GM}{a}$$

$$\Rightarrow \qquad r'=-\sqrt{\frac{2GM}{a}}\sqrt{\frac{a-r}{r}}\,.$$

(Remember, $r(t)$ is decreasing, which implies that $r'(t)<0$). Separating variables and integrating, we obtain

$$\int\sqrt{\frac{r}{a-r}}\,dr=\int\left(-\sqrt{\frac{2GM}{a}}\right)dt$$

$$\Rightarrow \qquad a\left(\arctan\sqrt{\frac{r}{a-r}}-\frac{\sqrt{r(a-r)}}{a}\right)=-\sqrt{\frac{2GM}{a}}\,t+C_2\,.$$

We apply the initial condition $r(0)=a$ once again to find the constant C_2. But this time we have to be careful because the argument of " arctan " function becomes infinite at $r=a$. So, we take the limit of both sides rather than making simple substitution.

$$\lim_{t\to+0} a\left(\arctan\sqrt{\frac{r(t)}{a-r(t)}}-\frac{\sqrt{r(t)\left(a-r(t)\right)}}{a}\right)=a\left(\lim_{t\to+0}\arctan\sqrt{\frac{r(t)}{a-r(t)}}-\lim_{t\to+0}\frac{\sqrt{r(t)\left(a-r(t)\right)}}{a}\right)$$

$$= a\left(\frac{\pi}{2} - 0\right) = a\frac{\pi}{2},$$

and

$$\lim_{t \to +0}\left(-\sqrt{\frac{2GM}{a}}\,t + C_2\right) = -\sqrt{\frac{2GM}{a}} \cdot 0 + C_2 = C_2 .$$

Thus $C_2 = a\pi/2$ and $r(t)$ satisfies

$$a\left(\arctan\sqrt{\frac{r(t)}{a - r(t)}} - \frac{\sqrt{r(t)(a - r(t))}}{a}\right) = -\sqrt{\frac{2GM}{a}}\,t + a\frac{\pi}{2} .$$

At the moment $t = T_0$ when Earth splashes into the sun we have $r(T_0) = 0$. Substituting this condition into the last equation yields

$$a\left(\arctan\sqrt{\frac{0}{a - 0}} - \frac{\sqrt{0 \cdot (a - 0)}}{a}\right) = -\sqrt{\frac{2GM}{a}}\,T_0 + a\frac{\pi}{2}$$

$$\Rightarrow \qquad 0 = -\sqrt{\frac{2GM}{a}}\,T_0 + a\frac{\pi}{2}$$

$$\Rightarrow \qquad T_0 = a\frac{\pi}{2}\sqrt{\frac{a}{2GM}} = \frac{\pi}{2\sqrt{2}}\sqrt{\frac{a^3}{GM}} .$$

Then the required ratio is

$$\frac{T_0}{T} = \frac{\pi}{2\sqrt{2}}\sqrt{\frac{a^3}{GM}} \Bigg/ 2\pi\sqrt{\frac{a^3}{GM}} = \frac{1}{4\sqrt{2}} .$$

EXERCISES 4.11: A Closer Look at Free Mechanical Vibrations, page 238

9. Substituting the values $m = 2$, $k = 40$, and $b = 8\sqrt{5}$ into equation (12) on page 233 in the text and using the initial conditions, we obtain the initial value problem

$$2\frac{d^2y}{dt^2} + 8\sqrt{5}\frac{dy}{dt} + 40y = 0; \qquad y(0) = 0.1\,\text{m}, \quad y'(0) = 2\,\text{m/sec} .$$

The initial conditions are positive to reflect the fact that we have taken down to be positive in our coordinate system. The auxiliary equation for this system is

$$2r^2 + 8\sqrt{5}r + 40 = 0 \qquad \text{or} \qquad r^2 + 4\sqrt{5}r + 20 = 0 .$$

This equation has a double root at $r=-2\sqrt{5}$. Therefore, this system is critically damped and the equation of motion has the form

$$y(t)=\left(C_1+C_2t\right)e^{-2\sqrt{5}\,t}$$

To find the constants C_1 and C_2, we use the initial conditions $y(0)=0.1\,\text{m}$ and $y'(0)=2\,\text{m/sec}$. Thus, we have

$$y(0)=0.1=C_1,$$

$$y'(0)=2=C_2-2\sqrt{5}C_1 \qquad \Rightarrow \qquad C_2=2+0.2\sqrt{5}.$$

From this we obtain

$$y(t)=\left[0.1+\left(2+0.2\sqrt{5}\right)t\right]e^{-2\sqrt{5}t}.$$

The maximum displacement of the mass is found by determining the first time the velocity of the mass becomes zero. Therefore, we have

$$y'(t)=0=\left(2+0.2\sqrt{5}\right)e^{-2\sqrt{5}t}-2\sqrt{5}\left[0.1+\left(2+0.2\sqrt{5}\right)t\right]e^{-2\sqrt{5}t},$$

which gives

$$t=\frac{2}{2\sqrt{5}(2+0.2\sqrt{5})}=\frac{1}{\sqrt{5}(2+0.2\sqrt{5})}.$$

Thus the maximum displacement is

$$y\left(\frac{1}{\sqrt{5}(2+0.2\sqrt{5})}\right)=\left[0.10+\left(2+0.2\sqrt{5}\right)\left(\frac{1}{\sqrt{5}(2+0.2\sqrt{5})}\right)\right]e^{\frac{-2\sqrt{5}}{\sqrt{5}\left(2+0.2\sqrt{5}\right)}}\approx 0.242\text{ m}.$$

13. In Example 3, the solution was found to be

$$y(t)=\sqrt{\frac{7}{12}}\,e^{-2t}\sin\left(2\sqrt{3}\,t+\phi\right), \tag{10}$$

where $\phi=\pi+\arctan\dfrac{\sqrt{3}}{2}$. Therefore, we have

$$y'(t)=-\sqrt{\frac{7}{3}}\,e^{-2t}\sin\left(2\sqrt{3}\,t+\phi\right)+\sqrt{7}\,e^{-2t}\cos\left(2\sqrt{3}\,t+\phi\right).$$

Thus, to find the relative extrema for $y(t)$, we set

$$y'(t)=-\sqrt{\frac{7}{3}}\,e^{-2t}\sin\left(2\sqrt{3}\,t+\phi\right)+\sqrt{7}\,e^{-2t}\cos\left(2\sqrt{3}\,t+\phi\right)=0$$

$$\Rightarrow \qquad \frac{\sin\left(2\sqrt{3}\,t+\phi\right)}{\cos\left(2\sqrt{3}\,t+\phi\right)}=\frac{\sqrt{7}}{\sqrt{7/3}}=\sqrt{3}$$

$$\Rightarrow \qquad \tan\left(2\sqrt{3}\,t+\phi\right)=\sqrt{3}\,.$$

Since $\tan\theta=\sqrt{3}$ when $\theta=\dfrac{\pi}{3}+n\pi$, where n is an integer, we see that the relative extrema will occur at the points t_n, where

$$2\sqrt{3}\,t_n+\phi=\frac{\pi}{3}+n\pi \qquad \Rightarrow \qquad t_n=\frac{\dfrac{\pi}{3}+n\pi-\phi}{2\sqrt{3}}\,.$$

By substituting $\pi+\arctan\left(\dfrac{\sqrt{3}}{2}\right)$ for ϕ in the last equation above and by requiring that t be greater than zero, we obtain

$$t_n=\frac{\dfrac{\pi}{3}+(n-1)\pi-\arctan\left(\dfrac{\sqrt{3}}{2}\right)}{2\sqrt{3}}, \qquad n=1,2,3,\ldots$$

We see that the solution curve given by equation (10) above will touch the exponential curves $y(t)=\pm\sqrt{\dfrac{7}{12}}\,e^{-2t}$ when we have

$$\sqrt{\frac{7}{12}}\,e^{-2t}\sin\left(2\sqrt{3}\,t+\phi\right)=\pm\sqrt{\frac{7}{12}}\,e^{-2t},$$

where $\phi=\pi+\arctan\left(\dfrac{\sqrt{3}}{2}\right)$. This will occur when $\sin\left(2\sqrt{3}\,t+\phi\right)=\pm 1$. Since $\sin\theta=\pm1$ when $\theta=\dfrac{\pi}{2}+m\pi$ for any integer m, we see that the times T_m, when the solution touches the exponential curves, satisfy

$$2\sqrt{3}T_m+\phi=\frac{\pi}{2}+m\pi \qquad \Rightarrow \qquad T_m=\frac{\dfrac{\pi}{2}+m\pi-\phi}{2\sqrt{3}},$$

where $\phi=\pi+\arctan\left(\dfrac{\sqrt{3}}{2}\right)$ and m is an integer. Again requiring that t be positive we see that $y(t)$ touches the exponential curve when

$$T_m=\frac{\dfrac{\pi}{2}+(m-1)\pi-\arctan\left(\dfrac{\sqrt{3}}{2}\right)}{2\sqrt{3}}, \qquad m=1,2,3,\ldots.$$

From these facts it follows that for $y(t)$ to be an extremum and touch the curves $y=\pm\sqrt{\frac{7}{12}}\,e^{-2t}$ at the same time, there must be integers m and n such that $t_n=T_m$. That is, there must be integers m and n such that

$$\frac{\frac{\pi}{3}+n\pi-\arctan\left(\frac{\sqrt{3}}{2}\right)}{2\sqrt{3}}=\frac{\frac{\pi}{2}+m\pi-\arctan\left(\frac{\sqrt{3}}{2}\right)}{2\sqrt{3}}$$

$$\Rightarrow \qquad \frac{\pi}{3}+n\pi=\frac{\pi}{2}+m\pi$$

$$\Rightarrow \qquad n-m=\frac{1}{2}-\frac{1}{3}=\frac{1}{6}.$$

But, since m and n are integers, their difference is an integer and never $\frac{1}{6}$. Thus, the extrema of $y(t)$ do not occur on the exponential curves.

REVIEW PROBLEMS: page 250

41. Comparing the given homogeneous equations with mass-spring oscillator equation (13) on page 221 of the text,

$$[\text{inertia}]y''+[\text{damping}]y'+[\text{stiffness}]y=0,$$

we see that in equations (a) through (d) the damping coefficient is 0. So, the behavior, of solutions, as $t\to+\infty$, depends on the sign of the stiffness coefficient "k".

(a) "k" $=t^4>0$. This implies that all the solutions remain bounded as $t\to+\infty$.

(b) "k" $=-t^4<0$. The stiffness of the system is negative and increases unboundedly as $t\to+\infty$. It reinforces the displacement, $y(t)$, with magnitude increasing with time. Hence some solutions grow rapidly with time.

(c) "k" $=y^6>0$. Similarly to (a), we conclude that all the solutions are bounded.

(d) "k" $=y^7$. The function $f(y)=y^7$ is positive for positive y and negative if y is negative. Hence, we can expect that some of the solutions (say, ones satisfying negative initial conditions) are unbounded.

(e) "k" $=3+\sin t$. Since $|\sin t|\le 1$ for any t, we conclude that

$$\text{“}k\text{”}\ge 3+(-1)=2>0,$$

and all the solutions are bounded as $t\to+\infty$.

(f) Here there is positive damping "b" $=t^2$ increasing with time, which results an increasing drain of energy from the system, and positive constant stiffness $k=1$. Thus all the solutions are bounded.

(g) Negative damping "b" $=-t^2$, increases (in absolute value) with time, which imparts energy to the system instead of draining it. Note that the stiffness $k=-1$ is also negative. Thus we should expect that some of the solutions increase unboundedly as $t \to +\infty$.

CHAPTER 5: Introduction to Systems and Phase Plane Analysis

EXERCISES 5.2: Introduction to the Phase Plane, page 274

3. In this problem, $f(x,y)=x-y$, $g(x,y)=x^2+y^2-1$. To find the critical point set, we solve the system

$$\begin{aligned} x-y&=0, \\ x^2+y^2-1&=0 \end{aligned} \quad \Rightarrow \quad \begin{aligned} x&=y, \\ x^2+y^2&=1. \end{aligned}$$

Eliminating y yields

$$2x^2=1 \quad \Rightarrow \quad x=\pm\frac{1}{\sqrt{2}}.$$

Substituting x into the first equation, we find the corresponding value for y. Thus the critical points of the given system are $\left(1/\sqrt{2}\,,1/\sqrt{2}\right)$ and $\left(-1/\sqrt{2},-1/\sqrt{2}\right)$.

6. We see by Definition 1 on page 266 of the text that we must solve the system of equations given by

$$\begin{aligned} &y^2-3y+2=0, \\ &(x-1)(y-2)=0. \end{aligned}$$

By factoring the first equation above, we find that this system becomes

$$\begin{aligned} &(y-1)(y-2)=0, \\ &(x-1)(y-2)=0. \end{aligned}$$

Thus, we observe that if $y=2$ and x is any constant, then the system of differential equations given in this problem will be satisfied. Therefore, one family of critical points is given by the line $y=2$. If $y\neq 2$, then the system of equations above simplifies to $y-1=0$, and $x-1=0$. Hence, another critical point is the point (1, 1).

7. Here $f(x,y)=y-1, g(x,y)=e^{x+y}$. Thus the phase plane equation becomes

$$\frac{dy}{dx}=\frac{e^{x+y}}{y-1}=\frac{e^x e^y}{y-1}.$$

Separating variables yields

$$(y-1)\,e^{-y}dy=e^x dx$$

$$\Rightarrow \qquad \int(y-1)\,e^{-y}dy=\int e^x dx$$

$$\Rightarrow \qquad -ye^{-y} + c = e^x \qquad \text{or} \qquad e^x + ye^{-y} = c\,.$$

13. We will first find the critical points of this system. Therefore, we solve the system

$$(y-x)(y-1)=0\,,$$
$$(x-y)(x-1)=0\,.$$

Notice that both of these equations will be satisfied if $y=x$. Thus, $x=c$ and $y=c$, for any fixed constant c, will be a solution to the given system of differential equations and one family of critical points is the line $y=x$. We also see that we have a critical point at the point (1, 1). (This critical point is, of course, also on the line $y=x$.)

Next we will find the integral curves. Therefore, we must solve the first order differential equation given by

$$\frac{dy}{dx} = \frac{\dfrac{dy}{dt}}{\dfrac{dx}{dt}} = \frac{(x-y)(x-1)}{(y-x)(y-1)}$$

$$\Rightarrow \qquad \frac{dy}{dx} = \frac{1-x}{y-1}\,.$$

We can solve this last differential equation by the method of separation of variables. Thus, we have

$$\int (y-1)\,dy = \int (1-x)\,dx$$

$$\Rightarrow \qquad \frac{y^2}{2} - y = x - \frac{x^2}{2} + C$$

$$\Rightarrow \qquad x^2 - 2x + y^2 - 2y = 2C\,.$$

By completing the square, we obtain

$$(x-1)^2 + (y-1)^2 = c\,,$$

where $c = 2C+2$. Therefore, the integral curves are concentric circles with centers at the point (1, 1), including the critical point for the system of differential equations. The trajectories associated with the constants c = 1, 4, and 9, are sketched in Figure B.31 in the answers of the text.

Finally we will determine the flow along the trajectories. Notice that the variable t imparts a flow to the trajectories of a solution to a system of differential equations in the same manner as the parameter t imparts a direction to a curve written in parametric form. We will find this flow by determining the regions in the xy-plane where $x(t)$ is increasing (moving from left to right on each trajectory) and the regions where $x(t)$ is decreasing (moving from right to left on each trajectory). Therefore, we will use four cases to study the equation $\frac{dx}{dt}=(y-x)(y-1)$, the first equation in our system.

Case 1: $y>x$ and $y<1$. (This region is above the line $y=x$ but below the line $y=1$.) In this case, $y-x>0$ but $y-1<0$. Thus, $\frac{dx}{dt}=(y-x)(y-1)<0$. Hence, $x(t)$ will be decreasing here. Therefore, the flow along the trajectories will be from right to left and so the movement is clockwise.

Case 2: $y>x$ and $y>1$. (This region is above the lines $y=x$ and $y=1$.) In this case, we see that $y-x>0$ and $y-1>0$. Hence, $\frac{dx}{dt}=(y-x)(y-1)>0$. Thus, $x(t)$ will be increasing and the flow along the trajectories in this region will still be clockwise.

Case 3: $y<x$ and $y<1$. (This region is below the lines $y=x$ and $y=1$.) In this case, $y-x<0$ and $y-1<0$. Thus, $\frac{dx}{dt}>0$ and so $x(t)$ is increasing. Thus, the movement is from left to right and so the flow along the trajectories will be counterclockwise.

Case 4: $y<x$ and $y>1$. (This region is below the line $y=x$ and above the line $y=1$.) In this case, $y-x<0$ and $y-1>0$. Thus, $\frac{dx}{dt}<0$ and so $x(t)$ will be decreasing here. Therefore, the flow is from right to left and, thus, counterclockwise here also.

Therefore, above the line $y=x$ the flow is clockwise and below that line the flow is counterclockwise. See Figure B.31 in the answers of the text.

15.[†] From Definition 1 on page 266 of the text, we must solve the system of equations given by

$$2x+y+3=0,$$
$$-3x-2y-4=0.$$

By eliminating y in the first equation we obtain

$$x+2=0$$

and by eliminating x in the first equation we obtain

$$-y+1=0.$$

Thus, we observe that $x=-2$ and $y=1$ will satisfy both equations. Therefore $(-2, 1)$ is a critical point.

[†] In the 1st printing of the text, there was a misprint in the writing of the first equation of the system. The correct version reads $dx/dt=2x+y+3$.

From Figure B.32 in the answers of the text we see that all solutions passing near the point (–2, 1) do not stay close to it therefore the critical point (–2, 1) is unstable.

21. First we convert the given equation into a system of first order equations involving the functions $y(t)$ and $v(t)$ by using the substitution

$$v(t) = y'(t) \qquad \Rightarrow \qquad v'(t) = y''(t).$$

Therefore, this equation becomes the system

$$\frac{dy}{dt} = v,$$

$$\frac{dv}{dt} = -y - y^5 = -y\left(1 + y^4\right).$$

To find the critical points, we solve the system of equations given by $v = 0$ and $-y\left(1 + y^4\right) = 0$. This system is satisfied only when $v = 0$ and $y = 0$. Thus, the only critical point is the point (0, 0). To find the integral curves, we solve the first order equation given by

$$\frac{dv}{dy} = \frac{\frac{dv}{dt}}{\frac{dy}{dt}} = \frac{-y - y^5}{v}.$$

This is a separable equation and can be written as

$$v\,dv = \left(-y - y^5\right)dy \qquad \Rightarrow \qquad \frac{v^2}{2} = \frac{-y^2}{2} - \frac{y^6}{6} + C$$

$$\Rightarrow \qquad 3v^2 + 3y^2 + y^6 = c, \quad (c = 6C),$$

where we have integrated to obtain the second equation above. Therefore, the integral curves for this system are given by the equations $3v^2 + 3y^2 + y^6 = c$ for each positive constant c.

To determine the flow along the trajectories, we will examine the equation $\frac{dy}{dt} = v$. Thus, we see that

$$\frac{dy}{dt} > 0 \text{ when } v > 0, \qquad \text{and} \qquad \frac{dy}{dt} < 0 \text{ when } v < 0.$$

Therefore, y will be increasing when $v > 0$ and decreasing when $v < 0$. Hence, above the y-axis the flow will be from left to right and below the x-axis the flow will be from right to left. Thus, the flow on these trajectories will be clockwise (Figure B.35 in the answers of the text). Thus (0, 0) is a center (stable).

33. Since $S(t)$ and $I(t)$ represent population and we cannot have a negative population, we are only interested in the first quadrant of the SI-plane.

(a) In order to find the trajectory corresponding to the initial conditions $I(0)=1$ and $S(0)=700$, we must solve the first order equation

$$\frac{dI}{dS}=\frac{dI/dt}{dS/dt}=\frac{aSI-bI}{-aSI}=-\frac{aS-b}{aS}$$

$$\Rightarrow \qquad \frac{dI}{dS}=-1+\frac{b}{a}\frac{1}{S}. \qquad (1)$$

By integrating both sides of equation (1) with respect to *S*, we obtain the integral curves given by

$$I(S)=-S+\frac{b}{a}\ln S+c.$$

A sketch of this curve for $a=0.003$ and $b=0.5$ is shown in Figure B.39 in the answers of the text.

(b) From the sketch in Figure B.39 in the answers of the text we see that the peak number of infected people is 295.

(c) The peak number of infected people occurs when $\frac{dI}{dS}=0$. From equation (1) we have

$$\frac{dI}{dS}=0=-1+\frac{b}{a}\frac{1}{S}.$$

Solving for *S* we obtain

$$S=\frac{b}{a}=\frac{0.5}{0.003}\approx 167 \text{ people.}$$

37. **(a)** Denoting $y'=v$, we have $y''=v'$, and (with $m=\mu=k=1$) (16) can be written as a system

$$y'=v,$$

$$v'=-y+\begin{cases} y, & \text{if } |y|<1, v=0, \\ \text{sign}(y), & \text{if } |y|\ge 1, v=0, \\ -\text{sign}(v), & \text{if } v\ne 0 \end{cases} = \begin{cases} 0, & \text{if } |y|<1, v=0, \\ -y+\text{sign}(y), & \text{if } |y|\ge 1, v=0, \\ -y-\text{sign}(v), & \text{if } v\ne 0. \end{cases} \qquad (2)$$

(b) The condition $v\ne 0$ corresponds to the third case in (2), i.e., the system has the form

$$y'=v,$$

$$v'=-y-\text{sign}(v).$$

The phase plane equation for this system is

$$\frac{dv}{dy}=\frac{dv/dt}{dy/dt}=\frac{-y-\text{sign}(v)}{v}.$$

We consider two cases.

1) $v>0$. In this case $\text{sign}(v)=1$ and we have

$$\frac{dv}{dy}=\frac{-y-1}{v} \qquad \Rightarrow \qquad v\,dv=-(y+1)\,dy$$

$$\Rightarrow \qquad \int v\,dv=-\int (y+1)\,dy$$

$$\Rightarrow \qquad \frac{1}{2}v^2=-\frac{1}{2}(y+1)^2+C \qquad \Rightarrow \qquad v^2+(y+1)^2=c,$$

where $c=2C$.

2) $v<0$. In this case $\text{sign}(v)=-1$ and we have

$$\frac{dv}{dy}=\frac{-y+1}{v} \qquad \Rightarrow \qquad v\,dv=-(y-1)\,dy$$

$$\Rightarrow \qquad \int v\,dv=-\int (y-1)\,dy$$

$$\Rightarrow \qquad \frac{1}{2}v^2=-\frac{1}{2}(y-1)^2+C \qquad \Rightarrow \qquad v^2+(y-1)^2=c.$$

(c) The equation $v^2+(y+1)^2=c$ defines a circle in the *yv*-plane centered at $(-1,0)$ and of the radius $\sqrt{c}$ if $c\ge 0$, and it is the empty set if $c<0$. The condition $v>0$ means that we have to take only the half of these circles lying in the upper half plane. Moreover, the first equation, $y'=v$, implies that trajectories flow from left to right. Similarly, in the lower half plane, $v<0$, we have concentric semicircles $v^2+(y-1)^2=c$, $c\ge 0$, centered at $(1,0)$ and flowing from right to left.

(d) For the system found in (a),

$$f(y,v)=v,$$

$$g(y,v)=\begin{cases} 0, & \text{if } |y|<1, v=0, \\ -y+\text{sign}(y), & \text{if } |y|\ge 1, v=0, \\ -y-\text{sign}(v), & \text{if } v\ne 0. \end{cases}$$

Since

$$f(y,v)=0 \qquad \Leftrightarrow \qquad v=0$$

and

$$g(y,0)=\begin{cases} 0, & \text{if } |y|<1, \\ -y+\text{sign}(y), & \text{if } |y|\ge 1, \end{cases}$$

we consider two cases. If $|y|<1$, then $g(y,0)\equiv 0$. This means that any point of the interval $-1<y<1$ is a critical point. If $|y|\ge 1$, then $g(y,0)=-y+\text{sign}(y)$ which is 0 if $y=\pm 1$. Thus the critical point set is the segment $v=0$, $-1\le y\le 1$.

(e) According to (c), the mass released at $(7.5,0)$ goes in the lower half plane from right to left along a semicircle centered at $(1,0)$. The radius of this semicircle is $7.5-1=6.5$, and its other end is

$(1-6.5, 0)=(-5.5, 0)$. From this point, the mass goes from left to right in the upper half plane along the semicircle centered at $(-1, 0)$ and of the radius $-1-(-5.5)=4.5$, and comes to the point $(-1+4.5, 0)=(3.5, 0)$. Then the mass again goes from right to left in the lower half plane along the semicircle centered at $(1, 0)$ and of the radius $3.5-1=2.5$, and comes to the point $(1-2.5, 0)=(-1.5, 0)$. From this point, the mass goes in the upper half plane from left to right along the semicircle centered at $(-1, 0)$ and of the radius $-1-(-1.5)=0.5$, and comes to the point $(-1+0.5, 0)=(-0.5, 0)$. Here it comes to rest because $|-0.5|<1$, and there is not a lower semicircle starting at this point. See the colored curve in Figure B.42 of the text.

EXERCISES 5.3: Elimination Method for Systems, page 284

5. Writing this system in operator notation yields the system

$$\begin{aligned} &(D-1)[x]+D[y]=5, \\ &D[x]+(D+1)[y]=1. \end{aligned} \qquad (3)$$

We will first eliminate the function $x(t)$, although we could proceed just as easily by eliminating the function $y(t)$. Thus, we apply the operator D to the first equation and the operator $-(D-1)$ to the second equation to obtain

$$\begin{aligned} &D(D-1)[x]+D^2[y]=D[5]=0, \\ &-(D-1)D[x]-(D-1)(D+1)[y]=-(D-1)[1]=1. \end{aligned}$$

Adding these two equations yields

$$\left(D(D-1)-(D-1)D\right)[x]+\left(D^2-(D^2-1)\right)[y]=1$$

$$\Rightarrow \qquad 0\cdot x+1\cdot y=1 \qquad \Rightarrow \qquad y(t)=1.$$

To find the function $x(t)$, we will eliminate y from the system given in (3). Therefore, we "multiply" the first equation in (3) by $(D+1)$ and the second by $-D$ to obtain the system

$$(D+1)(D-1)[x]+(D+1)D[y]=(D+1)[5]=5,$$

$$-D^2[x]-D(D+1)[y]=-D[1]=0.$$

By adding these two equations we obtain

$$\left((D^2-1)-D^2\right)[x]=5$$

$$\Rightarrow \qquad -x=5 \qquad \Rightarrow \qquad x(t)=-5.$$

Therefore, this system of linear differential equation is solved by the functions

$$x(t)=-5 \quad \text{and} \quad y(t)=1.$$

9. Expressed in operator notation, this system becomes

$$(D+2)[x]+D[y]=0,$$
$$(D-1)[x]+(D-1)[y]=\sin t.$$

In order to eliminate the function $y(t)$, we will apply the operator $(D-1)$ to the first equation above and the operator $-D$ to the second. Thus, we have

$$(D-1)(D+2)[x]+(D-1)D[y]=(D-1)[0]=0\ ,$$

$$-D(D-1)[x]-D(D-1)[y]=-D[\sin t]=-\cos t.$$

Adding these two equations yields the differential equation involving the single function $x(t)$ given by

$$\left((D^2+D-2)-(D^2-D)\right)[x]=-\cos t$$

$$\Rightarrow \qquad 2(D-1)[x]=-\cos t. \qquad (4)$$

This is a linear first order differential equation with constant coefficients and so can be solved by the methods of Chapter 2. (See section 2.3.) However, we will use the methods of Chapter 4. We see that the auxiliary equation associated with the corresponding homogeneous equation is given by $2(r-1)=0$, which has the root $r=1$. Thus, a general solution to the corresponding homogeneous equation is

$$x_h(t)=c_1e^t.$$

We will use the method of undetermined coefficients to find a particular solution to the nonhomogeneous equation. To this end, we note that according to Table 4.1 on page 208 of the text, a particular solution to this differential equation will have the form

$$x_p(t)=A\cos t+B\sin t \qquad \Rightarrow \qquad x_p'(t)=-A\sin t+B\cos t.$$

Substituting these expressions into the nonhomogeneous equation given in (4) yields

$$2x_p'-2x_p=2(-A\sin t+B\cos t)-2(A\cos t+B\sin t)$$

$$=(2B-2A)\cos t+(-2A-2B)\sin t=-\cos t.$$

By equating coefficients we obtain

$$2B-2A=-1 \qquad \text{and} \qquad -2A-2B=0.$$

By solving these two equations simultaneously for A and B, we see that

$$A=\frac{1}{4} \qquad \text{and} \qquad B=-\frac{1}{4}.$$

Thus, a particular solution to the nonhomogeneous equation given in (4) will be

$$x_p(t)=\frac{1}{4}\cos t-\frac{1}{4}\sin t$$

and a general solution to the nonhomogeneous equation (4) will be

$$x(t) = x_h(t) + x_p(t) = c_1 e^t + \frac{1}{4}\cos t - \frac{1}{4}\sin t .$$

We now must find a function $y(t)$. To do this, we subtract the second of the two differential equations in the system from the first to obtain

$$3x + y = -\sin t \qquad \Rightarrow \qquad y = -3x - \sin t .$$

Therefore, we see that

$$y(t) = -3\left[c_1\, e^t + \frac{1}{4}\cos t \;-\frac{1}{4}\sin t \right] - \sin t$$

$$\Rightarrow \qquad y(t) = -3c_1\, e^t - \frac{3}{4}\cos t - \frac{1}{4}\sin t .$$

Hence this system of differential equations has the general solution

$$x(t) = c_1 e^t + \frac{1}{4}\cos t - \frac{1}{4}\sin t \qquad \text{and} \qquad y(t) = -3c_1\, e^t - \frac{3}{4}\cos t - \frac{1}{4}\sin t .$$

17. Expressed in operator notation, this system becomes

$$(D^2 + 5)[x] - 4[y] = 0,$$
$$-[x] + (D^2 + 2)[y] = 0.$$

In order to eliminate the function $x(t)$, we apply the operator $(D^2 + 5)$ to the second equation. Thus, we have

$$(D^2 + 5)[x] - 4[y] = 0,$$
$$-(D^2 + 5)[x] + (D^2 + 5)(D^2 + 2)[y] = 0.$$

Adding these two equations together yields the differential equation involving the single function $y(t)$ given by

$$\left((D^2 + 5)(D^2 + 2) - 4\right)[y] = 0$$

$$\Rightarrow \qquad (D^4 + 7D^2 + 6)[y] = 0 .$$

The auxiliary equation for this homogeneous equation is given by $r^4 + 7r^2 + 6 = \left(r^2 + 1\right)\left(r^2 + 6\right) = 0$, which has roots $r = \pm i, \pm i\sqrt{6}$. Thus, a general solution is

$$y(t) = c_1 \sin t + c_2 \cos t + c_3 \sin\sqrt{6}t + c_4 \cos\sqrt{6}t .$$

We must now find a function $x(t)$ that satisfies the system of differential equations given in the problem. To do this we solve the second equation of the system of differential equations for $x(t)$ to obtain

$$x(t) = (D^2 + 2)[y].$$

Substituting the expression we found for $y(t)$, we see that

$$x(t) = -c_1 \sin t - c_2 \cos t - 6c_3 \sin\sqrt{6}t - 6c_4 \cos\sqrt{6}t$$
$$+ 2\left(c_1 \sin t + c_2 \cos t + c_3 \sin\sqrt{6}t + c_4 \cos\sqrt{6}t\right)$$

$$\Rightarrow \qquad x(t) = c_1 \sin t + c_2 \cos t - 4c_3 \sin\sqrt{6}t - 4c_4 \cos\sqrt{6}t .$$

Hence this system of differential equations has the general solution

$$x(t) = c_1 \sin t + c_2 \cos t - 4c_3 \sin\sqrt{6}t - 4c_4 \cos\sqrt{6}t$$

and

$$y(t) = c_1 \sin t + c_2 \cos t + c_3 \sin\sqrt{6}t + c_4 \cos\sqrt{6}t .$$

21. To apply the elimination method, we first write the system using the operator notation:

$$\begin{aligned} D^2[x] - y &= 0, \\ -x + D^2[y] &= 0. \end{aligned} \tag{5}$$

Eliminating y by applying D^2 to the first equation and adding to the second equation gives

$$\left(D^2 D^2 - 1\right)[x] = 0,$$

which reduces to

$$\left(D^4 - 1\right)[x] = 0. \tag{6}$$

The corresponding auxiliary equation, $r^4 - 1 = 0$, has roots ± 1, $\pm i$. Thus, the general solution to (6) is given by

$$x(t) = C_1 e^t + C_2 e^{-t} + C_3 \cos t + C_4 \sin t . \tag{7}$$

Substituting $x(t)$ into the first equation in (5) yields

$$y(t) = x''(t) = C_1 e^t + C_2 e^{-t} - C_3 \cos t - C_4 \sin t . \tag{8}$$

We use initial conditions to determine constants $C_1, C_2, C_3,$ and C_4. Differentiating (7) and (8), we get

$$3 = x(0) = C_1 e^0 + C_2 e^{-0} + C_3 \cos 0 + C_4 \sin 0 = C_1 + C_2 + C_3 ,$$

$$1 = x'(0) = C_1 e^0 - C_2 e^{-0} - C_3 \sin 0 + C_4 \cos 0 = C_1 - C_2 + C_4 ,$$

$$1 = y(0) = C_1 e^0 + C_2 e^{-0} - C_3 \cos 0 - C_4 \sin 0 = C_1 + C_2 - C_3 ,$$

$$-1 = y'(0) = C_1 e^0 - C_2 e^{-0} + C_3 \sin 0 - C_4 \cos 0 = C_1 - C_2 - C_4$$

$$\Rightarrow \quad \begin{aligned} C_1 + C_2 + C_3 &= 3, \\ C_1 - C_2 + C_4 &= 1, \\ C_1 + C_2 - C_3 &= 1, \\ C_1 - C_2 - C_4 &= -1. \end{aligned}$$

Solving we obtain $C_1 = C_2 = C_3 = C_4 = 1$. So, the desired solution is

$$x(t) = e^t + e^{-t} + \cos t + \sin t,$$

$$y(t) = e^t + e^{-t} - \cos t - \sin t.$$

23. We will attempt to solve this system by first eliminating the function $y(t)$. Thus, we "multiply" the first equation by $(D+2)$ and the second by $-(D-1)$. Therefore, we obtain

$$(D+2)(D-1)[x] + (D+2)(D-1)D[y] = (D+2)\left[-3e^{-2t}\right] = 6e^{-2t} - 6e^{-2t} = 0,$$

$$-(D-1)(D+2)[x] - (D-1)(D+2)[y] = -(D-1)\left[3e^t\right] = -3e^t + 3e^t = 0.$$

Adding these two equations yields

$$0 \cdot x + 0 \cdot y = 0,$$

which will be true for any two functions $x(t)$ and $y(t)$. (But not every pair of functions will satisfy this system of differential equations.) Thus, this is a degenerate system, and has infinitely many linearly independent solutions. To see if we can find these solutions, we will examine the system more closely. Notice that we could write this system as

$$\begin{aligned} (D-1)[x+y] &= -3e^{-2t}, \\ (D+2)[x+y] &= 3e^t. \end{aligned}$$

Therefore, let's try the substitution $z(t) = x(t) + y(t)$. We want a function $z(t)$ that satisfies the two equations

$$z'(t) - z(t) = -3e^{-2t} \qquad \text{and} \qquad z'(t) + 2z(t) = 3e^t, \tag{9}$$

simultaneously. We start by solving the first equation given in (9). This is a linear differential equation with constant coefficients which has the associated auxiliary equation given by $r - 1 = 0$. Hence, the solution to the corresponding homogeneous equation is

$$z_h(t) = ce^t.$$

By the method of undetermined coefficients, we see that a particular solution will have the form

$$z_p(t) = Ae^{-2t} \qquad \Rightarrow \qquad z_p'(t) = -2Ae^{-2t}.$$

Substituting these expressions into the first differential equation given in (9) yields

$$z_p'(t) - z_p(t) = -2Ae^{-2t} - Ae^{-2t} = -3Ae^{-2t} = -3e^{-2t} \qquad \Rightarrow \qquad A = 1.$$

Thus, the first equation given in (9) has the general solution

$$z(t) = ce^{t} + e^{-2t}.$$

Now, substituting $z(t)$ into the second equation in (9) gives

$$ce^{t} - 2e^{-2t} + 2\left(ce^{t} + e^{-2t}\right) = 3e^{t}$$

$$\Rightarrow \qquad 3ce^{t} = 3e^{t}.$$

Hence, c must be 1. Therefore, $z(t) = e^{t} + e^{-2t}$ is the only solution that satisfies both differential equations given in (9) simultaneously. Thus, any two differentiable functions that satisfy the equation $x(t) + y(t) = e^{t} + e^{-2t}$ will satisfy the original system.

29. We begin by expressing the system in operator notation

$$(D-\lambda)[x] + y = 0,$$
$$-3x + (D-1)[y] = 0.$$

We eliminate y by applying $(D-1)$ to the first equation and subtracting the second equation from it. This gives

$$\{(D-1)(D-\lambda) - (-3)\}[x] = 0$$

$$\Rightarrow \qquad \left(D^2 - (\lambda+1)D + (\lambda+3)\right)[x] = 0. \tag{10}$$

Note that since the given system is homogeneous, $y(t)$ also satisfies this equation (compare (7) and (8) on page 281 of the text). So, can we investigate solutions $x(t)$ only. The auxiliary equation, $r^2 - (\lambda+1)r + (\lambda+3) = 0$, has roots

$$r_1 = \frac{(\lambda+1) - \sqrt{\Delta}}{2}, \qquad r_2 = \frac{(\lambda+1) + \sqrt{\Delta}}{2},$$

where the discriminant $\Delta := (\lambda+1)^2 - 4(\lambda+3)$. We consider two cases:

i) If $\lambda + 3 < 0$, i.e. $\lambda < -3$, then $\Delta > (\lambda+1)^2$ and the root

$$r_2 > \frac{(\lambda+1) + |\lambda+1|}{2} = 0.$$

Therefore, the solution $x(t) = e^{r_2 t}$ is unbounded as $t \to +\infty$.

ii) If $\lambda + 3 \ge 0$, i.e. $\lambda \ge -3$, then $\Delta \le (\lambda+1)^2$. If $\Delta < 0$, then a fundamental solution set to (10) is

$$\left\{ e^{(\lambda+1)t/2} \cos\left(\frac{\sqrt{-\Delta} t}{2}\right), e^{(\lambda+1)t/2} \sin\left(\frac{\sqrt{-\Delta} t}{2}\right) \right\}. \tag{11}$$

If $\Delta \ge 0$, then $\sqrt{\Delta} \le |\lambda+1|$ and a fundamental solution set is

$$\begin{cases} \left\{e^{r_1 t}, e^{r_2 t}\right\}, & \text{if } \Delta > 0, \\ \left\{e^{r_1 t}, te^{r_1 t}\right\}, & \text{if } \Delta = 0, \end{cases} \tag{12}$$

where both roots r_1, r_2 are non-positive if and only if $\lambda \le -1$. For $\lambda = -1$ we have $\Delta = (-1+1)^2 - 4(-1+3) < 0$, and we have a particular case of the fundamental solution set (11) (without exponential term) consisting of bounded functions. Finally, if $\lambda < -1$, then $r_1 < 0$, $r_2 \le 0$ and all the functions listed in (11), (12) are bounded.

Any solution $x(t)$ is a linear combination of fundamental solutions and, therefore, all solutions $x(t)$ are bounded if and only if $-3 \le \lambda \le -1$.

31. Solving this problem, we follow the arguments described in Section 5.1, page 261 of the text, i.e., $x(t)$, the mass of salt in the tank A, and $y(t)$, the mass of salt in the tank B, satisfy the system

$$\begin{aligned} \frac{dx}{dt} &= \text{input}_A - \text{output}_A \\ \frac{dy}{dt} &= \text{input}_B - \text{output}_B \end{aligned} \tag{13}$$

with initial conditions $x(0) = 0$, $y(0) = 200$. It is important to notice that the volume of each tank stays at 100 L because the net flow rate into each tank is the same as the net outflow. Next we observe that " input_A " consists of the salt coming from outside, which is

$$2\,\text{kg/L} \cdot 6\,\text{L/min} = 12\,\text{kg/min},$$

and the salt coming from the tank B, which is given by

$$\frac{y(t)}{100}\,\text{kg/L} \cdot 1\,\text{L/min} = \frac{y(t)}{100}\,\text{kg/min}.$$

Thus,

$$\text{input}_A = \left(12 + \frac{y(t)}{100}\right)\text{kg/min}.$$

" output_A " consists of two flows: one is going out of the system and the other one is going to the tank B. So,

$$\text{output}_A = \frac{x(t)}{100}\,\text{kg/L} \cdot (4+3)\,\text{L/min} = \frac{7x(t)}{100}\,\text{kg/min},$$

and the first equation in (13) becomes

$$\frac{dx}{dt} = 12 + \frac{y}{100} - \frac{7x}{100}.$$

Similarly, the second equation in (13) can be written as

$$\frac{dy}{dt}=\frac{3x}{100}-\frac{3y}{100}.$$

Rewriting this system in the operator form, we obtain

$$\begin{aligned}&(D+0.07)[x]-0.01\,y=12,\\ &-0.03x+(D+0.03)[y]=0.\end{aligned} \qquad (14)$$

Eliminating y yields

$$\{(D+0.07)(D+0.03)-(-0.01)(-0.03)\}[x]=(D+0.03)[12]=0.36\ ,$$

which simplifies to

$$\left(D^2+0.1D+0.0018\right)[x]=0.36. \qquad (15)$$

The auxiliary equation, $r^2+0.1r+0.0018=0$, has roots

$$r_1=-\frac{1}{20}-\sqrt{\frac{1}{400}-\frac{18}{10000}}=-\frac{1}{20}-\frac{\sqrt{7}}{100}=\frac{-5-\sqrt{7}}{100}\approx-0.0765\,,$$

$$r_2=\frac{-5+\sqrt{7}}{100}\approx-0.0235\,.$$

Therefore, the general solution the corresponding homogeneous equation is

$$x_h(t)=C_1e^{r_1t}+C_2e^{r_2t}\,.$$

Since the nonhomogeneous term in (15) is a constant (0.36), we are looking for a particular solution of the form $x_p(t)=A=\text{const}$. Substituting into (15) yields

$$0.0018A=0.36 \qquad \Rightarrow \qquad A=200,$$

and the general solution, $x(t)$, is

$$x(t)=x_h(t)+x_p(t)=C_1e^{r_1t}+C_2e^{r_2t}+200\,.$$

From the first equation in (14) we find

$$\begin{aligned}y(t)&=100\cdot\left((D+0.07I)[x]-12\right)=100\frac{dx}{dt}+7x(t)-1200\\ &=100\left\{r_1C_1e^{r_1t}+r_2C_2\,e^{r_2t}\right\}+7\left\{C_1e^{r_1t}+C_2e^{r_2t}+200\right\}-1200\\ &=\left(2-\sqrt{7}\right)C_1e^{r_1t}+\left(2+\sqrt{7}\right)C_2e^{r_2t}+200\,.\end{aligned}$$

The initial conditions imply

$$\begin{aligned}0&=x(0)=C_1+C_2+200,\\ 200&=y(0)=\left(2-\sqrt{7}\right)C_1+\left(2+\sqrt{7}\right)C_2+200\end{aligned}$$

$$\Rightarrow \quad \begin{aligned} &C_1 + C_2 = -200, \\ &\left(2-\sqrt{7}\right)C_1 + \left(2+\sqrt{7}\right)C_2 = 0 \end{aligned}$$

$$\Rightarrow \quad C_1 = -\left(100 + \frac{200}{\sqrt{7}}\right), \qquad C_2 = -\left(100 - \frac{200}{\sqrt{7}}\right).$$

Thus the solution to the problem is

$$x(t) = -\left(100 + \frac{200}{\sqrt{7}}\right)e^{r_1 t} - \left(100 - \frac{200}{\sqrt{7}}\right)e^{r_2 t} + 200\,\text{kg},$$

$$y(t) = \frac{300}{\sqrt{7}}e^{r_1 t} - \frac{300}{\sqrt{7}}e^{r_2 t} + 200\,\text{kg}.$$

35. Let $x(t)$ and $y(t)$ denote the temperatures at time t in zones A and B, respectively. Therefore, the rate of change of temperature in zone A will be $x'(t)$ and in zone B will be $y'(t)$. We can apply Newton's law of cooling to help us express these rates of change in an alternate manner. Thus, we observe that the rate of change of the temperature in zone A due to the outside temperature is $k_1[100 - x(t)]$ and due to the temperature in zone B is $k_2[y(t) - x(t)]$. Since the time constant for heat transfer between zone A and the outside is 2 hrs $\left(= \frac{1}{k_1}\right)$, we see that $k_1 = \frac{1}{2}$. Similarly, we see that $\frac{1}{k_2} = 4$ which implies that $k_2 = \frac{1}{4}$. Therefore, since there is no heating or cooling source in zone A, we can write the equation for the rate of change of the temperature in the attic as

$$x'(t) = \frac{1}{2}\left(100 - x(t)\right) + \frac{1}{4}\left(y(t) - x(t)\right).$$

In the same way, we see that the rate of change of the temperature in zone B due to the temperature of the attic is $k_3[x(t) - y(t)]$, where $\frac{1}{k_3} = 4$; and the rate of change of the temperature in this zone due to the outside temperature is $k_4[100 - y(t)]$, where $\frac{1}{k_4} = 4$. In this zone, however, we must consider the cooling due to the air conditioner. Since the heat capacity of zone B is $\frac{1}{2}$ °F per thousand Btu and the air conditioner has the cooling capacity of 24 thousand Btu per hr, we see that the air conditioner removes heat from this zone at the rate of $\left(\frac{1}{2} \times 24\right)^{\circ} = 12^{\circ}$ F/hr. (Since heat is *removed* from the house, this rate will be negative.) By combining these observations, we see that the rate of change of the temperature in zone B is given by

$$y'(t) = -12 + \frac{1}{4}\left(x(t) - y(t)\right) + \frac{1}{4}\left(100 - y(t)\right).$$

By simplifying these equations, we observe that this cooling problem satisfies the system

$$4x'(t)+3x(t)-y(t)=200\,,$$
$$-x(t)+4y'(t)+2y(t)=52\,.$$

In operator notation, this system becomes

$$(4D+3)[x]-[y]=200,$$
$$-[x]+(4D+2)[y]=52.$$

Since we are interested in the temperature in the attic, $x(t)$, we will eliminate the function $y(t)$ from the system above by applying $(4D+2)$ to the first equation and adding the resulting equations to obtain

$$((4D+2)(4D+3)-1)[x]=(4D+2)[200]+52=452$$

$$\Rightarrow \qquad \left(16D^2+20D+5\right)[x]=452\,. \qquad (16)$$

This last equation is a linear equation with constant coefficients whose corresponding homogeneous equation has the associated auxiliary equation $16r^2+20r+5=0$. By the quadratic formula, the roots to this auxiliary equation are

$$r_1=\frac{-5+\sqrt{5}}{8}\approx-0.345\,, \quad \text{and} \quad r_2=\frac{-5-\sqrt{5}}{8}\approx-0.905\,.$$

Therefore, the homogeneous equation associated with this equation has a general solution given by

$$x_h(t)=c_1e^{r_1t}+c_2e^{r_2t}\,,$$

where r_1 and r_2 are given above. By the method of undetermined coefficients, we observe that a particular solution to equation (16) will have the form

$$x_p(t)=A$$

$$\Rightarrow \qquad x_p'(t)=0$$

$$\Rightarrow \qquad x_p''(t)=0\,.$$

Substituting these expressions into equation (16) yields

$$16x_p''+20x_p'+5x_p=5A=452 \qquad \Rightarrow \qquad A=90.4\,.$$

Thus, a particular solution to the differential equation given in (16) is $x_p(t)=9.4$ and the general solution to this equation will be

$$x(t)=c_1e^{r_1t}+c_2e^{r_2t}+90.4\,,$$

where $r_1=\dfrac{-5+\sqrt{5}}{8}$ and $r_2=\dfrac{-5-\sqrt{5}}{8}$. To determine the maximum temperature of the attic, we will assume that zones A and B have sufficiently cool initial temperatures. (So that, for example,

c_1 and c_2 are negative.) Since r_1 and r_2 are negative, as t goes to infinity, $c_1e^{r_1t}$ and $c_2e^{r_2t}$ each go to zero. Therefore, the maximum temperature that can be attained in the attic will be

$$\lim_{t\to\infty} x(t) = 90.4°\text{ F}.$$

EXERCISES 5.4: Coupled Mass-Spring Systems, page 291

1. For the mass m_1 there is only one force acting on it; that is the force due to the spring with constant k_1. This equals $-k_1(x-y)$. Hence, we get

$$m_1x'' = -k_1(x-y).$$

For the mass m_2 there are two forces acting on it: the force due to the spring with constant k_2 is $-k_2y$; and the force due to the spring with constant k_1 is $k_1(y-x)$. So we get

$$m_2y'' = k_1(x-y) - k_2y.$$

So the system is

$$m_1x'' = k_1(y-x),$$

$$m_2y'' = -k_1(y-x) - k_2y,$$

or in operator form

$$\left(m_1D^2 + k_1\right)[x] - k_1y = 0,$$

$$-k_1x + \left(m_2D^2 + \left(k_1 + k_2\right)\right)[y] = 0.$$

With $m_1 = 1$, $m_2 = 2$, $k_1 = 4$, and $k_2 = \frac{10}{3}$, we get

$$\begin{aligned} &\left(D^2 + 4\right)[x] - 4y = 0, \\ &-4x + \left(2D^2 + \frac{22}{3}\right)[y] = 0, \end{aligned} \qquad (17)$$

with initial conditions:

$$x(0) = -1, \quad x'(0) = 0, \quad y(0) = 0, \quad y'(0) = 0.$$

Multiplying the second equation of the system given in (17) by 4, applying $\left(2D^2 + \frac{22}{3}\right)$ to the first equation of this system, and then adding, we get

$$\left(D^2 + 4\right)\left(2D^2 + \frac{22}{3}\right)[x] - 16x = 0,$$

$$\Rightarrow \quad \left(2D^4 + \frac{46}{3}D^2 + \frac{40}{3}\right)[x] = 0,$$

$$\Rightarrow \quad (3D^4 + 23D^2 + 20)[x] = 0.$$

The characteristic equation is

$$3r^4 + 23r^2 + 20 = 0,$$

which is a quadratic in r^2. So

$$r^2 = \frac{-23 \pm \sqrt{529 - 240}}{6} = \frac{-23 \pm 17}{6}.$$

Since $\frac{-20}{3}$ and -1 are negative, the roots of the characteristic equation are $\pm i\beta_1$ and $\pm i\beta_2$, where

$$\beta_1 = \sqrt{\frac{20}{3}}, \quad \beta_2 = 1.$$

Hence

$$x(t) = c_1 \cos \beta_1 t + c_2 \sin \beta_1 t + c_3 \cos \beta_2 t + c_4 \sin \beta_2 t.$$

Solving the first equation of the system given in (17) for y, we get

$$y(t) = \frac{1}{4}(D^2 + 4I)[x] = \frac{1}{4}\left[(-\beta_1^2 + 4)c_1 \cos \beta_1 t + (-\beta_1^2 + 4)c_2 \sin \beta_1 t\right.$$

$$\left. + (-\beta_2^2 + 4)c_3 \cos \beta_2 t + (-\beta_2^2 + 4)c_4 \sin \beta_2 t\right].$$

Next we substitute into the initial conditions. Setting $x(0) = -1$, $x'(0) = 0$ yields

$$-1 = c_1 + c_3,$$

$$0 = c_2\beta_1 + c_4\beta_2.$$

From the initial conditions $y(0) = 0$, $y'(0) = 0$, we get

$$0 = \frac{1}{4}\left[(-\beta_1^2 + 4)c_1 + (-\beta_2^2 + 4)c_3\right],$$

$$0 = \frac{1}{4}\left[\beta_1(-\beta_1^2 + 4)c_2 + \beta_2(-\beta_2^2 + 4)c_4\right].$$

The solution to the above system is

$$c_2 = c_4 = 0, \qquad c_1 = -\frac{9}{17}, \qquad c_3 = -\frac{8}{17},$$

which yields the solutions

$$x(t) = -\frac{9}{17}\cos\sqrt{\frac{20}{3}}t - \frac{8}{17}\cos t\,,$$

$$y(t) = \frac{6}{17}\cos\sqrt{\frac{20}{3}}t - \frac{6}{17}\cos t\,.$$

5. This spring system is similar to the system in Example 2 on page 290 of the text, except the middle spring has been replaced by a dashpot. We proceed as in Example 1. Let x and y represent the displacement of masses m_1 and m_2 to the right of their respective equilibrium positions. The mass m_1 has a force F_1 acting on its left side due to the left spring and a force F_2 acting on its right side due to the dashpot. Applying Hooke's law, we see that

$$F_1 = -k_1 x\,.$$

Assuming as we did in Section 4.1 that the damping force due to the dashpot is proportional to the magnitude of the velocity, but opposite in direction, we have

$$F_2 = b(y' - x'),$$

where b is the damping constant. Notice that velocity of the arm of the dashpot is the difference between the velocities of mass m_2 and mass m_1. The mass m_2 has a force F_3 acting on its left side due to the dashpot and a force F_4 acting on its right side due to the right spring. Using similar arguments, we find

$$F_3 = -b(y' - x') \qquad \text{and} \qquad F_4 = -k_2 y\,.$$

Applying Newton's second law to each mass gives

$$m_1 x''(t) = F_1 + F_2 = -k_1 x(t) + b[y'(t) - x'(t)]\,,$$

$$m_2 y''(t) = F_3 + F_4 = -b[y'(t) - x'(t)] - k_2 y(t)\,.$$

Plugging in the constants $m_1 = m_2 = 1$, $k_1 = k_2 = 1$, and $b = 1$, and simplifying yields

$$\begin{aligned} x''(t) + x'(t) + x(t) - y'(t) &= 0, \\ -x'(t) + y''(t) + y'(t) + y(t) &= 0. \end{aligned} \tag{18}$$

The initial conditions for the system will be $y(0) = 0$ (m_2 is held in its equilibrium position), $x(0) = -2$ (m_1 is pushed to the left 2 ft), and $x'(0) = y'(0) = 0$ (the masses are simply released at time $t = 0$ with no additional velocity). In operator notation this system becomes

$$\begin{aligned} &(D^2 + D + 1)[x] - D[y] = 0, \\ &-D[x] + (D^2 + D + 1)[y] = 0, \end{aligned}$$

By multiplying the first equation above by D and the second by $(D^2 + D + 1)$ and adding the resulting equations, we can eliminate the function $y(t)$. Thus, we have

$$\left\{\left(D^2+D+1\right)^2-D^2\right\}[x]=0$$

$$\Rightarrow \quad \left\{\left[\left(D^2+D+1\right)-D\right]\cdot\left[\left(D^2+D+1\right)+D\right]\right\}[x]=0$$

$$\Rightarrow \quad \left\{\left(D^2+1\right)\left(D+1\right)^2\right\}[x]=0\,.$$

This last equation is a fourth order linear differential equation with constant coefficients whose associated auxiliary equation has roots $r=-1$, -1, i, and $-i$. Therefore, the solution to this differential equation is

$$x(t)=c_1e^{-t}+c_2te^{-t}+c_3\cos t+c_4\sin t$$

$$\Rightarrow \quad x'(t)=\left(-c_1+c_2\right)e^{-t}-c_2te^{-t}-c_3\sin t+c_4\cos t$$

$$\Rightarrow \quad x''(t)=\left(c_1-2c_2\right)e^{-t}+c_2te^{-t}-c_3\cos t-c_4\sin t\,.$$

To find $y(t)$, note that by the first equation of the system given in (18), we have

$$y'(t)=x''(t)+x'(t)+x(t)\,.$$

Substituting the function $x(t)$ into this equation yields

$$y'(t)=\left(c_1-2c_2\right)e^{-t}+c_2te^{-t}-c_3\cos t-c_4\sin t+\left(-c_1+c_2\right)e^{-t}-c_2te^{-t}-c_3\sin t+c_4\cos t$$
$$+c_1e^{-t}+c_2te^{-t}+c_3\cos t+c_4\sin t$$

$$\Rightarrow \quad y'(t)=\left(c_1-c_2\right)e^{-t}+c_2te^{-t}-c_3\sin t+c_4\cos t\,.$$

By integrating both sides of this equation with respect to t, we obtain

$$y(t)=-\left(c_1-c_2\right)e^{-t}-c_2te^{-t}-c_2e^{-t}+c_3\cos t+c_4\sin t+c_5\,,$$

where we have integrated c_2te^{-t} by parts. Simplifying yields

$$y(t)=-c_1e^{-t}-c_2te^{-t}+c_3\cos t+c_4\sin t+c_5\,.$$

To determine the five constants, we will use the four initial conditions and the second equation in system (18). (We used the first equation to determine y). Substituting into the second equation in (18) gives

$$-\left(\left(-c_1+c_2\right)e^{-t}-c_2te^{-t}-c_3\sin t+c_4\cos t\right)+\left(\left(-c_1+2c_2\right)e^{-t}-c_2te^{-t}-c_3\cos t-c_4\sin t\right)$$
$$+\left(\left(c_1-c_2\right)e^{-t}+c_2te^{-t}-c_3\sin t+c_4\cos t\right)+\left(-c_1e^{-t}-c_2te^{-t}+c_3\cos t+c_4\sin t+c_5\right)=0,$$

which reduces to $c_5=0$. Using the initial conditions and the fact that $c_5=0$, we see that

$$x(0)=c_1+c_3=-2\,, \qquad x'(0)=\left(-c_1+c_2\right)+c_4=0\,,$$

$$y(0)=-c_1+c_3=0\,, \qquad y'(0)=\left(c_1-c_2\right)+c_4=0\,.$$

By solving these equations simultaneously, we find

$$c_1 = -1, \quad c_2 = -1, \quad c_3 = -1, \quad \text{and} \quad c_4 = 0.$$

Therefore, the solution to this spring-mass-dashpot system is

$$x(t) = -e^{-t} - te^{-t} - \cos t, \quad y(t) = e^{-t} + te^{-t} - \cos t.$$

9. Writing the equations of this system in operator form we obtain

$$\left(mD^2 + \left(\frac{mg}{l} + k\right)\right)[x_1] - kx_2 = 0,$$

$$-kx_1 + \left(mD^2 + \left(\frac{mg}{l} + k\right)\right)[x_2] = 0.$$

Applying $\left(mD^2 + (mg/l + k)\right)$ to the first equation, multiplying the second equation by k, and then adding, results in

$$\left[\left(mD^2 + \left(\frac{mg}{l} + k\right)\right)^2 - k^2\right][x_1] = 0.$$

This equation has the auxiliary equation

$$\left(mr^2 + \frac{mg}{l} + k\right)^2 - k^2 = \left(mr^2 + \frac{mg}{l}\right)\left(mr^2 + \frac{mg}{l} + 2k\right) = 0$$

with roots $\pm i\sqrt{\frac{g}{l}}$ and $\pm i\sqrt{\frac{g}{l} + \frac{2k}{m}}$. As discussed on page 291 of the text $\sqrt{\frac{g}{l}}$ and $\sqrt{\frac{g}{l} + \frac{2k}{m}}$ are the normal angular frequencies. To find the normal frequencies we divide each one by 2π and obtain $\left(\frac{1}{2\pi}\right)\sqrt{\frac{g}{l}}$ and $\left(\frac{1}{2\pi}\right)\sqrt{\frac{g}{l} + \frac{2k}{m}}$.

EXERCISES 5.5: Electrical Circuits, page 300

3. In this problem $L = 4$, $R = 120$, $C = (2200)^{-1}$, and $E(t) = 10\cos 20t$. Therefore, we see that $\frac{1}{C} = 2200$ and $E'(t) = -200\sin 20t$. By substituting these values into equation (4) on page 295 of the text, we obtain the equation

$$4\frac{d^2I}{dt^2} + 120\frac{dI}{dt} + 2200I = -200\sin 20t.$$

Or, by simplifying, we have

$$\frac{d^2 I}{dt^2}+30\frac{dI}{dt}+550I=-50\sin 20t\,. \tag{19}$$

The auxiliary equation associated with the homogeneous equation corresponding to (19) above is $r^2+30\mathrm{r}+550=0$. This equation has roots $r=-15\pm 5\sqrt{13}i$. Therefore, the transient current, that is $I_h(t)$, is given by

$$I_h(t)=\mathrm{e}^{-15t}\left[C_1\cos\left(5\sqrt{13}t\right)+C_2\sin\left(5\sqrt{13}t\right)\right].$$

By the method of undetermined coefficients, a particular solution, $I_p(t)$, of equation (19) will be of the form $I_p(t)=t^s\left[A\cos 20t+B\sin 20t\right]$. Since neither $y(t)=\cos 20t$ nor $y(t)=\sin 20t$ is a solution to the homogeneous equation (that is the system is not at resonance), we can let $s=0$ in $I_p(t)$. Thus, we see that $I_p(t)$, the steady-state current, has the form

$$I_p(t)=A\cos 20t+B\sin 20t\,.$$

To find the steady-state current, we must, therefore, find *A* and *B*. To accomplish this, we observe that

$$I_p'(t)=-20A\sin 20t+20B\cos 20t$$

$$\Rightarrow\qquad I_p''(t)=-400A\cos 20t-400B\sin 20t\,.$$

Plugging these expressions into equation (19) yields

$$I_p''(t)+30I_p'(t)+550I_p=-400A\cos 20t-400B\sin 20t-600A\sin 20t+600B\cos 20t$$

$$+550A\cos 20t+550B\sin 20t=-50\sin 20t$$

$$\Rightarrow\qquad (150A+600B)\cos 20t+(150B-600A)\sin 20t=-50\sin 20t\,.$$

By equating coefficients we obtain the system of equations

$$15A+60B=0\,,$$

$$-60A+15B=-5\,.$$

By solving these equations simultaneously for *A* and *B*, we obtain $A=\frac{4}{51}$ and $B=\frac{-1}{51}$. Thus, we have the steady-state current given by

$$I_p(t)=\frac{4}{51}\cos 20t-\frac{1}{51}\sin 20t\,.$$

As was observed on page 299 of the text, there is a correlation between the RLC series circuits and mechanical vibration. Therefore, we can discuss the resonance frequency of the RLC series circuit. To do so we associate the variable *L* with *m*, *R* with *b*, and $1/C$ with *k*. Thus, we see that the resonance frequency for an RLC series circuit is given by $\gamma_r/2\pi$, where

$$\gamma_r = \sqrt{\frac{1}{CL} - \frac{R^2}{2L^2}},$$

provided that $R^2 < 2L/C$. For this problem

$$R^2 = 14{,}400 < 2L/C = 17{,}600.$$

Therefore, we can find the resonance frequency of this circuit. To do so we first find

$$\gamma_r = \sqrt{\frac{1}{CL} - \frac{R^2}{2L^2}} = \sqrt{\frac{2200}{4} - \frac{14400}{32}} = 10.$$

Hence the resonance frequency of this circuit is $\frac{10}{2\pi} = \frac{5}{\pi}$.

7. This spring system satisfies the differential equation

$$7\frac{d^2x}{dt^2} + 2\frac{dx}{dt} + 3x = 10\cos 10t.$$

Since we want to find an *RLC* series circuit analog for the spring system with $R = 10$ ohms, we must find L, $1/C$, and $E(t)$ so that the differential equation

$$L\frac{d^2q}{dt^2} + 10\frac{dq}{dt} + \frac{1}{C}q = E(t)$$

corresponds to the one above. Thus, we want $E(t) = 50\cos 10t$ volts, $L = 35$ henrys, and $C = 1/15$ farads.

11. For this electric network, there are three loops. Loop 1 is through a 10V battery, a 10Ω resistor, and a 20H inductor. Loop 2 is through a 10V battery, a 10Ω resistor, a 5Ω resistor, and a $\frac{1}{30}$F capacitor. Loop 3 is through a 5Ω resistor, a $\frac{1}{30}$F capacitor, and a 20H inductor. Therefore, applying Kirchhoff's second law to this network yields the three equations given by

Loop 1 $\quad 10I_1 + 20\frac{dI_2}{dt} = 10,$

Loop 2 $\quad 10I_1 + 5I_3 + 30q_3 = 10,$

Loop 3 $\quad 5I_3 + 30q_3 - 20\frac{dI_2}{dt} = 0.$

Since the equation for loop 2 minus the equation for loop 1 yields the remaining equation, we will use the first and second equations above for our calculations. By examining a junction point, we see that we also have the equation $I_1 = I_2 + I_3$. Thus, we have $\frac{dI_1}{dt} = \frac{dI_2}{dt} + \frac{dI_3}{dt}$. We begin by dividing the equation for loop 1 by 10 and the equation for loop 2 by 5. Differentiating the equation for loop 2 yields the system

$$I_1 + 2\frac{dI_2}{dt} = 1,$$

$$2\frac{dI_1}{dt} + \frac{dI_3}{dt} + 6I_3 = 0,$$

where $I_3 = \frac{dq_3}{dt}$. Since $I_1 = I_2 + I_3$ and $\frac{dI_1}{dt} = \frac{dI_2}{dt} + \frac{dI_3}{dt}$, we can rewrite the system using operator notation in the form

$$(2D+1)[I_2] + I_3 = 1,$$

$$(2D)[I_2] + (3D+6)[I_3] = 0.$$

If we "multiply" the first equation above by $(3D+6)$ and then subtract the second equation, we obtain

$$\{(3D+6)(2D+1) - 2D\}[I_2] = 6 \qquad \Rightarrow \qquad (6D^2 + 13D + 6)[I_2] = 6.$$

This last differential equation is a linear equation with constant coefficients whose corresponding homogeneous equation has the associated auxiliary equation $6r^2 + 13r + 6 = 0$. The roots to this auxiliary are $r = \frac{-3}{2}, \frac{-2}{3}$. Therefore, the solution to the homogeneous equation corresponding to the equation above is given by

$$I_{2h}(t) = c_1 e^{-3t/2} + c_2 e^{-2t/3}.$$

By the method of undetermined coefficients, according to Table 4.1 on page 208 of the text, the form of a particular solution to the differential equation above will be $I_{2p}(t) = A$. By substituting this function into the differential equation, we see that a particular solution is given by

$$I_{2p}(t) = 1.$$

Thus, the current, I_2, will satisfy the equation

$$I_2(t) = c_1 e^{-3t/2} + c_2 e^{-2t/3} + 1.$$

As we noticed above, I_3 can now be found from the first equation

$$I_3(t) = -(2D+1)[I_2] + 1 = -2\left[\frac{-3}{2}c_1 e^{-3t/2} - \frac{2}{3}c_2 e^{-2t/3}\right] - \left[c_1 e^{-3t/2} + c_2 e^{-2t/3} + 1\right] + 1$$

$$\Rightarrow \qquad I_3(t) = 2c_1 e^{-3t/2} + \frac{1}{3}c_2 e^{-2t/3}.$$

To find I_1, we will use the equation $I_1 = I_2 + I_3$. Therefore, we have

$$I_1(t) = c_1 e^{-3t/2} + c_2 e^{-2t/3} + 1 + 2c_1 e^{-3t/2} + \frac{1}{3}c_2 e^{-2t/3}$$

$$\Rightarrow \qquad I_1(t) = 3c_1 e^{-3t/2} + \frac{4}{3}c_2 e^{-2t/3} + 1.$$

We will use the initial condition $I_2(0) = I_3(0) = 0$ to find the constants c_1 and c_2. Thus, we have

$$I_2(0) = c_1 + c_2 + 1 = 0 \qquad \text{and} \qquad I_3(0) = 2c_1 + \frac{1}{3}c_2 = 0.$$

Solving these two equations simultaneously yields $c_1 = \frac{1}{5}$ and $c_2 = \frac{-6}{5}$. Therefore, the equations for the currents for this electric network are given by

$$I_1(t) = \frac{3}{5}e^{-3t/2} - \frac{8}{5}e^{-2t/3} + 1,$$

$$I_2(t) = \frac{1}{5}e^{-3t/2} - \frac{6}{5}e^{-2t/3} + 1,$$

$$I_3(t) = \frac{2}{5}e^{-3t/2} - \frac{2}{5}e^{-2t/3}.$$

EXERCISES 5.6: Numerical Methods for Higher – Order Equations and Systems, page 311

5. First we express the given system as

$$x'' = x' - y + 2t,$$
$$y'' = x - y - 1.$$

Setting $x_1 = x$, $x_2 = x'$, $x_3 = y$, $x_4 = y'$ we obtain

$$\begin{aligned} x_1' &= x' = x_2, & & & x_1' &= x_2, \\ x_2' &= x'' = x_2 - x_3 + 2t, & &\Rightarrow & x_2' &= x_2 - x_3 + 2t, \\ x_3' &= y' = x_4, & & & x_3' &= x_4, \\ x_4' &= y'' = x_1 - x_3 - 1 & & & x_4' &= x_1 - x_3 - 1 \end{aligned}$$

with initial conditions

$$x_1(3) = 5, \quad x_2(3) = 2, \quad x_3(3) = 1, \quad x_4(3) = -1.$$

10. To see how the improved Euler's method can be extended let's recall, from Section 3.5, the improved Euler's method on page 129 of the text. For the initial value problem

$$x' = f(x,t), \qquad x(t_0) = x_0,$$

the recursive formulas for the improved Euler's method are

$$t_{n+1} = t_n + h,$$

$$x_{n+1} = x_n + \frac{h}{2}\left(f(t_n, x_n) + f(t_n + h, x_n + hf(t_n, x_n))\right),$$

where h is the step size. Now suppose we want to approximate the solution $x_1(t)$, $x_2(t)$ to the system

$$x_1' = f_1(t, x_1, x_2), \quad \text{and} \quad x_2' = f_2(t, x_1, x_2),$$

that satisfies the initial conditions

$$x_1(t_0) = a_1, \; x_2(t_0) = a_2.$$

Let $x_{n,1}$ and $x_{n,2}$, and denote approximations to $x_1(t_n)$ and $x_2(t_n)$, respectively, where $t_n = t_0 + nh$ for $n = 0, 1, 2, \ldots$. The recursive formulas for the improved Euler's method are obtained by forming the vector analogue of the scalar formula. We obtain

$$t_{n+1} = t_n + h,$$

$$x_{n+1,1} = x_{n,1} + \frac{h}{2}\left(f_1(t_n, x_{n,1}, x_{n,2}) + f_1(t_n + h, x_{n,1} + hf_1(t_n, x_{n,1}, x_{n,2}), x_{n,2} + hf_2(t_n, x_{n,1}, x_{n,2}))\right),$$

$$x_{n+1,2} = x_{n,2} + \frac{h}{2}\left(f_2(t_n, x_{n,1}, x_{n,2}) + f_2(t_n + h, x_{n,1} + hf_1(t_n, x_{n,1}, x_{n,2}), x_{n,2} + hf_2(t_n, x_{n,1}, x_{n,2}))\right).$$

The approach can be used more generally for systems of m equations in normal form. Suppose we want to approximate the solution $x_1(t)$, $x_2(t)$, ..., $x_m(t)$ to the system

$$x_1' = f_1(t, x_1, x_2, \ldots, x_m),$$

$$x_2' = f_2(t, x_1, x_2, \ldots, x_m),$$

$$\vdots$$

$$x_m' = f_m(t, x_1, x_2, \ldots, x_m),$$

with the initial conditions

$$x_1(t_0) = a_1, \quad x_2(t_0) = a_2, \quad \ldots, \quad x_m(t_0) = a_m.$$

We adapt the recursive formulas above to obtain

$$t_{n+1} = t_n + h, \qquad n = 0, 1, 2, \ldots;$$

$$x_{n+1,1} = x_{n,1} + \frac{h}{2}\big(f_1(t_n, x_{n,1}, x_{n,2}, \ldots, x_{n,m}) + f_1(t_n + h, x_{n,1} + hf_1(t_n, x_{n,1}, x_{n,2}, \ldots, x_{n,m}),$$
$$x_{n,2} + hf_2(t_n, x_{n,1}, x_{n,2}, \ldots, x_{n,m}), \ldots, x_{n,m} + hf_m(t_n, x_{n,1}, x_{n,2}, \ldots, x_{n,m}))\big),$$

$$x_{n+1,2} = x_{n,2} + \frac{h}{2}\big(f_2(t_n, x_{n,1}, x_{n,2}, \ldots, x_{n,m}) + f_2(t_n + h, x_{n,1} + hf_1(t_n, x_{n,1}, x_{n,2}, \ldots, x_{n,m}),$$
$$x_{n,2} + hf_2(t_n, x_{n,1}, x_{n,2}, \ldots, x_{n,m}), \ldots, x_{n,m} + hf_m(t_n, x_{n,1}, x_{n,2}, \ldots, x_{n,m}))\big),$$

$$\vdots$$

$$x_{n+1,m} = x_{n,m} + \frac{h}{2}\big(f_{m1}(t_n, x_{n,1}, x_{n,2}, \ldots, x_{n,m}) + f_m(t_n + h, x_{n,1} + hf_1(t_n, x_{n,1}, x_{n,2}, \ldots, x_{n,m}),$$
$$x_{n,2} + hf_2(t_n, x_{n,1}, x_{n,2}, \ldots, x_{n,m}), \ldots, x_{n,m} + hf_m(t_n, x_{n,1}, x_{n,2}, \ldots, x_{n,m}))\big).$$

21. Let $x_1 := x$ and $x_2 := y$. For starting values we take $t_0 = 0$, $x_{0,1} = 1$, and $x_{0,2} = 0.5$, which are determined by the initial conditions. Here $h = 0.25$, and

$$f_1(t, x_1, x_2) = \cos(x_1 + x_2),$$

$$f_2(t, x_1, x_2) = (x_1^2 + t)^{-1} + t.$$

Now using the definitions of $t_n, x_{n,i}, k_{1,i}, k_{2,i}, k_{3,i}$ and $k_{4,i}$ on pages 303-304 of the text, we have

$$k_{1,1} = hf_1(t_n, x_{n,1}, x_{n,2}) = (0.25)\cos(x_{n,1} + x_{n,2}),$$

$$k_{1,2} = hf_2(t_n, x_{n,1}, x_{n,2}) = (0.25)\left(\frac{1}{x_{n,1}^2 + t_n} + t_n\right),$$

$$k_{2,1} = hf_1\left(t_n + \frac{h}{2}, x_{n,1} + \frac{k_{1,1}}{2}, x_{n,2} + \frac{k_{1,2}}{2}\right) = (0.25)\cos\left(x_{n,1} + \frac{k_{1,1}}{2} + x_{n,2} + \frac{k_{1,2}}{2}\right),$$

$$k_{2,2} = hf_2\left(t_n + \frac{h}{2}, x_{n,1} + \frac{k_{1,1}}{2}, x_{n,2} + \frac{k_{1,2}}{2}\right) = (0.25)\left(\frac{1}{\left(x_{n,1} + \frac{k_{1,1}}{2}\right)^2 + t_n + 0.125} + t_n + 0.125\right),$$

$$k_{3,1} = hf_1\left(t_n + \frac{h}{2}, x_{n,1} + \frac{k_{2,1}}{2}, x_{n,2} + \frac{k_{2,2}}{2}\right) = (0.25)\cos\left(x_{n,1} + \frac{k_{2,1}}{2} + x_{n,2} + \frac{k_{2,2}}{2}\right),$$

$$k_{3,2} = hf_2\left(t_n + \frac{h}{2}, x_{n,1} + \frac{k_{2,1}}{2}, x_{n,2} + \frac{k_{2,2}}{2}\right) = (0.25)\left(\frac{1}{\left(x_{n,1} + \frac{k_{2,1}}{2}\right)^2 + t_n + 0.125} + t_n + 0.125\right),$$

$$k_{4,1} = hf_1(t_n + h, x_{n,1} + k_{3,1}, x_{n,2} + k_{3,2}) = (0.25)\cos(x_{n,1} + k_{3,1} + x_{n,2} + k_{3,2}),$$

$$k_{4,2} = hf_2(t_n + h, x_{n,1} + k_{3,1}, x_{n,2} + k_{3,2}) = (0.25)\left(\frac{1}{(x_{n,1} + k_{3,1})^2 + t_n + 0.25} + t_n + 0.25\right).$$

Using these values, we find

$$t_{n+1} = t_n + h = t_n + 0.25,$$

$$x_{n+1,1} = x_{n,1} + \frac{1}{6}(k_{1,1} + 2k_{2,1} + 2k_{3,1} + k_{4,1}),$$

$$x_{n+1,2} = x_{n,2} + \frac{1}{6}(k_{1,2} + 2k_{2,2} + 2k_{3,2} + k_{4,2}).$$

In Table 5-A we have given the approximations $x_{n,1} = x_n$ and $x_{n,2} = y_n$ for $x_1(t_n) = x(t_n)$ and $x_2(t_n) = y(t_n)$ where $t_n = t_0 + n(0.25)$, for $n = n =$ 1, 2, 3, and 4.

Table 5-A: Approximations of the Solution to Problem 21.

n	t_n	$x(t_n) \approx x_{n,1}$	$y(t_n) \approx x_{n,2}$
1	0.25	0.98661	0.75492
2	0.50	0.91749	1.04259
3	0.75	0.79823	1.38147
4	1.00	0.63303	1.77948

22. For starting values we take $t_0 = 0$, $x_{0,1} = 10$, and $x_{0,2} = 15$, which are determined by the initial conditions. Here $h = 0.1$, and

$$f_1(t, x_1, x_2) = -(0.1)x_1x_2,$$

$$f_2(t, x_1, x_2) = -x_1.$$

Now, using the definitions of $t_n, x_{n,i}, k_{1,i}, k_{2,i}, k_{3,i}$, and $k_{4,i}$ on page 306 of the text, we have

$$k_{1,1} = hf_1(t_n, x_{n,1}, x_{n,2}) = -h(0.1)x_{n,1}x_{n,2},$$

$$k_{1,2} = hf_2(t_n, x_{n,1}, x_{n,2}) = -hx_{n,1},$$

$$k_{2,1} = hf_1\left(t_n + \frac{h}{2}, x_{n,1} + \frac{k_{1,1}}{2}, x_{n,2} + \frac{k_{1,2}}{2}\right) = -h(0.1)\left(x_{n,1} + \frac{k_{1,1}}{2}\right)\left(x_{n,2} + \frac{k_{1,2}}{2}\right),$$

$$k_{2,2} = hf_2\left(t_n + \frac{h}{2}, x_{n,1} + \frac{k_{1,1}}{2}, x_{n,2} + \frac{k_{1,2}}{2}\right) = -h\left(x_{n,1} + \frac{k_{1,1}}{2}\right),$$

$$k_{3,1} = hf_1\left(t_n + \frac{h}{2}, x_{n,1} + \frac{k_{2,1}}{2}, x_{n,2} + \frac{k_{2,2}}{2}\right) = -h(0.1)\left(x_{n,1} + \frac{k_{2,1}}{2}\right)\left(x_{n,2} + \frac{k_{2,2}}{2}\right),$$

$$k_{3,2} = hf_2\left(t_n + \frac{h}{2}, x_{n,1} + \frac{k_{2,1}}{2}, x_{n,2} + \frac{k_{2,2}}{2}\right) = -h\left(x_{n,1} + \frac{k_{2,1}}{2}\right),$$

$$k_{4,1} = hf_1(t_n + h, x_{n,1} + k_{3,1}, x_{n,2} + k_{3,2}) = -h(0.1)(x_{n,1} + k_{3,1})(x_{n,2} + k_{3,2}),$$

$$k_{4,2} = hf_2(t_n + h, x_{n,1} + k_{3,1}, x_{n,2} + k_{3,2}) = -h(x_{n,1} + k_{3,1}).$$

Using these values, we find

$$t_{n+1} = t_n + h = t_n + 0.1,$$

$$x_{n+1,1} = x_{n,1} + \frac{1}{6}(k_{1,1} + 2k_{2,1} + 2k_{3,1} + k_{4,1}),$$

$$x_{n+1,2} = x_{n,2} + \frac{1}{6}(k_{1,2} + 2k_{2,2} + 2k_{3,2} + k_{4,2}).$$

In Table 5-B we give approximate values for t_n, $x_{n,1}$, and $x_{n,2}$.

Table 5-B: Approximations of the Solutions to Problem 22.

t_n	$x_{n,1} \approx$	$x_{n,2} \approx$
0	10	15
1	3.124	9.353
2	1.381	7.254
3	0.707	6.256
4	0.389	5.726
5	0.223	5.428

From Table 5-B we see that the strength of the guerrilla troops, x_1, approaches zero, therefore with the combat effectiveness coefficients of 0.1 for guerrilla troops and 1 for conventional troops the conventional troops win.

27. Let $x_1 = y$ and $x_2 = y'$ to give the initial value problem

$$x_1' = f_1(t, x_1, x_2) = x_2 \qquad x_1(0) = a,$$

$$x_2' = f_2(t, x_1, x_2) = -x_1(1 + rx_1^2) \qquad x_2(0) = 0.$$

Now, using the definitions of $t_n, x_{n,i}, k_{1,i}, k_{2,i}, k_{3,i}$, and $k_{4,i}$ on page 306 of the text, we have

$$k_{1,1} = hf_1(t_n, x_{n,1}, x_{n,2}) = hx_{n,2},$$

$$k_{1,2} = hf_2(t_n, x_{n,1}, x_{n,2}) = -hx_{n,1}(1 + rx_{n,1}^2),$$

$$k_{2,1} = hf_1\left(t_n + \frac{h}{2}, x_{n,1} + \frac{k_{1,1}}{2}, x_{n,2} + \frac{k_{1,2}}{2}\right) = h\left(x_{n,2} + \frac{k_{1,2}}{2}\right),$$

$$k_{2,2} = hf_2\left(t_n + \frac{h}{2}, x_{n,1} + \frac{k_{1,1}}{2}, x_{n,2} + \frac{k_{1,2}}{2}\right) = -h\left(x_{n,1} + \frac{k_{1,1}}{2}\right)\left(1 + r\left(x_{n,1} + \frac{k_{1,1}}{2}\right)^2\right),$$

$$k_{3,1} = hf_1\left(t_n + \frac{h}{2}, x_{n,1} + \frac{k_{2,1}}{2}, x_{n,2} + \frac{k_{2,2}}{2}\right) = h\left(x_{n,2} + \frac{k_{2,2}}{2}\right),$$

$$k_{3,2} = hf_2\left(t_n + \frac{h}{2}, x_{n,1} + \frac{k_{2,1}}{2}, x_{n,2} + \frac{k_{2,2}}{2}\right) = -h\left(x_{n,1} + \frac{k_{2,1}}{2}\right)\left(1 + r\left(x_{n,1} + \frac{k_{2,1}}{2}\right)^2\right),$$

$$k_{4,1} = hf_1(t_n + h, x_{n,1} + k_{3,1}, x_{n,2} + k_{3,2}) = h(x_{n,2} + k_{3,2}),$$

$$k_{4,2} = hf_2(t_n + h, x_{n,1} + k_{3,1}, x_{n,2} + k_{3,2}) = -h(x_{n,1} + k_{3,1})(1 + r(x_{n,1} + k_{3,1})^2).$$

Using these values, we find

$$t_{n+1} = t_n + h = t_n + 0.1,$$

$$x_{n+1,1} = x_{n,1} + \frac{1}{6}(k_{1,1} + 2k_{2,1} + 2k_{3,1} + k_{4,1}),$$

$$x_{n+1,2} = x_{n,2} + \frac{1}{6}(k_{1,2} + 2k_{2,2} + 2k_{3,2} + k_{4,2}).$$

In Table 5-C we give the approximate period for $r = 1$ and 2 with $a = 1$, 2 and 3, from this we see that the period varies as r is varied or as a is varied.

Table 5-C: Approximate period of the solution to Problem 27.

r	$a=1$	$a=2$	$a=3$
1	4.8	3.3	2.3
2	4.0	2.4	1.7

29. With $x_1 = y$, $x_2 = y'$ and $x_3 = y''$, the initial value problem can be expressed as the system

$$x_1' = x_2, \qquad x_1(0) = 1,$$

$$x_2' = x_3, \qquad x_2(0) = 1,$$

$$x_3' = t - x_3 - x_1^2, \qquad x_3(0) = 1.$$

Here

$$f_1(t, x_1, x_2, x_3) = x_2,$$

$$f_2(t, x_1, x_2, x_3) = x_3,$$

$$f_3(t, x_1, x_2, x_3) = t - x_3 - x_1^2.$$

Since we are computing the approximations for $c = 1$, the initial value for h in Step 1 of the algorithm on page 306 of the text is

$$h = (1-0)2^{-0} = 1.$$

The equations in Step 3 are

$$k_{1,1} = hf_1(t, x_1, x_2, x_3) = hx_2,$$

$$k_{1,2} = hf_2(t, x_1, x_2, x_3) = hx_3,$$

$$k_{1,3} = hf_3(t, x_1, x_2, x_3) = h(t - x_3 - x_1^2),$$

$$k_{2,1} = hf_1\left(t + \frac{h}{2}, x_1 + \frac{1}{2}k_{1,1}, x_2 + \frac{1}{2}k_{1,2}, x_3 + \frac{1}{2}k_{1,3}\right) = h\left(x_2 + \frac{1}{2}k_{1,2}\right),$$

$$k_{2,2} = hf_2\left(t + \frac{h}{2}, x_1 + \frac{1}{2}k_{1,1}, x_2 + \frac{1}{2}k_{1,2}, x_3 + \frac{1}{2}k_{1,3}\right) = h\left(x_3 + \frac{1}{2}k_{1,3}\right),$$

$$k_{2,3} = hf_3\left(t+\frac{h}{2}, x_1+\frac{1}{2}k_{1,1}, x_2+\frac{1}{2}k_{1,2}, x_3+\frac{1}{2}k_{1,3}\right) = h\left[t+\frac{h}{2}-x_3-\frac{1}{2}k_{1,3}-\left(x_1+\frac{1}{2}k_{1,1}\right)^2\right],$$

$$k_{3,1} = hf_1\left(t+\frac{h}{2}, x_1+\frac{1}{2}k_{2,1}, x_2+\frac{1}{2}k_{2,2}, x_3+\frac{1}{2}k_{2,3}\right) = h\left(x_2+\frac{1}{2}k_{2,2}\right),$$

$$k_{3,2} = hf_2\left(t+\frac{h}{2}, x_1+\frac{1}{2}k_{2,1}, x_2+\frac{1}{2}k_{2,2}, x_3+\frac{1}{2}k_{2,3}\right) = h\left(x_3+\frac{1}{2}k_{2,3}\right),$$

$$k_{3,3} = hf_3\left(t+\frac{h}{2}, x_1+\frac{1}{2}k_{2,1}, x_2+\frac{1}{2}k_{2,2}, x_3+\frac{1}{2}k_{2,3}\right) = h\left[t+\frac{h}{2}-x_3-\frac{1}{2}k_{2,3}-\left(x_1+\frac{1}{2}k_{2,1}\right)^2\right],$$

$$k_{4,1} = hf_1\left(t+h, x_1+k_{3,1}, x_2+k_{3,2}, x_3+\frac{1}{2}k_{3,3}\right) = h(x_2+k_{3,2}),$$

$$k_{4,2} = hf_2(t+h, x_1+k_{3,1}, x_2+k_{3,2}, x_3+k_{3,3}) = h(x_3+k_{3,3}),$$

$$k_{4,3} = hf_3(t+h, x_1+k_{3,1}, x_2+k_{3,2}, x_3+k_{3,3}) = h[t+h-x_3-k_{3,3}-(x_1+k_{3,1})^2]$$

Using the starting values $t_0 = 0$, $a_1 = 1$, $a_2 = 0$, and $a_3 = 1$, we obtain the first approximations

$x_1(1;1) = 1.29167$,

$x_2(1;1) = 0.28125$,

$x_3(1;1) = 0.03125$.

Repeating the algorithm with $h = 2^{-1}, 2^{-2}$, and 2^{-3}, we obtain the approximations in Table 5-D.

Table 5-D: Approximations of the Solution to Problem 29.

n	h	$y(1) \approx x_1(1;2^{-n})$	$x_2(1;2^{-n})$	$x_3(1;2^{-n})$
0	1.0	1.29167	0.28125	0.03125
1	0.5	1.26039	0.34509	–0.06642
2	0.25	1.25960	0.34696	–0.06957
3	0.125	1.25958	0.34704	–0.06971

We stopped at $n = 3$ since

$$\left|\frac{x_1(1;2^{-3})-x_1(1;2^{-2})}{x_1(1;2^{-3})}\right|=\left|\frac{1.25958-1.25960}{1.25958}\right|=0.00002<0.01,$$

$$\left|\frac{x_2(1;2^{-3})-x_2(1;2^{-2})}{x_2(1;2^{-3})}\right|=\left|\frac{0.34704-0.34696}{0.34704}\right|=0.00023<0.01, \qquad \text{and}$$

$$\left|\frac{x_3(1;2^{-3})-x_3(1;2^{-2})}{x_3(1;2^{-3})}\right|=\left|\frac{-0.06971+0.06957}{-0.06971}\right|=0.00201<0.01.$$

Hence

$$y(1)\approx x_1(1;2^{-3})=1.25958,$$

with tolerance 0.01.

EXERCISES 5.7: Dynamical Systems, Poincaré Maps, and Chaos, page 323

1. Let $\omega=\frac{3}{2}$. Using system (3) on page 316 of the text with $A=F=1$, $\phi=0$, and $\omega=\frac{3}{2}$, we define the Poincaré map

$$x_n=\sin(3\pi n)+\frac{1}{9/4-4/4}=\sin(3\pi n)+\frac{4}{5}=\frac{4}{5},$$

$$v_n=\frac{3}{2}\cos(3\pi n)=(-1)^n\frac{3}{2},$$

for $n=0$, 1, 2, … . Calculating the first few values of (x_n,v_n), we find that they alternate between $\left(\frac{4}{5},\frac{3}{2}\right)$ and $\left(\frac{4}{5},-\frac{3}{2}\right)$. Consequently, we can deduce that there is a subharmonic solution of period 4π. Let $\omega=\frac{3}{5}$. Using system (3) on page 316 of the text with $A=F=1$, $\phi=0$, and $\omega=\frac{3}{5}$, we define the Poincaré map

$$x_n=\sin\left(\frac{6\pi n}{5}\right)+\frac{1}{9/25-1}=\sin\left(\frac{6\pi n}{5}\right)-1.5625,$$

$$v_n=\frac{3}{5}\cos\left(\frac{6\pi n}{5}\right)=(0.6)\cos\left(\frac{6\pi n}{5}\right),$$

for $n=0$, 1, 2, … . Calculating the first few values of (x_n,v_n), we find that the Poincaré map cycles through the points

(–1.5625, 0.6), $\qquad n=0$, 5, 10, …,

$(-2.1503, -0.4854)$, $n = 1, 6, 11, \ldots$,

$(-0.6114, 0.1854)$, $n = 2, 7, 12, \ldots$,

$(-2.5136, 0.1854)$, $n = 3, 8, 13, \ldots$,

$(-0.9747, -0.4854)$, $n = 4, 9, 14, \ldots$.

Consequently, we can deduce that there is a subharmonic solution of period 10π.

5. We want to construct the Poincaré map using $t = 2\pi n$ for $x(t)$ given in equation (5) on page 318 of the text with $A = F = 1$, $\phi = 0$, $\omega = \frac{1}{3}$, and $b = 0.22$. Since $\tan\theta = \frac{\omega^2 - 1}{b} = -4.040404$, we take $\theta = -1.328172$ and get

$$x_n = x(2\pi n) = e^{-0.22\pi n}\sin(0.629321\pi n) - (1.092050)\sin(1.328172),$$

$$v_n = x'(2\pi n) = -0.11e^{-0.22\pi n}\sin(0.629321\pi n) + (1.258642)e^{-0.22\pi n}\cos(0.629321\pi n)$$
$$+ (1.09250)\cos(1.328172).$$

In Table 5-E we have listed the first 21 values of the Poincaré map. As n get large, we see that

$$x_n \approx -(1.092050)\sin(1.328172) \approx -1.060065,$$

$$v_n \approx (1.092050)\sin(1.328172) \approx 0.262366.$$

Hence, as $n \to \infty$ the Poincaré map approaches the point $(-1.060065, 0.262366)$.

Table 5-E. Poincaré map for Problem 5.

n	x_n	v_n	n	x_n	v_n
0	−1.060065	1.521008	11	−1.059944	0.261743
1	−0.599847	−0.037456	12	−1.060312	0.262444
2	−1.242301	0.065170	13	−1.059997	0.262491
3	−1.103418	0.415707	14	−1.060030	0.262297
4	−0.997156	0.251142	15	−1.060096	0.262362
5	−1.074094	0.228322	16	−1.060061	0.262385
6	−1.070300	0.278664	17	−1.060058	0.262360
7	−1.052491	0.264458	18	−1.060068	0.262364
8	−1.060495	0.257447	19	−1.060065	0.262369
9	−1.061795	0.263789	20	−1.060064	0.262366
10	−1.059271	0.263037			

9. **(a)** When $x_0 = \frac{1}{7}$, the doubling modulo 1 map gives

$$x_1 = \frac{2}{7}(\text{mod}\,1) = \frac{2}{7}, \qquad x_2 = \frac{4}{7}(\text{mod}\,1) = \frac{4}{7},$$

$$x_3 = \frac{8}{7}(\text{mod}\,1) = \frac{1}{7}, \qquad x_4 = \frac{2}{7}(\text{mod}\,1) = \frac{2}{7},$$

$$x_5 = \frac{4}{7}(\text{mod}\,1) = \frac{4}{7}, \qquad x_6 = \frac{8}{7}(\text{mod}\,1) = \frac{1}{7},$$

$$x_7 = \frac{2}{7}(\text{mod}\,1) = \frac{2}{7}, \qquad \text{etc.}$$

This is the sequence $\left\{\frac{1}{7}, \frac{2}{7}, \frac{4}{7}, \frac{1}{7}, \ldots\right\}$. For $x_0 = \frac{k}{7}$, $k = 2, \ldots, 6$, we obtain

$$\left\{\frac{2}{7}, \frac{4}{7}, \frac{1}{7}, \frac{2}{7}, \ldots\right\}, \qquad \left\{\frac{3}{7}, \frac{6}{7}, \frac{5}{7}, \frac{3}{7}, \ldots\right\},$$

$$\left\{\frac{4}{7}, \frac{1}{7}, \frac{2}{7}, \frac{4}{7}, \ldots\right\}, \qquad \left\{\frac{5}{7}, \frac{3}{7}, \frac{6}{7}, \frac{5}{7}, \ldots\right\},$$

$$\left\{\frac{6}{7}, \frac{5}{7}, \frac{3}{7}, \frac{6}{7}, \ldots\right\}.$$

These sequences fall into two classes. The first has the repeating sequence $\overline{\frac{1}{7}, \frac{2}{7}, \frac{4}{7}}$ and the second has the repeating sequence $\overline{\frac{3}{7}, \frac{6}{7}, \frac{5}{7}}$.

(c) To see what happens, when $x_0 = \frac{k}{2^j}$, let's consider the special case when $x_0 = \frac{3}{2^2} = \frac{3}{4} = \frac{3}{4}$. Then,

$$x_1 = 2\left(\frac{3}{4}\right)(\text{mod}\,1) = \frac{3}{2}(\text{mod}\,1) = \frac{1}{2},$$

$$x_2 = 2\left(\frac{1}{2}\right)(\text{mod}\,1) = 1(\text{mod}\,1) = 0,$$

$$x_3 = 0,$$

$$x_4 = 0,$$

etc.

Observe that

$$x_2 = 2^2\left(\frac{3}{2^2}\right)(\text{mod}\,1) = 3(\text{mod}\,1) = 0.$$

In general,

$$x_j = 2^j\left(\frac{k}{2^j}\right)(\text{mod}\,1) = k(\text{mod}\,1) = 0.$$

Consequently, $x_n = 0$ for $n \ge j$.

REVIEW PROBLEMS: page 327

3. With the notation used in (1) on page 263 of the text,

$$f(x,y) = 4 - 4y$$
$$g(x,y) = -4x\ ,$$

and the phase plane equation (see equation (2) on page 264 of the text) can be written as

$$\frac{dy}{dx} = \frac{g(x,y)}{f(x,y)} = \frac{-4x}{4-4y} = \frac{x}{y-1}.$$

This equation is separable. Separating variables yields

$$(y-1)\,dy = x\,dx \qquad \Rightarrow \qquad \int (y-1)\,dy = \int x\,dx \qquad \Rightarrow \qquad (y-1)^2 + C = x^2$$

or $x^2 - (y-1)^2 = C$, where C is an arbitrary constant. We find the critical points by solving the system

$$\begin{aligned} f(x,y) &= 4 - 4y = 0, \\ g(x,y) &= -4x = 0 \end{aligned} \qquad \Rightarrow \qquad \begin{aligned} y &= 1, \\ x &= 0. \end{aligned}$$

So, (0,1) is the unique critical point. For $y > 1$,

$$\frac{dx}{dt} = 4(1-y) < 0,$$

which implies that trajectories flow to the left. Similarly, for $y < 1$, trajectories flow to the right. Comparing the phase plane diagram with those given on Figure 5.7 on page 270 of the text, we conclude that the critical point (0,1) is a saddle (unstable) point.

7. Expressing the system in the operator notation gives

$$D[x] + \left(D^2 + 1\right)[y] = 0,$$
$$D^2[x] + D[y] = 0.$$

Eliminating x by applying D to the first equation and subtracting the second equation from it yields

$$\{D(D^2+1)-D\}[y]=0 \qquad \Rightarrow \qquad D^3[y]=0 .$$

Thus on integrating 3 times we get

$$y(t)=C_3+C_2t+C_1t^2 .$$

We substitute this solution into the first equation of given system to get

$$x'=-(y''+y)=-\left[(2C_1)+\left(C_3+C_2t+C_1t^2\right)\right. \;=-\left[\left(C_3+2C_1\right)+C_2t+C_1t^2\right].$$

Integrating we obtain

$$x(t)=-\int\left[\left(C_3+2C_1\right)+C_2\,t+C_1t^2\right]dt=C_4-\left(C_3+2C_1\right)t-\frac{1}{2}C_2\,t^2-\frac{1}{3}C_1t^3 .$$

Thus the general solution of the given system is

$$x(t)=C_4-\left(C_3+2C_1\right)t-\frac{1}{2}C_2\,t^2-\frac{1}{3}C_1t^3 ,$$

$$y(t)=C_3+C_2t+C_1t^2 .$$

11. Differentiating the second equation, we obtain

$$y''=z' .$$

We eliminate z from the first and the third equations by substituting y' for z and y'' for z' into them:

$$\begin{aligned} x'&=y'-y, \\ y''&=y'-x \end{aligned} \qquad \Rightarrow \qquad \begin{aligned} x'-y'+y&=0, \\ y''-y'+x&=0, \end{aligned} \tag{20}$$

or, in operator notation,

$$\begin{aligned} D[x]-(D-1)[y]&=0, \\ [x]+(D^2-D)[y]&=0. \end{aligned}$$

Applying D to the first equation and adding the result to the second equation, we eliminate y:

$$\left(D^2[x]-D(D-1)[y]\right)+\left([x]+(D^2-D)[y]\right)=0$$

$$\Rightarrow \qquad \left(D^2+1\right)[x]=0 .$$

This equation is the simple harmonic equation, and its general solution is given by

$$x(t)=C_1\cos t+C_2\sin t .$$

Substituting $x(t)$ into the first equation of the system (20) yields

$$y'-y=-C_1\sin t+C_2\cos t . \tag{21}$$

The general solution to the corresponding homogeneous equation, $y' - y = 0$, is

$$y_h(t) = C_3 e^t .$$

To find a particular solution to (21) we observe that its right-hand side is of Type III from the Table 4.1 on page 208 of the text. Hence we look for a particular solution of (21) of the form

$$y_p(t) = C_4 \cos t + C_5 \sin t .$$

Differentiating, we obtain

$$y_p'(t) = -C_4 \sin t + C_5 \cos t .$$

Thus the equation (21) becomes

$$-C_1 \sin t + C_2 \cos t = y_p' - y_p = (-C_4 \sin t + C_5 \cos t) - (C_4 \cos t + C_5 \sin t)$$
$$= (C_5 - C_4)\cos t - (C_5 + C_4)\sin t .$$

Equating the coefficients yields

$$C_5 - C_4 = C_2 ,$$
$$C_5 + C_4 = C_1$$

$$\Rightarrow \quad \text{(by adding the equations)} \quad 2C_5 = C_1 + C_2 \quad \Rightarrow \quad C_5 = \frac{C_1 + C_2}{2} .$$

From the second equation in (22), we find

$$C_4 = C_1 - C_5 = \frac{C_1 - C_2}{2} .$$

Therefore, the general solution to the equation (21) is

$$y(t) = y_h(t) + y_p(t) = C_3 e^t + \frac{C_1 - C_2}{2} \cos t + \frac{C_1 + C_2}{2} \sin t .$$

Finally, we find $z(t)$ from the second equation:

$$z(t) = y'(t) = \left(C_3 e^t + \frac{C_1 - C_2}{2} \cos t + \frac{C_1 + C_2}{2} \sin t \right)'$$

$$= C_3 e^t - \frac{C_1 - C_2}{2} \sin t + \frac{C_1 + C_2}{2} \cos t .$$

Hence, the general solution to the given system is

$$x(t) = C_1 \cos t + C_2 \sin t ,$$

$$y(t) = C_3 e^t + \frac{C_1 - C_2}{2} \cos t + \frac{C_1 + C_2}{2} \sin t ,$$

$$z(t) = C_3 e^t - \frac{C_1 - C_2}{2} \sin t + \frac{C_1 + C_2}{2} \cos t .$$

To find constants C_1, C_2, and C_3, we use the initial conditions. So we get

$$0 = x(0) = C_1 \cos 0 + C_2 \sin 0 = C_1 ,$$

$$0 = y(0) = C_3 e^0 + \frac{C_1 - C_2}{2} \cos 0 + \frac{C_1 + C_2}{2} \sin 0 = C_3 + \frac{C_1 - C_2}{2} ,$$

$$2 = z(0) = C_3 e^0 - \frac{C_1 - C_2}{2} \sin 0 + \frac{C_1 + C_2}{2} \cos 0 = C_3 + \frac{C_1 + C_2}{2} ,$$

or

$$C_1 = 0 ,$$

$$C_1 - C_2 + 2C_3 = 0 ,$$

$$C_1 + C_2 + 2C_3 = 4 .$$

Solving we obtain $C_1 = 0$, $C_2 = 2$, $C_3 = 1$ and so

$$x(t) = 2 \sin t , \qquad y(t) = e^t - \cos t + \sin t , \qquad z(t) = e^t + \cos t + \sin t .$$

17. We first rewrite the given differential equation in an equivalent form as

$$y''' = \frac{1}{3}\left(5 + e^t y - 2y'\right).$$

Denoting $x_1(t) := y(t)$, $x_2(t) := y'(t)$, and $x_3(t) := y''(t)$, we conclude that

$$x_1' = y' = x_2 ,$$

$$x_2' = (y')' = y'' = x_3 ,$$

$$x_3' = (y'')' = y''' = \frac{1}{3}\left(5 + e^t y - 2y'\right) = \frac{1}{3}\left(5 + e^t x_1 - 2x_2\right),$$

i.e.,

$$x_1' = x_2 ,$$

$$x_2' = x_3 ,$$

$$x_3' = \frac{1}{3}\left(5 + e^t x_1 - 2x_2\right).$$

CHAPTER 6: Theory of Higher Order Linear Differential Equations

EXERCISES 6.1: Basic Theory of Linear Differential Equations, page 345

3. For this problem, $p_1(x) = -1$, $p_2(x) = \sqrt{x-1}$, and $g(x) = \tan x$. Note that $p_1(x)$ is continuous everywhere, $p_2(x)$ is continuous for $x \geq 1$, and $g(x)$ is continuous everywhere except at odd multiples of $\pi/2$. Therefore, these three functions are continuous simultaneously on the intervals

$$\left[1, \frac{\pi}{2}\right), \left(\frac{\pi}{2}, \frac{3\pi}{2}\right), \left(\frac{3\pi}{2}, \frac{5\pi}{2}\right), \ldots.$$

Because 5, the initial point, is in the interval $(3\pi/2, 5\pi/2)$, Theorem 1 guarantees that we have a unique solution to the initial value problem on this interval.

9. Let $y_1 = \sin^2 x$, $y_2 = \cos^2 x$, and $y_3 = 1$. We want to find c_1, c_2, and c_3, not all zero, such that

$$c_1 y_1 + c_2 y_2 + c_3 y_3 = c_1 \sin^2 x + c_2 \cos^2 x + c_3 \cdot 1 = 0,$$

for all x in the interval $(-\infty, \infty)$. Since $\sin^2 x + \cos^2 x = 1$ for all real numbers x, we can choose $c_1 = 1$, $c_2 = 1$, and $c_3 = -1$. Thus, these functions are linearly dependent. The Wronskian of these functions is given by

$$W[y_1, y_2, y_3](x) = \begin{vmatrix} \cos^2 x & \sin^2 x & 1 \\ -2\sin x \cos x & 2\sin x \cos x & 0 \\ 2\sin^2 x - 2\cos^2 x & 2\cos^2 x - 2\sin^2 x & 0 \end{vmatrix} = 0.$$

11. Let $y_1 = x^{-1}$, $y_2 = x^{1/2}$, and $y_3 = x$. We want to find constants c_1, c_2, and c_3 such that

$$c_1 y_1 + c_2 y_2 + c_3 y_3 = c_1 x^{-1} + c_2 x^{1/2} + c_3 x = 0 ,$$

for all x on the interval $(0, \infty)$. This equation must hold if $x = 1$, 4, or 9 (or any other values for x in the interval $(0, \infty)$). By plugging these values for x into the equation above, we see that c_1, c_2, and c_3 must satisfy the three equations

$$c_1 + c_2 + c_3 = 0,$$
$$\frac{c_1}{4} + 2c_2 + 4c_3 = 0,$$
$$\frac{c_1}{9} + 3c_2 + 9c_3 = 0.$$

Solving these three equations simultaneously yields $c_1 = c_2 = c_3 = 0$. Thus, the only way for $c_1 x^{-1} + c_2 x^{1/2} + c_3 x = 0$ for all x on the interval $(0, \infty)$, is for $c_1 = c_2 = c_3 = 0$. Therefore, these three functions are linearly independent on $(0, \infty)$. The Wronskian of these functions is given by

$$W[y_1, y_2, y_3](x) = \begin{vmatrix} x^{-1} & x^{1/2} & x \\ -x^{-2} & \frac{1}{2}x^{-1/2} & 1 \\ 2x^{-3} & -\frac{1}{4}x^{-3/2} & 0 \end{vmatrix} = \frac{3}{2}x^{-5/2}.$$

19. **(a)** Since $\{e^x,\ e^{-x}\cos 2x,\ e^{-x}\sin 2x$ is a fundamental solution set for the associated homogeneous differential equation and since $y_p = x^2$ is a solution to the nonhomogeneous equation, by the superposition principle, we have a general solution given by

$$y(x) = C_1 e^x + C_2 e^{-x}\cos 2x + C_3 e^{-x}\sin 2x + x^2.$$

(b) To find the solution that satisfies the initial conditions, we must differentiate the general solution $y(x)$ twice with respect to x. Thus, we have

$$y'(x) = C_1 e^x - C_2 e^{-x}\cos 2x - 2C_2 e^{-x}\sin 2x - C_3 e^{-x}\sin 2x + 2C_3 e^{-x}\cos 2x + 2x$$

$$= C_1 e^x + (-C_2 + 2C_3)e^{-x}\cos 2x + (-2C_2 - C_3)e^{-x}\sin 2x + 2x$$

$$\Rightarrow \quad y''(x) = C_1 e^x + (C_2 - 2C_3)e^{-x}\cos 2x - 2(-C_2 + 2C_3)e^{-x}\sin 2x$$

$$-(-2C_2 - C_3)e^{-x}\sin 2x + 2(-2C_2 - C_3)e^{-x}\cos 2x + 2$$

$$= C_1 e^x + (-3C_2 - 4C_3)e^{-x}\cos 2x + (4C_2 - 3C_3)e^{-x}\sin 2x + 2.$$

Plugging the initial conditions into these formulas, yields the equations

$$y(0) = C_1 + C_2 = -1,$$
$$y'(0) = C_1 - C_2 + 2C_3 = 1,$$
$$y''(0) = C_1 - 3C_2 - 4C_3 + 2 = -3.$$

By solving these equations simultaneously, we obtain $C_1 = -1$, $C_2 = 0$, and $C_3 = 1$. Therefore, the solution to the initial value problem is given by

$$y(x) = -e^x + e^{-x}\sin 2x + x^2.$$

EXERCISES 6.2: Homogeneous Linear Equations with Constant Coefficients, page 352

3. The auxiliary equation for this problem is $6r^3 + 7r^2 - r - 2 = 0$. By inspection we see that $r = -1$ is a root to this equation and so we can factor it as follows

$$6r^3 + 7r^2 - r - 2 = (r+1)(6r^2 + r - 2) = (r+1)(3r+2)(2r-1) = 0.$$

Thus, we see that the roots to the auxiliary equation are $r = -1, \frac{-2}{3}$, and $\frac{1}{2}$. These roots are real and non-repeating. Therefore, a general solution to this problem is given by

$$z(x) = c_1 e^{-x} + c_2 e^{-2x/3} + c_3 e^{x/2}.$$

13. The auxiliary equation for this problem is $r^4 + 4r^2 + 4 = 0$. This can be factored as $(r^2+2)^2 = 0$. Therefore, this equation has roots $r = \sqrt{2}\,i, -\sqrt{2}\,i, \sqrt{2}\,i, -\sqrt{2}\,i$, which we see are repeated and complex. Therefore, a general solution to this problem is given by

$$y(x) = c_1 \cos(\sqrt{2}\,x) + c_2 x \cos(\sqrt{2}\,x) + c_3 \sin(\sqrt{2}\,x) + c_4 x \sin(\sqrt{2}\,x).$$

15. The roots to this auxiliary equation, $(r-1)^2(r+3)(r^2+2r+5)^2 = 0$, are

$$r = 1, 1, -3,\ (-1 \pm 2i),\ (-1 \pm 2i),$$

where we note that 1 and $(-1 \pm 2i)$ are repeated roots. Therefore, a general solution to the differential equation with the given auxiliary equation is

$$y(x) = c_1 e^{x} + c_2 x e^{x} + c_3 e^{-3x} + (c_4 + c_5 x) e^{-x} \cos 2x + (c_6 + c_7 x) e^{-x} \sin 2x.$$

31. **(b)** If we let $y = x^r$, then we see that

$$y' = r x^{r-1}$$

$$\Rightarrow \quad y'' = r(r-1)x^{r-2} = (r^2 - r)x^{r-2}$$

$$\Rightarrow \quad y''' = r(r-1)(r-2)x^{r-3} = (r^3 - 3r^2 + 2r)x^{r-3}$$

$$\Rightarrow \quad y^{(4)} = r(r-1)(r-2)(r-3)x^{r-4} = (r^4 - 6r^3 + 11r^2 - 6r)x^{r-4}.$$

Thus, if $y = x^r$ is a solution to this fourth order Cauchy–Euler equation, then we must have

$$x^4(r^4 - 6r^3 + 11r^2 - 6r)x^{r-4} + 6x^3(r^3 - 3r^2 + 2r)x^{r} + 2x^2(r^2 - r)x^{r-2} - 4xrx^{r-1} + 4x^r = 0$$

$$\Rightarrow \quad (r^4 - 6r^3 + 11r^2 - 6r)x^r + 6(r^3 - 3r^2 + 2r)x^r + 2(r^2 - r)x^r - 4rx^r + 4x^r = 0$$

$$\Rightarrow \quad (r^4 - 5r^2 + 4)x^r = 0. \tag{1}$$

Therefore, in order for $y = x^r$ to be a solution to the equation with $x > 0$, we must have $r^4 - 5r^2 + 4 = 0$. Factoring this equation yields

$$\left(r^4 - 5r^2 + 4\right) = \left(r^2 - 4\right)\left(r^2 - 1\right) = (r - 2)(r + 2)(r - 1)(r + 1).$$

Equation (1) will equal zero and, therefore, the differential equation will be satisfied, if $r = \pm 1, \pm 2$. Thus, four solutions to the differential equation are $y = x$, $y = x^{-1}$, $y = x^2$, and $y = x^{-2}$. Since these functions are linearly independent, they form a fundamental solution set.

(c) Substituting $y = x^r$ into this differential equation yields

$$\left(r^3 - 3r^2 + 2r\right)x^r - 2\left(r^2 - r\right)x^r + 13rx^r - 13x^r = 0$$

$$\Rightarrow \qquad \left(r^3 - 5r^2 + 17r - 13\right)x^r = 0.$$

Thus, in order for $y = x^r$ to be a solution to this differential equation with $x > 0$, we must have $r^3 - 5r^2 + 17r - 13 = 0$. By inspection we find that $r = 1$ is a root to this equation. Therefore, we can factor this equation as follows

$$(r - 1)\left(r^2 - 4r + 13\right) = 0.$$

We find the remaining roots by using the quadratic formula. Thus, we obtain the roots $r = 1, 2 \pm 3i$. From the root $r = 1$, we obtain the solution $y = x$. From the roots $r = 2 \pm 3i$, by applying the hint given in the problem, we see that a solution is given by

$$y(x) = x^{2+3i} = x^2\{\cos(3\ln x) + i\sin(3\ln x)\}.$$

Therefore, by Lemma 2 on page 193 of the text, we find that two real-valued solutions to this differential equation are $y(x) = x^2\cos(3\ln x)$ and $y(x) = x^2\sin(3\ln x)$. Since these functions and the function $y(x) = x$ are linearly independent, we obtain the fundamental solution set

$$\left\{x,\ x^2\cos(3\ln x),\ x^2\sin(3\ln x)\right\}.$$

EXERCISES 6.3: Undetermined Coefficients and the Annihilator Method, page 359

1. The corresponding homogeneous equation for this problem is $y''' - 2y'' - 5y' + 6y = 0$ which has the associated auxiliary equation given by $r^3 - 2r^2 - 5r + 6 = 0$. By inspection we see that $r = 1$ is a root to this equation. Therefore, this equation can be factored as follows

$$(r - 1)\left(r^2 - r - 6\right) = (r - 1)(r - 3)(r + 2).$$

Thus, the roots to the auxiliary equation are given by $r = 1, 3$, and –2, and a general solution to the homogeneous equation is

$$y_h(x) = Ae^x + Be^{3x} + Ce^{-2x}.$$

The nonhomogeneous term, $g(x)=e^x+x^2$, is, according to Table 4.1, the sum of terms of Types I and II. Therefore, this equation has a particular solution of the form

$$y_p(x)=x^{s_1}c_1e^x+x^{s_2}\left(c_2+c_3x+c_4x^2\right).$$

Since e^x is a solution to the associated homogeneous equation and xe^x is not, we set $s_1=1$. Since none of the terms x^2, x, or 1 is a solution to the associated homogeneous equation, we set $s_2=0$. Thus, the form of a particular solution is

$$y_p(x)=c_1xe^x+c_2+c_3x+c_4x^2.$$

3. The associated homogeneous equation for this equation is $y'''+3y''-4y=0$. This equation has the corresponding auxiliary equation $r^3+3r^2-4=0$, which, by inspection, has $r=1$ as one of its roots. Thus, the auxiliary equation can be factored as follows

$$(r-1)\left(r^2+4r+4\right)=(r-1)(r+2)^2.$$

From this we see that the roots to the auxiliary equation are $r=1,-2,-2$. Therefore, a general solution to the homogeneous equation is

$$y_h(x)=Ae^x+Be^{-2x}+Cxe^{-2x}.$$

The nonhomogeneous term $g(x)=e^{-2x}$ is, according to Table 4.1, of Type II. Therefore, a particular solution to the original differential equation has the form $y_p(x)=x^sc_1e^{-2x}$. Since both e^{-2x} and xe^{-2x} solutions to the associated homogeneous equation, we set $s=2$. (Note that this means that $r=-2$ will be a root of multiplicity three of the auxiliary equation associated with the operator equation $A[L[y]](x)=0$, where A is an annihilator of the nonhomogeneous term $g(x)=e^{-2x}$ and L is the linear operator $L:=D^3+3D^2-4$.) Thus, the form of a particular solution to this equation is

$$y_p(x)=c_1x^2e^{-2x}.$$

5. In the solution to Problem 1, we determined that a general solution to the homogeneous differential equation associated with this problem is

$$y_h(x)=Ae^x+Be^{3x}+Ce^{-2x},$$

and that a particular solution has the form

$$y_p(x)=c_1xe^x+c_2+c_3x+c_4x^2.$$

By differentiating $y_p(x)$, we find

$$y_p'(x)=c_1xe^x+c_1e^x+c_3+2c_4x$$

$$\Rightarrow \qquad y_p''(x)=c_1xe^x+2c_1e^x+2c_4$$

$$\Rightarrow \qquad y_p'''(x) = c_1xe^x + 3c_1e^x .$$

Substituting these expressions into the original differential equation, we obtain

$$y_p'''(x) - 2y_p''(x) - 5y_p'(x) + 6y_p(x) = c_1xe^x + 3c_1e^x - 2c_1xe^x - 4c_1e^x - 4c_4$$
$$- 5c_1xe^x - 5c_1e^x - 5c_3 - 10c_4x + 6c_1xe^x + 6c_2 + 6c_3x + 6c_4x^2 = e^x + x^2$$

$$\Rightarrow \qquad -6c_1e^x + (-4c_4 - 5c_3 + 6c_2) + (-10c_4 + 6c_3)x + 6c_4x^2 = e^x + x^2 .$$

Equating coefficients yields

$$-6c_1 = 1 \qquad \Rightarrow \qquad c_1 = \frac{-1}{6}, \qquad 6c_4 = 1 \qquad \Rightarrow \qquad c_4 = \frac{1}{6},$$

$$-10c_4 + 6c_3 = 0 \qquad \Rightarrow \qquad c_3 = \frac{10c_4}{6} = \frac{10}{36} = \frac{5}{18},$$

$$-4c_4 - 5c_3 + 6c_2 = 0 \qquad \Rightarrow \qquad c_2 = \frac{4c_4 + 5c_3}{6} = \frac{\frac{4}{6} + \frac{25}{18}}{6} = \frac{37}{108}.$$

Thus, a general solution to the nonhomogeneous equation is given by

$$y(x) = y_h(x) + y_p(x) = Ae^x + Be^{3x} + Ce^{-2x} - \frac{1}{6}xe^x + \frac{1}{6}x^2 + \frac{5}{18}x + \frac{37}{108}.$$

7. In Problem 3, a general solution to the associated homogeneous equation was found to be

$$y_h(x) = Ae^x + Be^{-2x} + Cxe^{-2x},$$

and the form of a particular solution to the nonhomogeneous equation was

$$y_p(x) = c_1x^2e^{-2x} .$$

Differentiating $y_p(x)$ yields

$$y_p'(x) = 2c_1xe^{-2x} - 2c_1x^2e^{-2x} = 2c_1\left(x - x^2\right)e^{-2x}$$

$$\Rightarrow \quad y_p''(x) = -4c_1\left(x - x^2\right)e^{-2x} + 2c_1(1 - 2x)e^{-2x} = 2c_1\left(2x^2 - 4x + 1\right)e^{-2x}$$

$$\Rightarrow \quad y_p'''(x) = -4c_1\left(2x^2 - 4x + 1\right)e^{-2x} + 2c_1(4x - 4)e^{-2x} = 4c_1\left(-2x^2 + 6x - 3\right)e^{-2x} .$$

By substituting these expressions into the nonhomogeneous equation, we obtain

$$y_p'''(x) + 3y_p''(x) - 4y_p(x) = 4c_1\left(-2x^2 + 6x - 3\right)e^{-2x} + 6c_1\left(2x^2 - 4x + 1\right)e^{-2x} - 4c_1x^2e^{-2x} = e^{-2x}$$

$$\Rightarrow \qquad -6c_1e^{-2x} = e^{-2x} .$$

By equating coefficients, we see that $c_1 = \frac{-1}{6}$. Thus, a general solution to the nonhomogeneous differential equation is given by

$$y(x) = y_h(x) + y_p(x) = Ae^x + Be^{-2x} + Cxe^{-2x} - \frac{1}{6}x^2e^{-2x}.$$

15. The operator $(D-2)$ annihilates the function $f_1(x) := e^{2x}$ and the operator $(D-1)$ annihilates the function $f_2(x) := -6e^x$. Thus, the composition of these two operators, namely $A := (D-2)(D-1)$, annihilates both of these functions and so, by linearity, it annihilates their sum.

17. This function has the same form as the functions given in equation (9) on page 356 of the text. Here we see that $\alpha = -1$, $\beta = 2$, and $m - 1 = 2$. Thus, the operator

$$A := \left[(D - \{-1\})^2 + 2^2\right]^3 = \left[(D+1)^2 + 4\right]^3$$

annihilates this function.

23. The function $g(x) = e^{3x} - x^2$ is annihilated by the operator

$$A := D^3(D-3).$$

Applying the operator A to both sides of the differential equation given in this problem yields

$$A[y'' - 5y' + 6y] = A\left[e^{3x} - x^2\right] = 0$$

$$\Rightarrow \quad D^3(D-3)(D^2 - 5D + 6)[y] = D^3(D-3)^2(D-2)[y] = 0.$$

This last equation has the associated auxiliary equation

$$r^3(r-3)^2(r-2) = 0,$$

which has roots $r = 0, 0, 0, 3, 3, 2$. Thus, a general solution to the differential equation associated with this auxiliary equation is

$$y(x) = c_1e^{2x} + c_2e^{3x} + c_3xe^{3x} + c_4x^2 + c_5x + c_6.$$

The homogeneous equation, $y'' - 5y' + 6y = 0$, associated with the original problem has as its corresponding auxiliary equation

$$r^2 - 5r + 6 = (r-2)(r-3) = 0.$$

Therefore, the solution to the homogeneous equation associated with the original problem is $y_h(x) = c_1e^{2x} + c_2e^{3x}$. Since a general solution to this original problem is given by

$$y(x) = y_h(x) + y_p(x) = c_1e^{2x} + c_2e^{3x} + y_p(x)$$

and since $y(x)$ must be of the form

$$y(x) = c_1e^{2x} + c_2e^{3x} + c_3xe^{3x} + c_4x^2 + c_5x + c_6,$$

we see that

$$y_p(x) = c_3 x e^{3x} + c_4 x^2 + c_5 x + c_6 .$$

EXERCISES 6.4: Method of Variation of Parameters, page 363

5. Since the nonhomogeneous term, $g(x) = \tan x$, is not a solution to a homogeneous linear differential equation with constant coefficients, we will find a particular solution by the method of variation of parameters. To do this, we must first find a fundamental solution set for the corresponding homogeneous equation, $y''' + y' = 0$. Its auxiliary equation is $r^3 + r = 0$, which factors as $r^3 + r = r(r^2 + r)$. Thus, the roots to this auxiliary equation are $r = 0, \pm i$. Therefore, a fundamental solution set to the homogeneous equation is $\{1, \cos x, \sin x\}$. We now seek a particular solution to the nonhomogeneous equation of the form

$$y_p(x) = v_1(x) + v_2(x)\cos x + v_3(x)\sin x .$$

To accomplish this, we must find the four determinants $W[1, \cos x, \sin x](x)$, $W_1(x)$, $W_2(x)$, $W_3(x)$. That is, we calculate

$$W[1, \cos x, \sin x](x) = \begin{vmatrix} 1 & \cos x & \sin x \\ 0 & -\sin x & \cos x \\ 0 & -\cos x & -\sin x \end{vmatrix} = \sin^2 x + \cos^2 x = 1,$$

$$W_1(x) = \begin{vmatrix} 0 & \cos x & \sin x \\ 0 & -\sin x & \cos x \\ 1 & -\cos x & -\sin x \end{vmatrix} = (-1)^{3-1} \begin{vmatrix} \cos x & \sin x \\ -\sin x & \cos x \end{vmatrix} = (-1)^2 \left(\cos^2 x + \sin^2 x\right) = 1,$$

$$W_2(x) = \begin{vmatrix} 1 & 0 & \sin x \\ 0 & 0 & \cos x \\ 0 & 1 & -\sin x \end{vmatrix} = (-1)^{3-2} \begin{vmatrix} 1 & \sin x \\ 0 & \cos x \end{vmatrix} = -\cos x,$$

$$W_3(x) = \begin{vmatrix} 1 & \cos x & 0 \\ 0 & -\sin x & 0 \\ 0 & -\cos x & 1 \end{vmatrix} = (-1)^{3-3} \begin{vmatrix} 1 & \cos x \\ 0 & -\sin x \end{vmatrix} = -\sin x.$$

By using formula (11) on page 361 of the text, we can now find $v_1(x)$, $v_2(x)$, and $v_3(x)$. Since $g(x) = \tan x$, we have (assuming that all constants of integration are zero)

$$v_1(x) = \int \frac{g(x) W_1(x)}{W[1, \cos x, \sin x]}\, dx = \int \tan x \, dx = \ln(\sec x),$$

$$v_2(x)=\int\frac{g(x)W_2(x)}{W[1,\cos x,\sin x]}dx=\int(\tan x)(-\cos x)\,dx=-\int\sin x\,dx=\cos x$$

$$v_3(x)=\int\frac{g(x)W_3(x)}{W[1,\cos x,\sin x]}dx=\int(\tan x)(-\sin x)\,dx=-\int\frac{\sin^2 x}{\cos x}dx$$

$$=-\int\frac{1-\cos^2 x}{\cos x}dx=\int(\cos x-\sec x)\,dx=\sin x-\ln(\sec x+\tan x)$$

Therefore, we have

$$y_p(x)=v_1(x)+v_2(x)\cos x+v_3(x)\sin x$$

$$=\ln(\sec x)+\cos^2 x+\sin^2 x-(\sin x)\ln(\sec x+\tan x)$$

$$\Rightarrow\qquad y_p(x)=\ln(\sec x)-(\sin x)\ln(\sec x+\tan x)+1.$$

Since $y\equiv 1$ is a solution to the homogeneous equation, we may choose

$$y_p(x)=\ln(\sec x)-(\sin x)\ln(\sec x+\tan x).$$

Note: We left the absolute value signs off $\ln(\sec x)$ and $\ln(\sec x+\tan x)$ because of the stated domain: $0<x<\pi/2$.

9. To find a particular solution to the nonhomogeneous equation, we will use the method of variation of parameters. We must first calculate the four determinants $W\left[e^x,e^{-x},e^{2x}\right](x)$, $W_1(x)$, $W_2(x)$, $W_3(x)$. Thus, we have

$$W\left[e^x,e^{-x},e^{2x}\right](x)=\begin{vmatrix} e^x & e^{-x} & e^{2x}\\ e^x & -e^{-x} & 2e^{2x}\\ e^x & e^{-x} & 4e^{2x}\end{vmatrix}=-4e^{2x}+2e^{2x}+e^{2x}+e^{2x}-2e^{2x}-4e^{2x}=-6e^{2x},$$

$$W_1(x)=\begin{vmatrix} 0 & e^{-x} & e^{2x}\\ 0 & -e^{-x} & 2e^{2x}\\ 1 & e^{-x} & 4e^{2x}\end{vmatrix}=(-1)^{3-1}\begin{vmatrix} e^{-x} & e^{2x}\\ -e^{-x} & 2e^{2x}\end{vmatrix}=2e^x+e^x=3e^x,$$

$$W_2(x)=\begin{vmatrix} e^x & 0 & e^{2x}\\ e^x & 0 & 2e^{2x}\\ e^x & 1 & 4e^{2x}\end{vmatrix}=(-1)^{3-2}\begin{vmatrix} e^x & e^{2x}\\ e^x & 2e^{2x}\end{vmatrix}=-(2e^{3x}-e^{3x})=-e^{3x},$$

$$W_3(x)=\begin{vmatrix} e^x & e^{-x} & 0\\ e^x & -e^{-x} & 0\\ e^x & e^{-x} & 1\end{vmatrix}=(-1)^{3-3}\begin{vmatrix} e^x & e^{-x}\\ e^x & -e^{-x}\end{vmatrix}=(-1-1)=-2.$$

Therefore, according to formula (12) on page 362 of the text, a particular solution, $y_p(x)$, will be given by

$$y_p(x) = e^x \int \frac{3e^x g(x)}{-6e^{2x}}\,dx + e^{-x} \int \frac{-e^{3x} g(x)}{-e^{2x}}\,dx + e^{2x} \int \frac{-2g(x)}{-6e^{2x}}\,dx$$

$$\Rightarrow \qquad y_p(x) = \frac{-1}{2} e^x \int e^{-x} g(x)\,dx + \frac{1}{6} e^{-x} \int e^x g(x)\,dx + \frac{1}{3} e^{2x} \int e^{-2x} g(x)\,dx .$$

CHAPTER 7: Laplace Transforms

EXERCISES 7.2: Definition of the Laplace Transform, page 380

1. Using Definition 1 on page 373, we compute

$$\mathcal{L}\{t\} = \int_0^\infty e^{-st}\, t\, dt = \lim_{N\to\infty} \int_0^N e^{-st}\, t\, dt$$

Using integration by parts, we find

$$\mathcal{L}\{t\} = \lim_{N\to\infty}\left[\frac{-t}{s}e^{-st}\Big|_0^N + \frac{1}{s}\int_0^N e^{-st}\,dt\right] = \lim_{N\to\infty}\left[\frac{-t}{s}e^{-st}\Big|_0^N - \frac{1}{s^2}e^{-st}\Big|_0^N\right]$$

$$= \lim_{N\to\infty}\left[\frac{-N}{s}e^{-sN} - \frac{1}{s^2}e^{-sN} + \frac{1}{s^2}\right]$$

$$\Rightarrow \qquad \mathcal{L}\{t\} = 0 + 0 + \frac{1}{s^2} = \frac{1}{s^2} \qquad \text{for } s > 0.$$

9. As in Example 4 on page 375 of the text, we first break the integral into separate parts. Thus,

$$\int_0^\infty e^{-st} f(t)\, dt = \int_0^2 e^{-2t}\cdot 0\, dt + \int_2^\infty e^{-st} t\, dt = 0 + \int_2^\infty e^{-st} t\, dt.$$

Using integration by parts, we obtain for $s > 0$

$$\int_2^\infty e^{-st} t\, dt = \lim_{N\to\infty}\left\{\left[\frac{-te^{-st}}{s} - \frac{e^{-st}}{s^2}\right]\Bigg|_2^N\right\}$$

$$= \lim_{N\to\infty}\left[\frac{-Ne^{-sN}}{s} - \frac{e^{-sN}}{s^2} + \frac{2e^{-2s}}{s} + \frac{e^{-2s}}{s^2}\right] = \frac{2e^{-2s}}{s} + \frac{e^{-2s}}{s^2} = \frac{e^{-2s}(2s+1)}{s^2}.$$

Notice that the above limit does not exist for $s < 0$, and when $s = 0$, the integral becomes $\int_2^\infty t\, dt$, which diverges.

17. By the linearity of the Laplace transform,

$$\mathcal{L}\left\{e^{3t}\sin 6t - t^3 + e^t\right\} = \mathcal{L}\left\{e^{3t}\sin 6t\right\} - \mathcal{L}\left\{t^3\right\} + \mathcal{L}\left\{e^t\right\}.$$

From Table 7.1 on page 380 of the text, we see that

$$\mathcal{L}\left\{e^{3t}\sin 6t\right\} = \frac{6}{(s-3)^2+36}, \qquad s > 3, \tag{1}$$

$$\mathcal{L}\{t^3\} = \frac{3!}{s^4}, \qquad s > 0, \tag{2}$$

and

$$\mathcal{L}\{e^t\} = \frac{1}{s-1}, \qquad s > 1. \tag{3}$$

Thus, we have

$$\mathcal{L}\{e^{3t}\sin 6t - t^3 + e^t\} = \frac{6}{(s-3)^2 + 36} - \frac{6}{s^4} + \frac{1}{s-1}.$$

Since (1), (2), and (3) all hold for $s > 3$, we see that our answer is valid for $s > 3$.

27. Since $\frac{1}{t} \to +\infty$ as $t \to 0^+$, $f(t)$ has an "infinite jump" at the origin, and is, therefore, neither continuous nor piecewise continuous on $[0, 10]$.

29. **(a)** First observe that $\left|t^3 \sin t\right| \le \left|t^3\right|$ for all t. Next, three applications of L'Hospital's rule show that as $t \to \infty$, we have $t^3 e^{-\alpha t} \to 0$ for all $\alpha > 0$. Thus for some $T > 0$, we have $\left|t^3\right| < e^{\alpha t}$ for all $t \ge T$, and so

$$\left|t^3 \sin t\right| \le \left|t^3\right| < e^{\alpha t}, \qquad t \ge T.$$

Therefore, $t^3 \sin t$ is of exponential order α, for any $\alpha > 0$.

(c) Since

$$\lim_{t\to\infty} \frac{e^{t^3}}{e^{\alpha t}} = \lim_{t\to\infty} e^{t\left(t^2-\alpha\right)} = +\infty,$$

we see that e^{t^3} grows faster than $e^{\alpha t}$ for any α. Thus e^{t^3} is *not* of exponential order.

EXERCISES 7.3: Properties of the Laplace Transform, page 386

3. By the linearity of the Laplace transform,

$$\mathcal{L}\{e^{-t}\cos 3t + e^{6t} - 1\} = \mathcal{L}\{e^{-t}\cos 3t\} + \mathcal{L}\{e^{6t}\} - \mathcal{L}\{1\}.$$

From Table 7.1 on page 380 of the text, we see that

$$\mathcal{L}\{e^{-t}\cos 3t\} = \frac{s+1}{(s+1)^2 + 9}, \qquad s > -1, \tag{4}$$

$$\mathcal{L}\{e^{6t}\}=\frac{1}{s-6}, \qquad s>6, \tag{5}$$

$$\mathcal{L}\{1\}=\frac{1}{s}, \qquad s>0. \tag{6}$$

Thus

$$\mathcal{L}\{e^{-t}\cos 3t+e^{6t}-1\}=\frac{s+1}{(s+1)^2+9}+\frac{1}{s-6}-\frac{1}{s}.$$

Since (4), (5), and (6) all hold for $s>6$, we see that our answer is valid for $s>6$. Note that (4) and (5) could also be obtained from the Laplace transforms for $\cos 3t$ and 1, respectively, by applying the translation property, property (1) on page 382 of the text.

7. Since $(t-1)^4=t^4-4t^3+6t^2-4t+1$, we have from the linearity of the Laplace transform that

$$\mathcal{L}\{(t-1)^4\}=\mathcal{L}\{t^4\}-4\mathcal{L}\{t^3\}+6\mathcal{L}\{t^2\}-4\mathcal{L}\{t\}+\mathcal{L}\{1\}.$$

From Table 7.1 on page 380 of the text, we get

$$\mathcal{L}\{t^4\}=\frac{4!}{s^5}=\frac{24}{s^5}, \qquad s>0,$$

$$\mathcal{L}\{t^3\}=\frac{3!}{s^4}=\frac{6}{s^4}, \qquad s>0,$$

$$\mathcal{L}\{t^2\}=\frac{2!}{s^3}=\frac{2}{s^3}, \qquad s>0,$$

$$\mathcal{L}\{t\}=\frac{1}{s^2}, \qquad s>0,$$

$$\mathcal{L}\{1\}=\frac{1}{s}, \qquad s>0.$$

Thus $\mathcal{L}\{(t-1)^4\}=\dfrac{24}{s^5}-\dfrac{24}{s^4}+\dfrac{12}{s^3}-\dfrac{4}{s^2}+\dfrac{1}{s}, \quad s>0.$

15. From the trigonometric identity $\cos^2 t=\frac{1}{2}(1+\cos 2t)$, we find that

$$\cos^3 t=\cos t\cos^2 t=\frac{1}{2}\cos t+\frac{1}{2}\cos t\cos 2t.$$

Next we write

$$\cos t\cos 2t=\frac{1}{2}[\cos(2t+t)+\cos(2t-t)]=\frac{1}{2}\cos 3t+\frac{1}{2}\cos t.$$

Thus,

$$\cos^3 t = \frac{1}{2}\cos t + \frac{1}{4}\cos 3t + \frac{1}{4}\cos t = \frac{3}{4}\cos t + \frac{1}{4}\cos 3t .$$

We now use the linearity of the Laplace transform and Table 7.1 on page 380 of the text to find that

$$\mathcal{L}\{\cos^3 t\} = \frac{3}{4}\mathcal{L}\{\cos t\} + \frac{1}{4}\mathcal{L}\{\cos 3t\} = \frac{3}{4}\frac{s}{s^2+1} + \frac{1}{4}\frac{s}{s^2+9}, \qquad s > 0 .$$

25. **(a)** By property (6) on page 385 of the text,

$$\mathcal{L}\{t\cos bt\} = -\frac{d}{ds}\mathcal{L}\{\cos bt\} = -\frac{d}{ds}\left[\frac{s}{s^2+b^2}\right] = \frac{s^2-b^2}{(s^2+b^2)^2}, \qquad s > 0 .$$

(b) Again using the same property,

$$\mathcal{L}\{t^2\cos bt\} = \frac{d^2}{ds^2}\mathcal{L}\{\cos bt\} = \frac{d^2}{ds^2}\left[\frac{s}{s^2+b^2}\right] = \frac{d}{ds}\left[\frac{b^2-s^2}{(s^2+b^2)^2}\right] \qquad \text{(from (a))}$$

$$= \frac{2s^3-6sb^2}{\left(s^2+b^2\right)^3}, \qquad s > 0 .$$

27. First observe that since $f(t)$ is piecewise continuous on $[0, \infty)$ and $f(t)/t$ approaches a finite limit as $t \to 0^+$, we conclude that $f(t)/t$ is also piecewise continuous on $[0, \infty)$. Next, since for $t > 1$ we have $|f(t)/t| \le |f(t)|$, we see that $f(t)/t$ is of exponential order α since $f(t)$ is. These observations and Theorem 2 on page 379 of the text show that $\mathcal{L}\{f(t)/t\}$ exists. When the results of Problem 26 are applied to $f(t)/t$, we see that

$$\lim_{N\to\infty} \mathcal{L}\left\{\frac{f(t)}{t}\right\}(N) = 0.$$

By property (6) on page 385 of the text, we have that

$$F(s) = \int_0^\infty e^{-st} f(t)dt = \int_0^\infty \frac{te^{-st} f(t)}{t}dt = -\frac{d}{ds}\mathcal{L}\left\{\frac{f(t)}{t}\right\}.$$

Thus,

$$\int_s^\infty F(u)du = -\int_s^\infty \frac{d}{du}\mathcal{L}\left\{\frac{f(t)}{t}\right\}(u)du = \int_\infty^s \frac{d}{du}\mathcal{L}\left\{\frac{f(t)}{t}\right\}(u)du$$

$$= \mathcal{L}\left\{\frac{f(t)}{t}\right\}(s) - \lim_{N\to\infty}\mathcal{L}\left\{\frac{f(t)}{t}\right\}(N) = \mathcal{L}\left\{\frac{f(t)}{t}\right\}(s).$$

29. From the linearity properties (2) and (3) on page 376 of the text we have

$$\mathcal{L}\{g(t)\} = \mathcal{L}\{y''(t)+6y'(t)+10y(t)\} = \mathcal{L}\{y''(t)\}+\mathcal{L}\{6y'(t)\}+\mathcal{L}\{10y(t)\}$$
$$= \mathcal{L}\{y''(t)\}+6\mathcal{L}\{y'(t)\}+10\mathcal{L}\{y(t)\}.$$

Next, applying properties (2) and (4) on pages 383 and 384 of the text gives

$$\mathcal{L}\{g(t)\}(s) = s^2\mathcal{L}\{y\}(s)-sy(0)-y'(0)+6\left[s\mathcal{L}\{y\}(s)-y(0)\right] + 10\mathcal{L}\{y\}(s).$$

Keeping in mind the fact that all initial conditions are zero the above equation becomes

$$G(s)=\left(s^2+6s+10\right)Y(s), \quad \text{where } Y(s)=\mathcal{L}\{y\}(s).$$

Therefore, the transfer function $H(s)$ is given by

$$H(s)=\frac{Y(s)}{G(s)}=\frac{1}{s^2+6s+10}.$$

EXERCISES 7.4: Inverse Laplace Transform, page 396

7. By completing the square in the denominator, we can rewrite $\dfrac{2s+16}{s^2+4s+13}$ as

$$\frac{2s+16}{(s^2+4s+4)+9}=\frac{2s+16}{(s+2)^2+3^2}=\frac{2(s+2)}{(s+2)^2+3^2}+\frac{4(3)}{(s+2)^2+3^2}.$$

Thus, by the linearity of the inverse Laplace transform,

$$\mathcal{L}^{-1}\left\{\frac{2s+16}{s^2+4s+13}\right\}=2\mathcal{L}^{-1}\left\{\frac{s+2}{(s+2)^2+3^2}\right\}+4\mathcal{L}^{-1}\left\{\frac{3}{(s+2)^2+3^2}\right\}.$$

From Table 7.1 on page 380 of the text, we find that

$$\mathcal{L}^{-1}\left\{\frac{2s+16}{s^2+4s+13}\right\}=2e^{-2t}\cos 3t+4e^{-2t}\sin 3t.$$

17. First we need to completely factor the denominator. Since $s^2+s-6=(s-2)(s+3)$, we have

$$\frac{3s+5}{s\left(s^2+s-6\right)}=\frac{3s+5}{s(s-2)(s+3)}.$$

Since the denominator has only nonrepeated linear factors, we can write

$$\frac{3s+5}{s(s-2)(s+3)}=\frac{A}{s}+\frac{B}{s-2}+\frac{C}{s+3}$$

for some choice of A, B and C. Clearing fractions gives us

$$3s+5=A(s-2)(s+3)+Bs(s+3)+Cs(s-2).$$

With $s=0$, this yields $5=A(-2)(3)$ so that $A=-5/6$. With $s=2$, we get $11=B(2)(5)$ so that $B=11/10$. Finally, $s=-3$ yields $-4=C(-3)(-5)$ so that $C=-4/15$. Thus,

$$\frac{3s+5}{s(s^2+s-6)}=\frac{-5}{6s}+\frac{11}{10(s-2)}-\frac{4}{15(s+3)}.$$

19. First observe that the quadratic s^2+2s+2 is irreducible because the discriminant $2^2-4\cdot 2\cdot 1=-4$ is negative. Since the denominator has one nonrepeated linear factor and one nonrepeated quadratic factor, we can write

$$\frac{1}{(s-3)(s^2+2s+2)}=\frac{1}{(s-3)[(s+1)^2+1]}=\frac{A}{s-3}+\frac{B(s+1)}{(s+1)^2+1}+\frac{C}{(s+1)^2+1},$$

where we have chosen a form which is more convenient for taking the inverse Laplace transform. Clearing fractions gives us

$$1=A[(s+1)^2+1]+B(s+1)(s-3)+C(s-3). \qquad (7)$$

With $s=3$, this yields $1=A(17)$ so that $A=1/17$. Equation (7) now becomes

$$1=\frac{1}{17}[(s+1)^2+1]+B(s+1)(s-3)+C(s-3). \qquad (8)$$

If we choose $s=-1$, the B will drop out of (8) giving us $1=\frac{1}{17}(1)+C(-4)$, so that $C=-4/17$. Thus (8) becomes

$$1=\frac{1}{17}[(s+1)^2+1]+B(s+1)(s-3)-\frac{4}{17}(s-3). \qquad (9)$$

Choosing $s=0$ in (9) and solving for B, yields $B=-1/17$. Thus

$$\frac{1}{(s-3)(s^2+2s+2)}=\frac{1}{17}\left[\frac{1}{s-3}-\frac{s+1}{(s+1)^2+1}-\frac{4}{(s+1)^2+1}\right].$$

25. Let $F=\dfrac{7s^2+23s+30}{(s-2)(s^2+2s+5)}$. Observing that the quadratic s^2+2s+5 is irreducible, the partial fractions decomposition for F has the form

$$F=\frac{A}{s-2}+\frac{B(s+1)}{(s+1)^2+2^2}+\frac{C\cdot 2}{(s+1)^2+2^2}.$$

After some algebra we find that

$$F = \frac{8}{s-2} - \frac{s+1}{(s+1)^2+2^2} + \frac{3(2)}{(s+1)^2+2^2}.$$

Thus by the linearity of the inverse Laplace transform,

$$\mathcal{L}^{-1}\{F\} = \mathcal{L}^{-1}\left\{\frac{8}{s-2}\right\} - \mathcal{L}^{-1}\left\{\frac{s+1}{(s+1)^2+2^2}\right\} + 3\mathcal{L}^{-1}\left\{\frac{2}{(s+1)^2+2^2}\right\}$$

$$= 8e^{2t} - e^{-t}\cos 2t + 3e^{-t}\sin 2t.$$

33. We are looking for $\mathcal{L}^{-1}\{F(s)\} = f(t)$. According to the formula given just before Problem 33 on page 397 of the text, $f(t) = \frac{-1}{t}\mathcal{L}^{-1}\left\{\frac{dF}{ds}\right\}$ (take $n=1$ in the formula). Since

$$F(s) = \ln\left(\frac{s+2}{s-5}\right) = \ln(s+2) - \ln(s-5),$$

we have $\frac{dF}{ds} = \frac{1}{s+2} - \frac{1}{s-5}$, and so by Table 7.1, $\mathcal{L}^{-1}\left\{\frac{dF}{ds}\right\} = e^{-2t} - e^{5t}$. Thus

$$\mathcal{L}^{-1}\{F(s)\} = f(t) = \frac{-1}{t}\left[e^{-2t} - e^{5t}\right] = \frac{e^{5t}}{t} - \frac{e^{-2t}}{t}.$$

EXERCISES 7.5: Solving Initial Value Problems, page 405

3. Taking the Laplace transform of $y'' + 6y' + 9y = 0$ and applying the linearity of the Laplace transform yields

$$\mathcal{L}\{y''\} + 6\mathcal{L}\{y'\} + 9\mathcal{L}\{y\} = \mathcal{L}\{0\}. \tag{10}$$

If we put $Y(s) = \mathcal{L}\{y\}(s)$ and apply properties (2) and (4) on pages 383 and 384 of the text, we get

$$\mathcal{L}\{y'\} = sY(s) + 1, \tag{11}$$

and

$$\mathcal{L}\{y''\} = s^2Y(s) + s - 6. \tag{12}$$

Combining (10), (11), and (12) and using the fact that $\mathcal{L}\{0\} = 0$, gives us the equation

$$s^2Y(s) + s - 6 + 6sY(s) + 6 + 9Y(s) = 0.$$

Solving for $Y(s)$ gives us

$$Y(s) = \frac{-s}{s^2+6s+9} = \frac{-s}{(s+3)^2} = \frac{3}{(s+3)^2} - \frac{1}{s+3}, \tag{13}$$

where the last equality comes from the partial fraction expansion of $\frac{-s}{(s+3)^2}$. Using Table 7.1 on page 380 of the text to find the inverse Laplace transform of (13) gives us

$$y(t) = 3te^{-3t} - e^{-3t}.$$

5. Let $W = \mathcal{L}\{w\}$. Then taking the Laplace transform of the equation and using linearity yields

$$\mathcal{L}\{w''\} + \mathcal{L}\{w\} = \mathcal{L}\{t^2\} + \mathcal{L}\{2\}.$$

Solving for W, we find

$$s^2W - s + 1 + W = \frac{2}{s^3} + \frac{2}{s}$$

$$\Rightarrow \qquad (s^2+1)W = s - 1 + \frac{2}{s^3} + \frac{2}{s} = s - 1 + \frac{2s^2+2}{s^3}$$

$$\Rightarrow \qquad W = \frac{s-1}{s^2+1} + \frac{2s^2+2}{(s^2+1)s^3} = \frac{s}{s^2+1} - \frac{1}{s^2+1} + \frac{2}{s^3}.$$

Now, taking the inverse Laplace transform, we obtain

$$w = \mathcal{L}^{-1}\left\{\frac{s}{s^2+1}\right\} - \mathcal{L}^{-1}\left\{\frac{1}{s^2+1}\right\} + \mathcal{L}^{-1}\left\{\frac{2}{s^3}\right\}$$

$$\Rightarrow \qquad w = \cos t - \sin t + t^2.$$

11. As in Example 3 on page 401 of the text, we first need to shift the initial conditions to 0. If we set $v(t) = y(t+2)$, Problem 11 becomes

$$v'' - v = t; \tag{14}$$

$$v(0) = 3, \qquad v'(0) = 0.$$

Taking the Laplace transform of (14) and applying the linearity of the Laplace transform gives us

$$\mathcal{L}\{v''\} - \mathcal{L}\{v\} = \mathcal{L}\{t\}. \tag{15}$$

If we put $V(s) = \mathcal{L}\{v\}$ and apply property (4) on page 384 of the text (with $n = 2$), we get

$$\mathcal{L}\{v''\} = s^2V(s) - 3s. \tag{16}$$

Combining (15), (16) and the fact that $\mathcal{L}\{t\} = \frac{1}{s^2}$ yields the equation

$$s^2V(s) - 3s - V(s) = \frac{1}{s^2}.$$

Solving for $V(s)$ gives us

$$V(s) = \frac{3s^3 + 1}{(s^2 - 1)s^2} = \frac{-1}{s^2} + \frac{1}{s+1} + \frac{2}{s-1}.$$

Table 7.1 on page 380 of the text now tells us that

$$v(t) = -t + e^{-t} + 2e^{t}.$$

Since $v(t) = y(t+2)$ we see that $y(t) = v(t-2)$ and so

$$y(t) = 2 - t + e^{2-t} + 2e^{t-2}.$$

15. Taking the Laplace transform of

$$y'' - 3y' + 2y = \cos t$$

and applying the linearity of the Laplace transform yields

$$\mathcal{L}\{y''\} - 3\mathcal{L}\{y'\} + 2\mathcal{L}\{y\} = \mathcal{L}\{\cos t\}. \tag{17}$$

If we put $Y(s) = \mathcal{L}\{y\}(s)$ and apply properties (2), page 383, and (4), page 384 of the text, we get

$$\mathcal{L}\{y'\} = sY(s) \tag{18}$$

and

$$\mathcal{L}\{y''\} = s^2Y(s) + 1. \tag{19}$$

Combining (17), (18), and (19) and the fact that $\mathcal{L}\{\cos t\} = \frac{s}{s^2+1}$, yields the equation

$$s^2Y(s) + 1 - 3sY(s) + 2Y(s) = \frac{s}{s^2+1}.$$

Solving for $Y(s)$, we get

$$Y(s) = \frac{-s^2 + s - 1}{(s^2+1)(s^2 - 3s + 2)} = \frac{-s^2 + s - 1}{(s^2+1)(s-1)(s-2)}.$$

25. Taking the Laplace transform of

$$y''' - y'' + y' - y = 0$$

and applying the linearity of the Laplace transform yields

$$\mathcal{L}\{y'''\} - \mathcal{L}\{y''\} + \mathcal{L}\{y'\} - \mathcal{L}\{y\} = \mathcal{L}\{0\}. \tag{20}$$

If we put $Y(s) = \mathcal{L}\{y\}$ and apply property (4) on page 384 of the text, we get

$$\mathcal{L}\{y'\} = sY(s) - 1, \tag{21}$$

$$\mathcal{L}\{y''\} = s^2Y(s) - s - 1, \tag{22}$$

and

$$\mathcal{L}\{y'''\} = s^3Y(s) - s^2 - s - 3. \tag{23}$$

Combining (20), (21), (22), and (23) and using the fact that $\mathcal{L}\{0\} = 0$, gives us the equation

$$s^3Y(s) - s^2 - s - 3 - s^2Y(s) + s + 1 + sY(s) - 1 - Y(s) = 0.$$

Solving for $Y(s)$ yields

$$Y(s) = \frac{s^2 + 3}{s^3 - s^2 + s - 1} = \frac{s^2 + 3}{(s-1)(s^2+1)}.$$

Expanding $\dfrac{s^2+3}{(s-1)(s^2+1)}$ by partial fractions gives us

$$Y(s) = \frac{2}{s-1} - \frac{s+1}{s^2+1} = \frac{2}{s-1} - \frac{s}{s^2+1} - \frac{1}{s^2+1}.$$

From Table 7.1 on page 380 of the text, we see that

$$y(t) = 2e^t - \cos t - \sin t.$$

35. Taking the Laplace transform of

$$y'' + 3ty' - 6y = 1,$$

and applying the linearity of the Laplace transform yields

$$\mathcal{L}\{y''\} + 3\mathcal{L}\{ty'\} - 6\mathcal{L}\{y\} = \mathcal{L}\{1\}. \tag{24}$$

If we put $Y(s) = \mathcal{L}\{y\}$ and apply property (4) on page 384 of the text, we get

$$\mathcal{L}\{y'\} = sY(s) \tag{25}$$

and

$$\mathcal{L}\{y''\} = s^2Y(s). \tag{26}$$

Applying property (6) on page 385 of the text to (25) with $f(t) = ty'(t)$ yields

$$\mathcal{L}\{ty'\} = -\frac{d}{ds}[sY(s)] = -Y(s) - sY'(s). \tag{27}$$

Combining (24), (26), and (27) we get the equation

$$-3sY'(s) - 3Y(s) + s^2Y(s) - 6Y(s) = \frac{1}{s}$$

$$\Rightarrow \qquad Y'(s)+\frac{9-s^2}{3s}Y(s)=\frac{-1}{3s^2}.$$

This is a first order linear differential equation in $Y(s)$, which when solved by the techniques of Section 2.3 in the text, gives us

$$Y(s)=\frac{1}{s^3}+\frac{Ce^{s^2/6}}{s^3}.$$

Just as in Example 4 on page 402 of the text, C must be zero in order to ensure that $\lim_{s\to\infty} Y(s)=0$. Thus,

$$Y(s)=\frac{1}{s^3},$$

and from Table 7.1 on page 380 of the text, we get

$$y(t)=\frac{1}{2}t^2.$$

41. As in Problem 40, the differential equation modeling the automatic pilot is

$$Iy''(t)=-ke(t)-\mu e'(t), \qquad (28)$$

but now the error is given by

$$e(t)=y(t)-at. \qquad (29)$$

Taking the Laplace transform of (28) and applying properties (2), (3) on page 376, and (4) on page 384 of the text yields

$$Is^2Y(s)=-kE(s)-\mu sE(s), \qquad (30)$$

where $Y(s)=\mathcal{L}\{y\}$, and $E(s)=\mathcal{L}\{e\}$. Notice that as in Example 5 on page 404 of the text, we have assumed that

$$y(0)=y'(0)=e(0)=0.$$

Taking the Laplace transform of (29) yields

$$E(s)=Y(s)-\frac{a}{s^2} \quad \text{or}$$

$$Y(s)=E(s)+\frac{a}{s^2}. \qquad (31)$$

Combining (30) and (31) gives us

$$Is^2\left[E(s)+\frac{a}{s^2}\right]=-kE(s)-\mu sE(s),$$

which when solved for $E(s)$ yields

$$E(s)=\frac{-Ia}{s^2I+s\mu+k}$$

$$=\frac{-a}{s^2+\frac{s\mu}{I}+\frac{k}{I}}=\frac{-a}{\left(s+\frac{\mu}{2I}\right)^2+\frac{k}{I}-\frac{\mu^2}{4I^2}}=\frac{-a}{\left(s+\frac{\mu}{2I}\right)^2+\frac{4kI-\mu^2}{4I^2}}$$

$$=\left[\frac{-2Ia}{\sqrt{4kI-\mu^2}}\right]\left[\frac{\frac{\sqrt{4kI-\mu^2}}{2I}}{\left(s+\frac{\mu}{2I}\right)^2+\frac{4kI-\mu^2}{4I^2}}\right].$$

Using Table 7.1 on page 380 of the text to invert this expression gives us

$$e(t)=\left[\frac{-2Ia}{\sqrt{4kI-\mu^2}}\right]e^{-\mu t/2I}\sin\left[\frac{\sqrt{4kI-\mu^2}}{2I}t\right].$$

Compare this with Example 5 of the text and observe, how for moderate damping with $\mu<2\sqrt{kI}$, the oscillations of Example 5 die out exponentially.

EXERCISES 7.6: Transforms of Discontinuous and Periodic Functions, page 417

3. The graph of $t^2u(t-2)$ is shown in Figure 7-A. To apply the shifting property, we observe that $g(t)=t^2$ and $a=2$. Hence

$$g(t+a) = g(t+2) = (t+2)^2 = t^2+4t+4.$$

Now the Laplace transform of $g(t+2)$ is

$$\mathcal{L}\{g(t+2)\}(s) = \frac{2}{s^3}+\frac{4}{s^2}+\frac{4}{s}.$$

Hence, by formula (8) on page 409 of the text, we have

$$\mathcal{L}\{t^2u(t-2)\}(s)=e^{-2s}\left(\frac{2}{s^3}+\frac{4}{s^2}+\frac{4}{s}\right)$$

$$=\frac{e^{-2s}\left(4s^2+4s+2\right)}{s^3}.$$

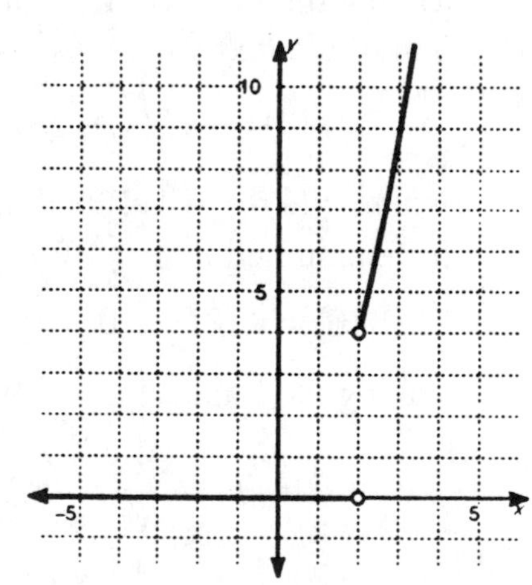

Figure 7-A. Graph of $t^2u(t-2)$.

7. Observe from the graph that $g(t)$ is given by

$$g(t)=\begin{cases}0 & t<1,\\ t & 1<t<2,\\ 1 & 2<t.\end{cases}$$

The function $g(t)$ equals zero until t reaches 1, at which point $g(t)$ jumps to the function t. We can express this jump by $tu(t-1)$. At $t=2$ the function $g(t)$ jumps from the function t to the value 1. This can be expressed by adding the term $(1-t)u(t-2)$. Hence

$$g(t)=tu(t-1)+(1-t)u(t-2).$$

Finally, taking the Laplace transform and using formula (8) on page 409, we find

$$\begin{aligned}\mathcal{L}\{tu(t-1)+(1-t)u(t-2)\}(s) &= \mathcal{L}\{tu(t-1)\}(s)+\mathcal{L}\{(1-t)u(t-2)\}(s)\\ &= e^{-s}\mathcal{L}\{t+1\}(s)-e^{-2s}\mathcal{L}\{t+1\}(s)\\ &= e^{-s}\left(\frac{1}{s^2}+\frac{1}{s}\right)-e^{-2s}\left(\frac{1}{s^2}+\frac{1}{s}\right)\\ &= \left(e^{-s}-e^{-2s}\right)\left(\frac{1}{s^2}+\frac{1}{s}\right)=\frac{(e^{-s}-e^{-2s})(s+1)}{s^2}.\end{aligned}$$

17. By partial fractions,

$$\frac{s-5}{(s+1)(s+2)}=-\frac{6}{s+1}+\frac{7}{s+2}$$

so that

$$\begin{aligned}\mathcal{L}^{-1}\left\{\frac{e^{-3s}(s-5)}{(s+1)(s+2)}\right\} &= -6\,\mathcal{L}^{-1}\left\{\frac{e^{-3s}}{s+1}\right\}+7\,\mathcal{L}^{-1}\left\{\frac{e^{-3s}}{s+2}\right\}\\ &= \left[-6e^{-(t-3)}+7e^{-2(t-3)}\right]u(t-3)=\left[7e^{6-2t}-6e^{3-t}\right]u(t-3).\end{aligned}$$

21. Since $f(t)=t$ on $0<t<2$, and has period 2, we see from Theorem 9 on page 413 of the text that

$$\mathcal{L}\{f\}=\frac{\int_0^2 e^{-st}t\,dt}{1-e^{-2s}}=\frac{\dfrac{-2e^{-2s}}{s}-\dfrac{e^{-2s}}{s^2}+\dfrac{1}{s^2}}{1-e^{-2s}}=\frac{1-2se^{-2s}-e^{-2s}}{s^2\left(1-e^{-2s}\right)}.$$

The graph of the function $f(t)$ is given in Figure B.45 in the answers of the text.

25. From the graph,

$$f(t)=u(t)-u(t-a),\qquad 0<t<2a,$$

and has period $2a$. From Theorem 9 on page 413 of the text, we have

$$\mathcal{L}\{f\}=\frac{\int_0^{2a} e^{-st}\left[u(t)-u(t-a)\right]dt}{1-e^{-2as}}=\frac{\int_0^{a} e^{-st}dt}{1-e^{-2as}}=\frac{\frac{1-e^{-as}}{s}}{1-e^{-2as}}=\frac{1-e^{-as}}{(1-e^{-as})(1+e^{-as})s}=\frac{1}{(1+e^{-as})s}.$$

27. Observe that if we let

$$f_{2a}(t)=\begin{cases} f(t) & 0<t<2a, \\ 0 & 2a<t, \end{cases}$$

denote the windowed version of f, then from formula (12) on page 413 of the text we have

$$\mathcal{L}\{f\}(s)=\mathcal{L}\{f_{2a}\}(s)/\left(1-e^{-2as}\right)=\mathcal{L}\{f_{2a}\}(s)/\left[\left(1-e^{-as}\right)\left(1+e^{-as}\right)\right].$$

Now

$$f_{2a}(t)=\frac{t}{a}+\left(2-\frac{2t}{a}\right)u(t-a)+\left(\frac{t}{a}-2\right)u(t-2a).$$

Hence,

$$\mathcal{L}\{f_{2a}\}(s)=\frac{1}{as^2}-\frac{2}{a}\mathcal{L}\{(t-a)u(t-a)\}(s)+\frac{1}{a}\mathcal{L}\{(t-2a)u(t-2a)\}(s)$$

$$=\frac{1}{as^2}-\frac{2}{a}e^{-as}\frac{1}{s^2}+\frac{1}{a}e^{-2as}\frac{1}{s^2}=\frac{1}{as^2}\left(1-2e^{-as}+e^{-2as}\right)=\frac{1}{as^2}\left(1-e^{-as}\right)^2.$$

So

$$\mathcal{L}\{f\}(s)=\frac{\frac{1}{as^2}\left(1-e^{-as}\right)^2}{\left(1-e^{-as}\right)\left(1+e^{-as}\right)}=\frac{1-e^{-as}}{as^2\left(1+e^{-as}\right)}.$$

33. By equation (4) on page 408 of the text,

$$\mathcal{L}\{u(t-2\pi)-u(t-4\pi)\}=\frac{e^{-2\pi s}}{s}-\frac{e^{-4\pi s}}{s}.$$

Thus, taking the Laplace transform of

$$y''+2y'+2y=u(t-2\pi)-u(t-4\pi)$$

and applying the initial conditions $y(0)=y'(0)=1$ gives us

$$s^2Y(s)-s-1+2sY(s)-2+2Y(s)=\frac{e^{-2\pi s}-e^{-4\pi s}}{s}.$$

Solving for $Y(s)$ yields

$$Y(s)=\frac{s^2+3s+e^{-2\pi s}-e^{-4\pi s}}{s(s^2+2s+2)}$$

$$= \left(\frac{s+1}{(s+1)^2+1}\right) + \left(\frac{2}{(s+1)^2+1}\right) + \left(\frac{e^{-2\pi s}}{s(s^2+2s+2)}\right) - \left(\frac{e^{-4\pi s}}{s(s^2+2s+2)}\right)$$

$$= \left(\frac{s+1}{(s+1)^2+1}\right) + \left(\frac{2}{(s+1)^2+1}\right) + \left(\frac{1}{2}\right)e^{-2\pi s}\left(\frac{1}{s} - \frac{s+2}{s^2+2s+2}\right)$$

$$- \left(\frac{1}{2}\right)e^{-4\pi s}\left(\frac{1}{s} - \frac{s+2}{s^2+2s+2}\right) \quad \text{(partial fractions)}$$

$$= \left(\frac{s+1}{(s+1)^2+1}\right) + \left(\frac{2}{(s+1)^2+1}\right) + \left(\frac{1}{2}\right)e^{-2\pi s}\left(\frac{1}{s} - \frac{s+1}{s^2+2s+2} - \frac{1}{s^2+2s+2}\right)$$

$$- \left(\frac{1}{2}\right)e^{-4\pi s}\left(\frac{1}{s} - \frac{s+1}{s^2+2s+2} - \frac{1}{s^2+2s+2}\right).$$

With the help of Theorem 8 and Table 7.1 on page 380 of the text, we invert this to find that

$$y(t) = e^{-t}\cos t + 2e^{-t}\sin t + \frac{1}{2}\left\{1 - e^{-(t-2\pi)}\left[\cos(t-2\pi) + \sin(t-2\pi)\right]\right\}u(t-2\pi)$$

$$- \frac{1}{2}\left\{1 - e^{-(t-4\pi)}\left[\cos(t-4\pi) + \sin(t-4\pi)\right]\right\}u(t-4\pi)$$

$$= e^{-t}\cos t + 2e^{-t}\sin t + \frac{1}{2}\left[1 - e^{(2\pi-t)}(\cos t + \sin t)\right]u(t-2\pi)$$

$$- \frac{1}{2}\left[1 - e^{(4\pi-t)}(\cos t + \sin t)\right]u(t-4\pi).$$

41. First observe that for $s > 0$, $T > 0$, we have $0 < e^{-Ts} < 1$ so that

$$\frac{1}{1-e^{-Ts}} = 1 + e^{-Ts} + e^{-2Ts} + e^{-3Ts} + \cdots$$

and the series converges for all $s > 0$. Thus,

$$\left[(s+\alpha)(1-e^{-Ts})\right]^{-1} = \frac{1}{s+\alpha}\frac{1}{1-e^{-Ts}} = \frac{1}{s+\alpha}\left(1 + e^{-Ts} + e^{-2Ts} + e^{-3Ts} + \cdots\right)$$

$$= \frac{1}{s+\alpha} + \frac{e^{-Ts}}{s+\alpha} + \frac{e^{-2Ts}}{s+\alpha} + \cdots,$$

and so

$$\mathcal{L}^{-1}\left\{\left[(s+\alpha)(1-e^{-Ts})\right]^{-1}\right\} = \mathcal{L}^{-1}\left\{\frac{1}{s+\alpha} + \frac{e^{-Ts}}{s+\alpha} + \frac{e^{-2Ts}}{s+\alpha} + \cdots\right\}.$$

Taking for granted that the linearity of the inverse Laplace transform extends to the infinite sum, $\dfrac{1}{s+\alpha}+\dfrac{e^{-Ts}}{s+\alpha}+\dfrac{e^{-2Ts}}{s+\alpha}+\cdots$, and ignoring convergence questions yields

$$\mathcal{L}^{-1}\left\{\left[(s+\alpha)\left(1-e^{-Ts}\right)\right]^{-1}\right\}=\mathcal{L}^{-1}\left\{\frac{1}{s+\alpha}\right\}+\mathcal{L}^{-1}\left\{\frac{e^{-Ts}}{s+\alpha}\right\}+\cdots$$

$$=e^{-\alpha t}+e^{-\alpha(t-T)}u(t-T)+e^{-2(t-2T)}u(t-2T)+\cdots$$

as claimed.

49. Recall that the Taylor's series for $\cos t$ about $t=0$ is

$$\cos t=1-\frac{t^2}{2!}+\frac{t^4}{4!}-\frac{t^6}{6!}+\cdots+(-1)^n\frac{t^{2n}}{(2n)!}+\cdots$$

so that

$$\frac{1-\cos t}{t}=\frac{t}{2!}-\frac{t^3}{4!}+\frac{t^5}{6!}+\cdots+(-1)^{n+1}\frac{t^{2n-1}}{(2n)!}+\cdots.$$

Thus

$$\mathcal{L}\left\{\frac{1-\cos t}{t}\right\}=\frac{1}{2!}\mathcal{L}\{t\}-\frac{1}{4!}\mathcal{L}\{t^3\}+\cdots+\frac{(-1)^{n+1}}{(2n)!}\mathcal{L}\{t^{2n-1}\}+\cdots$$

$$=\left(\frac{1}{2}\right)\left(\frac{1}{s^2}\right)-\left(\frac{1}{4}\right)\left(\frac{1}{s^4}\right)+\cdots+\left(\frac{(-1)^{n+1}}{2n}\right)\left(\frac{1}{s^{2n}}\right)+\cdots$$

$$=\sum_{n=1}^{\infty}\left(\frac{(-1)^{n+1}}{2n}\right)\left(\frac{1}{s^{2n}}\right)=\sum_{n=1}^{\infty}\frac{(-1)^{n+1}}{2ns^{2n}}.$$

To sum this series, recall that

$$\ln(1-x)=-\sum_{n=1}^{\infty}\frac{x^n}{n}.$$

Hence,

$$\ln\left(1+\frac{1}{s^2}\right)=-\sum_{n=1}^{\infty}\frac{(-1)^n}{ns^{2n}}=\sum_{n=1}^{\infty}\frac{(-1)^{n+1}}{ns^{2n}}.$$

Thus, we have

$$\frac{1}{2}\ln\left(1+\frac{1}{s^2}\right)=\sum_{n=1}^{\infty}\frac{(-1)^{n+1}}{2ns^{2n}}=\mathcal{L}\left\{\frac{1-\cos t}{t}\right\}.$$

This formula can also be obtained by using the result of Problem 27 on page 387 of the text.

55. Recall that

$$e^x = 1 + x + \frac{x^2}{2!} + \cdots + \frac{x^n}{n!} + \cdots .$$

Substituting $-\frac{1}{s}$ for x above yields

$$e^{-1/s} = 1 - \frac{1}{s} + \frac{1}{2!s^2} - \frac{1}{3!s^3} + \cdots + \frac{(-1)^n}{n!s^n} + \cdots .$$

Thus, we have

$$s^{-1/2}e^{-1/s} = \frac{1}{s^{1/2}} - \frac{1}{s^{3/2}} + \frac{1}{2!s^{5/2}} + \cdots + \frac{(-1)^n}{n!s^{n+(1/2)}} + \cdots = \sum_{n=0}^{\infty} \frac{(-1)^n}{n!s^{n+(1/2)}} .$$

By Problem 52 of this section,

$$\mathcal{L}^{-1}\left\{\frac{1}{s^{n+(1/2)}}\right\} = \frac{2^n t^{n-(1/2)}}{1\cdot 3\cdot 5\cdots(2n-1)\sqrt{\pi}},$$

so that

$$\mathcal{L}^{-1}\left\{s^{-1/2}e^{-1/s}\right\} = \mathcal{L}^{-1}\left\{\sum_{n=0}^{\infty} \frac{(-1)^n}{n!s^{n+(1/2)}}\right\}$$

$$= \sum_{n=0}^{\infty} \frac{(-1)^n}{n!}\mathcal{L}^{-1}\left\{\frac{1}{s^{n+(1/2)}}\right\} = \sum_{n=0}^{\infty} \frac{(-1)^n}{n!}\left(\frac{2^n t^{n-(1/2)}}{1\cdot 3\cdot 5\cdots(2n-1)\sqrt{\pi}}\right).$$

Multiplying the nth term by $\frac{2\cdot 4\cdots 2n}{2\cdot 4\cdots 2n}$, we obtain

$$\mathcal{L}^{-1}\left\{s^{-1/2}e^{-1/s}\right\} = \sum_{n=0}^{\infty} \frac{(-1)^n(2^n)^2 t^{n-(1/2)}}{(2n)!\sqrt{\pi}} = \sum_{n=0}^{\infty} \frac{(-1)^n(2^n)^2 t^n}{(2n)!\sqrt{\pi t}}$$

$$= \left(\frac{1}{\sqrt{\pi t}}\right)\sum_{n=0}^{\infty}\left(\frac{(-1)^n\left(2\sqrt{t}\right)^{2n}}{(2n)!}\right) = \left(\frac{1}{\sqrt{\pi t}}\right)\cos(2\sqrt{t}).$$

EXERCISES 7.7: Convolution, page 427

3. Taking the Laplace transform of $y'' + 4y' + 5y = g(t)$ and applying the initial conditions $y(0) = y'(0) = 1$ gives us

$$s^2Y(s) - s - 1 + 4sY(s) - 4 + 5Y(s) = G(s),$$

where $Y(s) = \mathcal{L}\{y\}$ and $G(s) = \mathcal{L}\{g\}$. Thus

$$Y(s) = \frac{G(s)}{s^2 + 4s + 5} + \frac{s + 5}{s^2 + 4s + 5}$$

$$= \frac{G(s)}{(s+2)^2 + 1} + \frac{s+2}{(s+2)^2 + 1} + \frac{3}{(s+2)^2 + 1}$$

Taking the inverse Laplace transform of $Y(s)$ with the help of the convolution theorem yields

$$Y(t) = \int_0^t e^{-2(t-v)} \sin(t - v) g(v) dv + e^{-2t} \cos t + 3e^{-2t} \sin t .$$

9. Since $\frac{s}{(s^2+1)^2} = \left(\frac{s}{s^2+1}\right)\left(\frac{1}{s^2+1}\right)$ the convolution theorem, Theorem 11 on page 422 of the text, tells us that

$$\mathcal{L}^{-1}\left\{\frac{s}{(s^2+1)^2}\right\} = \mathcal{L}^{-1}\left\{\left(\frac{s}{s^2+1}\right)\left(\frac{s}{s^2+1}\right)\right\} = \cos t * \sin t = \int_0^t \cos(t - v)\sin(v) dv .$$

Using the identity $\sin\alpha\cos\beta = \frac{1}{2}[\sin(\alpha + \beta) + \sin(\alpha - \beta)]$, we get

$$\mathcal{L}^{-1}\left\{\frac{s}{(s^2+1)^2}\right\} = \frac{1}{2}\int_0^t [\sin t + \sin(t - 2v)] dv$$

$$= \left(\frac{1}{2} v \sin t + \frac{1}{4}\cos(t - 2v)\right)\Bigg|_0^t = \frac{t}{2}\sin t .$$

11. Using the hint, we can write

$$\frac{s}{(s-1)(s+2)} = \frac{1}{s+2} + \frac{1}{(s-1)(s+2)},$$

so that by the convolution theorem, Theorem 11 on page 422 of the text,

$$\mathcal{L}^{-1}\left\{\frac{s}{(s-1)(s+2)}\right\} = \mathcal{L}^{-1}\left\{\frac{1}{s+2}\right\} + \mathcal{L}^{-1}\left\{\frac{1}{s-1}\frac{1}{s+2}\right\}$$

$$=e^{-2t}+e^{t}*e^{-2t}=e^{-2t}+\int_0^t e^{(t-v)}\,e^{-2v}dv$$

$$=e^{-2t}+\int_0^t e^{(t-3v)}\,dv=e^{-2t}-\frac{e^{-2t}}{3}+\frac{e^{t}}{3}=\frac{2}{3}e^{-2t}+\frac{1}{3}e^{t}.$$

21. As in Example 3 on page 424 of the text, we first rewrite the integro-differential equation as

$$y'(t)+y(t)-y(t)*\sin t=-\sin t\,,\qquad y(0)=1. \tag{32}$$

We now take the Laplace transform of (32) to obtain

$$sY(s)-1+Y(s)-\left(\frac{1}{s^2+1}\right)Y(s)=\frac{-1}{s^2+1},$$

where $Y=\mathcal{L}\{y\}$. Thus,

$$Y(s)=\frac{s^2}{s^3+s^2+s}=\frac{s}{s^2+s+1}=\frac{s}{(s+1/2)^2+3/4}$$

$$=\frac{s+1/2}{(s+1/2)^2+3/4}-\frac{\left(1/\sqrt{3}\right)\left(\sqrt{3}/2\right)}{(s+1/2)^2+3/4}.$$

Taking the inverse Laplace transform yields

$$Y(t)=e^{-t/2}\cos\left(\frac{\sqrt{3}t}{2}\right)-\left(\frac{1}{\sqrt{3}}\right)e^{-t/2}\sin\left(\frac{\sqrt{3}t}{2}\right).$$

23. Taking the Laplace transform of the differential equation, and assuming zero initial conditions, we obtain

$$s^2Y(s)+9Y(s)=G(s),$$

where $Y=\mathcal{L}\{y\}$ and $G=\mathcal{L}\{g\}$. Thus,

$$H(s)=\frac{Y(s)}{G(s)}=\frac{1}{s^2+9}.$$

The impulse response function is then

$$h(t)=\mathcal{L}^{-1}\{H(s)\}=\mathcal{L}^{-1}\left\{\frac{1}{s^2+9}\right\}=\frac{1}{3}\mathcal{L}^{-1}\left\{\frac{3}{s^2+3^2}\right\}=\frac{1}{3}\sin 3t.$$

To solve the initial value problem, we need to solution to the corresponding homogeneous problem. The auxiliary equation for the homogeneous equation is

$$r^2+9=0,$$

which has roots, $r=\pm 3i$. Thus a general solution is

$$C_1 \cos 3t + C_2 \sin 3t .$$

Applying the initial conditions

$$y(0) = 2 \qquad \text{and} \qquad y'(0) = -3 ,$$

we obtain

$$C_1 = 2 \qquad \text{and} \qquad C_2 = -1 .$$

So

$$y_k(t) = 2\cos 3t - \sin 3t .$$

Hence a formula for the solution to the initial value problem is

$$(h * g)(t) + y_k(t) = \frac{1}{3}\int_0^t (\sin 3(t-v))g(v)\,dv + 2\cos 3t - \sin 3t .$$

EXERCISES 7.8: Impulses and the Dirac Delta Function, page 434

3. By equation (3) on page 429 of the text,

$$\int_{-\infty}^{\infty} (\sin 3t)\,\delta\left[t - \frac{\pi}{2}\right] dt = \sin\left(\frac{3\pi}{2}\right) = -1 .$$

9. Using the definition of the Laplace transform, we see that since $\delta(t-1) = 0$ for $t < 1$,

$$\mathcal{L}\{t\delta(t-1)\} := \int_0^{\infty} e^{-st}\, t\,\delta(t-1)\,dt = \int_{-\infty}^{\infty} e^{-st}\, t\,\delta(t-1)\,dt = e^{-s} ,$$

by equation (3) on page 429 of the text.

11. Using the definition of the Laplace transform, we see that since $\delta(t-\pi) = 0$ for $t < \pi$,

$$\mathcal{L}\{\delta(t-\pi)\sin t\} := \int_0^{\infty} e^{-st}\,\delta(t-\pi)\sin t\,dt = \int_{-\infty}^{\infty} e^{-st}\,\delta(t-\pi)\sin t\,dt = e^{-\pi s}\sin\pi = 0 .$$

15. Taking the Laplace transform of $y'' + 2y' - 3y = \delta(t-1) - \delta(t-2)$ and applying the initial conditions $y(0) = 2$, $y'(0) = -2$, we obtain

$$s^2 Y(s) - 2s + 2 + 2sY(s) - 4 - 3Y(s) = e^{-s} - e^{-2s} ,$$

where $Y = \mathcal{L}\{y\}$ and we have used the fact that $\mathcal{L}\{\delta(t-a)\} = e^{-as}$. Solving for $Y(s)$ yields

$$Y(s) = \frac{2s + 2 + e^{-s} - e^{-2s}}{s^2 + 2s - 3}$$

$$= \left\{\frac{2s+2}{(s+3)(s-1)}\right\} + \left\{\frac{e^{-s}}{(s+3)(s-1)}\right\} - \left\{\frac{e^{-2s}}{(s+3)(s-1)}\right\}$$

$$= \frac{1}{s-1} + \frac{1}{s+3} + \frac{1}{4}\frac{e^{-s}}{s-1} - \frac{1}{4}\frac{e^{-s}}{s+3} - \frac{1}{4}\frac{e^{-2s}}{s-1} + \frac{1}{4}\frac{e^{-2s}}{s+3},$$

so that by Theorem 8 on page 409 of the text we get

$$y(t) = e^{t} + e^{-3t} + \frac{1}{4}\left(e^{t-1} - e^{3-3t}\right)u(t-1) - \frac{1}{4}\left(e^{t-2} - e^{6-3t}\right)u(t-2).$$

25. Taking the Laplace transform of $y'' + 4y' + 8y = \delta(t)$ with zero initial conditions yields

$$s^2Y(s) + 4sY(s) + 8Y(s) = 1.$$

Solving for $Y(s)$ yields

$$Y(s) = \frac{1}{s^2 + 4s + 8} = \frac{1}{(s+2)^2 + 4} = \frac{1}{2}\frac{2}{(s+2)^2 + 2^2}$$

so that $h(t) = y(t) = \frac{1}{2}e^{-2t}\sin 2t$. Notice that $H(s)$ for $y'' + 4y' + 8y = g(t)$ with $y(0) = y'(0) = 0$ is given by

$$H(s) = \frac{1}{s^2 + 4s + 8},$$

so that again

$$h(t) = \mathcal{L}^{-1}\{H(s)\} = \frac{1}{2}e^{-2t}\sin 2t.$$

31. By taking the Laplace transform of

$$ay'' + by' + cy = \delta(t), \qquad y(0) = y'(0) = 0,$$

and solving for $Y = \mathcal{L}\{y\}$, we find that the transfer function is given by

$$H(s) = \frac{1}{as^2 + bs + c}.$$

If the roots of $as^2 + bs + c$ are real and distinct, say r_1, r_2, then

$$H(s) = \frac{1}{(s-r_1)(s-r_2)} = \frac{A}{s-r_1} + \frac{B}{s-r_2}$$

for some constants A and B. Thus

$$h(t) = Ae^{r_1t} + Be^{r_2t}$$

and clearly $h(t)$ is bounded as $t \to \infty$ if and only if r_1 and r_2 are less than or equal to zero. If the roots of $as^2 + bs + c$ are complex, then by the quadratic formula, the roots are given by

$$\left(\frac{-b}{2a}\right) \pm \left(\frac{\sqrt{4ac - b^2}}{2a}\right)i$$

so that the real part of the roots is $\frac{-b}{2a}$. Now

$$H(s)=\frac{1}{as^2+bs+c}=\frac{1}{a}\left(\frac{1}{s^2+\frac{b}{a}s+\frac{c}{a}}\right)=\frac{1}{a}\left(\frac{1}{\left[s+\frac{b}{2a}\right]^2+\frac{c}{a}-\frac{b^2}{4a^2}}\right)$$

$$=\left(\frac{2}{\sqrt{4ac-b^2}}\right)\times\left(\frac{\frac{\sqrt{4ac-b^2}}{2a}}{\left[s+\frac{b}{2a}\right]^2+\frac{4ac-b^2}{4a^2}}\right)$$

so that

$$h(t)=\frac{2}{\sqrt{4ac-b^2}}e^{-(b/2a)t}\sin\left(\frac{\sqrt{4ac-b^2}}{2a}t\right),$$

and again it is clear that $h(t)$ is bounded if and only if $\frac{-b}{2a}$, the real part of the roots of as^2+bs+c, is less than or equal to zero.

EXERCISES 7.9: Solving Linear Systems with Laplace Transforms, page 438

7. We will first write this system without using operator notation. Thus, we have

$$\begin{aligned} x'-4x+6y&=9e^{-3t},\\ x-y'+y&=5e^{-3t}. \end{aligned} \tag{33}$$

By taking the Laplace transform of both sides of both of these differential equations (see the Laplace transform table on the inside back cover of the text) and using the linearity of the Laplace transform, we obtain

$$\begin{aligned} \mathcal{L}\{x'\}(s)-4X(s)+6Y(s)&=\frac{9}{s+3},\\ X(s)-\mathcal{L}\{y'\}(s)+Y(s)&=\frac{5}{s+3}, \end{aligned} \tag{34}$$

where $X(s)$ and $Y(s)$ are the Laplace transforms of $x(t)$ and $y(t)$, respectively. Using the initial conditions $x(0)=-9$ and $y(0)=4$, we can express $\mathcal{L}\{x'\}(s)$ in terms of $X(s)$ and $\mathcal{L}\{y'\}(s)$ in terms of $Y(s)$. Namely, we have

$$\mathcal{L}\{x'\}(s)=sX(s)-x(0)=sX(s)+9,$$

$$\mathcal{L}\{y'\}(s) = sY(s) - y(0) = sY(s) - 4.$$

Substituting these expressions into the system given in (34) and simplifying yields

$$(s-4)X(s) + 6Y(s) = -9 + \frac{9}{s+3} = \frac{-9s-18}{s+3},$$

$$X(s) + (-s+1)Y(s) = -4 + \frac{5}{s+3} = \frac{-4s-7}{s+3}.$$

By multiplying the second equation above by $-(s-4)$, adding the resulting equations, and simplifying, we obtain

$$\left(s^2 - 5s + 10\right)Y(s) = \frac{(4s+7)(s-4)}{s+3} + \frac{-9s-18}{s+3}$$

$$\Rightarrow \qquad \left(s^2 - 5s + 10\right)Y(s) = \frac{4s^2 - 18s - 46}{s+3}$$

$$\Rightarrow \qquad Y(s) = \frac{4s^2 - 18s - 46}{(s+3)(s^2 - 5s + 10)}.$$

The partial fraction expansion for *Y(s)* is

$$Y(s) = \left(\frac{1}{17}\right)\left(\frac{22}{s+3} + \frac{46s - 334}{s^2 - 5s + 10}\right).$$

Therefore, we have

$$y(t) = \mathcal{L}^{-1}\{Y(s)\}(t) = \left(\frac{1}{17}\right)\mathcal{L}^{-1}\left\{\frac{22}{s+3} + \frac{46s - 334}{s^2 - 5s + 10}\right\}(t).$$

By completing the square, we see that

$$\frac{46s - 334}{s^2 - 5s + 10} = \frac{46s - 334}{(s - 5/2)^2 + 15/4}$$

$$= 46\left(\frac{s - 5/2}{(s - 5/2)^2 + 15/4}\right) - \left(\frac{146\sqrt{15}}{85}\right)\left(\frac{\sqrt{15}/2}{(s - 5/2)^2 + 15/4}\right).$$

Therefore, we see that

$$y(t) = \frac{22}{17}\mathcal{L}^{-1}\left\{\frac{1}{s+3}\right\}(t) + \frac{46}{17}\mathcal{L}^{-1}\left\{\frac{s - 5/2}{(s - 5/2)^2 + 15/4}\right\}(t)$$

$$- \left(\frac{146\sqrt{15}}{85}\right)\mathcal{L}^{-1}\left\{\frac{\sqrt{15}/2}{(s - 5/2)^2 + 15/4}\right\}(t)$$

$$\Rightarrow \qquad y(t)=\frac{22}{17}e^{-3t}+\frac{46}{17}e^{5t/2}\cos\left(\frac{\sqrt{15}\,t}{2}\right)-\left(\frac{146\sqrt{15}}{85}\right)e^{5t/2}\sin\left(\frac{\sqrt{15}\,t}{2}\right). \qquad (35)$$

From the second equation in system (33) above, we notice that

$$x(t)=5e^{-3t}+y'(t)-y(t)=5e^{-3t}-\frac{66}{17}e^{-3t}+\frac{115}{17}e^{5t/2}\cos\left(\frac{\sqrt{15}\,t}{2}\right)$$

$$-\left(\frac{23\sqrt{15}}{17}+\frac{73\sqrt{15}}{17}\right)e^{5t/2}\sin\left(\frac{\sqrt{15}\,t}{2}\right)-\frac{219}{17}e^{5t/2}\cos\left(\frac{\sqrt{15}\,t}{2}\right)-\frac{22}{17}e^{-3t}$$

$$-\frac{46}{17}e^{5t/2}\cos\left(\frac{\sqrt{15}\,t}{2}\right)+\left(\frac{146\sqrt{15}}{85}\right)e^{5t/2}\sin\left(\frac{\sqrt{15}\,t}{2}\right)$$

$$\Rightarrow \qquad x(t)=\frac{-3}{17}e^{-3t}-\frac{150}{17}e^{5t/2}\cos\left(\frac{\sqrt{15}\,t}{2}\right)-\frac{334\sqrt{15}}{85}e^{5t/2}\sin\left(\frac{\sqrt{15}\,t}{2}\right). \qquad (36)$$

Thus, the solution to this initial value problem is $x(t)$ and $y(t)$ as given in equations (36) and (35), respectively.

9. Taking the Laplace transform of both sides of both of these differential equations and remembering that $\mathcal{L}\{0\}=0$ and $\mathcal{L}\{cf\}=c\mathcal{L}\{f\}$, yields the system

$$\mathcal{L}\{x''\}(s)+X(s)+2\mathcal{L}\{y'\}(s)=0,$$

$$-3\mathcal{L}\{x''\}(s)-3X(s)+2\mathcal{L}\{y''\}(s)+4Y(s)=0,$$

where $\mathcal{L}\{y\}(s)=Y(s)$ and $\mathcal{L}\{x\}(s)=X(s)$. Using the initial conditions $x(0)=2$, $x'(0)=-7$ $y(0)=4$, and $y'(0)=-9$ we see that

$$\mathcal{L}\{x''\}(s)=s^2X(s)-sx(0)-x'(0)=s^2X(s)-2s+7,$$

$$\mathcal{L}\{y'\}(s)=sY(s)-y(0)=sY(s)-4,$$

$$\mathcal{L}\{y''\}(s)=s^2Y(s)-sy(0)-y'(0)=s^2Y(s)-4s+9.$$

Substituting these expressions into the system given above yields

$$s^2X(s)-2s+7+X(s)+2sY(s)-8=0,$$

$$-3s^2X(s)+6s-21-3X(s)+2s^2Y(s)-8s+18+4Y(s)=0,$$

which simplifies to

$$\begin{aligned}&\left(s^2+1\right)X(s)+2sY(s)=2s+1,\\&-3\left(s^2+1\right)X(s)+2\left(s^2+2\right)Y(s)=2s+3.\end{aligned}\qquad(37)$$

Multiplying the first equation by 3 and adding the two resulting equations eliminates the function $X(s)$. Thus, we obtain

$$\left[2s^2+6s+4\right]Y(s)=8s+6 \qquad \Rightarrow \qquad Y(s)=\frac{4s+3}{(s+2)(s+1)},$$

where we have factored the expression $\left(s^2+3s+2\right)$. The partial fraction expansion for $Y(s)$ is

$$Y(s)=\frac{5}{s+2}-\frac{1}{s+1}.$$

Taking the inverse Laplace transform, we obtain

$$y(t)=\mathcal{L}^{-1}\{Y(s)\}(t)=5\mathcal{L}^{-1}\left\{\frac{1}{s+2}\right\}(t)-\mathcal{L}^{-1}\left\{\frac{1}{s+1}\right\}(t)=5e^{-2t}-e^{-t}.$$

To find the solution $x(t)$, we again examine the system given in (37) above. This time we will eliminate the function $Y(s)$ by multiplying the first equation by $\left(s^2+2\right)$ and the second by $-s$ and adding the resulting equations. Thus, we have

$$\left(s^2+3s+2\right)\left(s^2+1\right)X(s)=2s^3-s^2+s+2$$

$$\Rightarrow \qquad X(s)=\frac{2s^3-s^2+s+2}{(s+1)(s+2)(s^2+1)}.$$

Expressing $X(s)$ in a partial fraction expansion

$$\frac{2s^3-s^2+s+2}{(s+1)(s+2)(s^2+1)}=\frac{A}{s+1}+\frac{B}{s+2}+\frac{Cs+D}{s^2+1}$$

$$\Rightarrow \qquad X(s)=\frac{-1}{s+1}+\frac{4}{s+2}-\frac{s}{s^2+1}.$$

By taking the inverse Laplace transforms, we see that

$$x(t)=\mathcal{L}^{-1}\{X(s)\}(t)=\mathcal{L}^{-1}\left\{\frac{-1}{s+1}\right\}(t)+4\mathcal{L}^{-1}\left\{\frac{1}{s+2}\right\}(t)-\mathcal{L}^{-1}\left\{\frac{s}{s^2+1}\right\}(t)$$

$$\Rightarrow \qquad x(t)=-e^{-t}+4e^{-2t}-\cos t.$$

Hence, the solution to this initial value problem is

$$x(t)=-e^{-t}+4e^{-2t}-\cos t \qquad \text{and} \qquad y(t)=5e^{-2t}-e^{-t}.$$

19. We first take the Laplace transform of both sides of all three of these equations and use the initial conditions to obtain a system of equations for the Laplace transforms of the solution functions:

$$sX(s)+6=3X(s)+Y(s)-2Z(s),$$

$$sY(s)-2=-X(s)+2Y(s)+Z(s),$$

$$sZ(s)+12=4X(s)+Y(s)-3Z(s).$$

Simplifying yields

$$\begin{aligned}(s-3)X(s)-Y(s)+2Z(s)&=-6,\\ X(s)+(s-2)Y(s)-Z(s)&=2,\\ -4X(s)-Y(s)+(s+3)Z(s)&=-12.\end{aligned} \tag{38}$$

To solve this system, we will use substitution to eliminate the function $Y(s)$. Therefore, we solve for $Y(s)$ in the first equation in (38) to obtain

$$Y(s)=(s-3)X(s)+2Z(s)+6.$$

Substituting this expression into the two remaining equations in (38) and simplifying yields

$$\begin{aligned}\left(s^2-5s+7\right)X(s)+(2s-5)Z(s)&=-6s+14,\\ -(s+1)X(s)+(s+1)Z(s)&=-6.\end{aligned} \tag{39}$$

Next we will eliminate the function $X(s)$ from the system given in (39). To do this we can either multiply the first equation by $(s + 1)$ and the second by $\left(s^2-5s+7\right)$ and add, or we can solve the last equation given in (39) for $X(s)$ to obtain

$$X(s)=Z(s)+\frac{6}{s+1}, \tag{40}$$

and substitute this into the first equation in (39). By either method we see that

$$Z(s)=\frac{-12s^2+38s-28}{(s+1)(s^2-3s+2)}$$

$$\Rightarrow \qquad Z(s)=\frac{-12s^2+38s-28}{(s+1)(s-2)(s-1)}.$$

Now, $Z(s)$ has the partial fraction expansion

$$Z(s)=\frac{-13}{s+1}+\frac{1}{s-1}.$$

Therefore, by taking inverse Laplace transforms of both sides of this equation, we obtain

$$z(t)=\mathcal{L}^{-1}\{Z(s)\}(t)=-13\mathcal{L}^{-1}\left\{\frac{1}{s+1}\right\}(t)+\mathcal{L}^{-1}\left\{\frac{1}{s-1}\right\}(t)$$

$$\Rightarrow \quad z(t) = -13e^{-t} + e^{t}.$$

To find $X(s)$, we will use equation (40) and the expression found above for $Z(s)$. Thus, we have

$$X(s) = Z(s) + \frac{6}{s+1} = \frac{-13}{s+1} + \frac{1}{s-1} + \frac{6}{s+1}$$

$$\Rightarrow \quad X(s) = \frac{-7}{s+1} + \frac{1}{s-1}.$$

By taking the inverse Laplace transform, we obtain

$$x(t) = \mathcal{L}^{-1}\{X(s)\}(t) = -7\mathcal{L}^{-1}\left\{\frac{1}{s+1}\right\}(t) + \mathcal{L}^{-1}\left\{\frac{1}{s-1}\right\}(t)$$

$$\Rightarrow \quad x(t) = -7e^{-t} + e^{t}.$$

To find $y(t)$, we could substitute the expressions that we have already found for $X(s)$ and $Z(s)$ into the equation $Y(s) = (s - 3)X(s) + 2Z(s) + 6$, which we found above, or we could return to the original system of differential equations and use this to solve for $y(t)$. For the latter method, we solve the first equation in the original system for $y(t)$ to obtain

$$y(t) = x'(t) - 3x(t) + 2z(t) = 7e^{-t} + e^{t} + 21e^{-t} - 3e^{t} - 26e^{-t} + 2e^{t} = 2e^{-t}.$$

Therefore, the solution to the initial value problem is

$$x(t) = -7e^{-t} + e^{t}, \quad y(t) = 2e^{-t}, \quad \text{and} \quad z(t) = -13e^{-t} + e^{t}.$$

CHAPTER 8: Series Solutions of Differential Equations

EXERCISES 8.1: Introduction: The Taylor Polynomial Approximation, page 452

3. Using the initial condition, $y(0)=0$, we substitute $x=0$ and $y=0$ into the given equation and find $y'(0)$.

$$y'(0)=\sin(0)+e^{0}=1.$$

To determine $y''(0)$, we differentiate the given equation with respect to x and substitute $x=0$, $y=0$, and $y'=1$ in the formula obtained:

$$y''=\left(\sin y+e^{x}\right)'=(\sin y)'+\left(e^{x}\right)'=y'\cos y+e^{x},$$

$$y''(0)=1\cdot\cos(0)+e^{0}=2.$$

Similarly, differentiating $y''(x)$ and substituting, we obtain

$$y'''=\left(y'\cos y+e^{x}\right)'=(y'\cos y)'+\left(e^{x}\right)'=y''\cos y+(y')^{2}(-\sin y)+e^{x},$$

$$y'''(0)=y''(0)\cos y(0)+(y'(0))^{2}[-\sin y(0)]+e^{0}=2\cdot\cos 0+(1)^{2}(-\sin 0)+1=3.$$

Thus the first three nonzero terms in the Taylor polynomial approximations to the solution of the given initial value problem are

$$y(x)=y(0)+\frac{y'(0)}{1!}x+\frac{y''(0)}{2!}x^{2}+\frac{y'''(0)}{3!}x^{3}+\cdots$$

$$=0+\frac{1}{1}x+\frac{2}{2}x^{2}+\frac{3}{6}x^{3}+\cdots=x+x^{2}+\frac{1}{2}x^{3}+\cdots.$$

7. We use the initial conditions to find $y''(0)$. Writing the given equation in the form

$$y''(\theta)=-y(\theta)^{3}+\sin\theta$$

and substituting $\theta=0$, $y(0)=0$, we get

$$y''(0)=-y(0)^{3}+\sin 0=0.$$

Differentiating the given equation we obtain

$$y'''=\left(y''\right)'=-\left(y^{3}\right)'+(\sin\theta)'=-3y^{2}y'+\cos\theta$$

$$\Rightarrow\qquad y'''(0)=-3y(0)^{2}y'(0)+\cos 0=-3(0)^{2}(0)+1=1.$$

Similarly, we get

$$y^{(4)} = (y''')' = \left(-(y^3)' + \cos\theta\right)' = -3y^2y'' - 6y(y')^2 - \sin\theta$$

$$\Rightarrow \quad y^{(4)}(0) = -3y(0)^2 y''(0) - 6y(0)(y'(0))^2 - \sin 0 = 0 .$$

To simplify further computations we observe that since the Taylor expansion for $y(\theta)$ has the form

$$y(\theta) = \frac{1}{3!}\theta^3 + \cdots ,$$

then the Taylor expansion for $y(\theta)^3$ must begin with the term $(1/3!)^3\theta^9$, so that

$$\left(y(\theta)^3\right)^{(k)}\Big|_{\theta=0} = 0 \qquad \text{for} \quad k = 0, 1, \ldots, 8 .$$

Hence

$$y^{(5)} = -\left(y^3\right)^{(3)} - \cos\theta \qquad \Rightarrow \qquad y^{(5)}(0) = -\left(y^3\right)^{(3)}\Big|_{\theta=0} - \cos 0 = -1 ,$$

$$y^{(6)} = -\left(y^3\right)^{(4)} + \sin\theta \qquad \Rightarrow \qquad y^{(4)}(0) = -\left(y^3\right)^{(4)}\Big|_{\theta=0} + \sin 0 = 0 ,$$

$$y^{(7)} = -\left(y^3\right)^{(5)} + \cos\theta \qquad \Rightarrow \qquad y^{(7)}(0) = -\left(y^3\right)^{(5)}\Big|_{\theta=0} + \cos 0 = 1 .$$

Thus, the first three nonzero terms of the Taylor approximations are

$$\frac{1}{3!}\theta^3 - \frac{1}{5!}\theta^5 + \frac{1}{7!}\theta^7 + \cdots = \frac{1}{6}\theta^3 - \frac{1}{120}\theta^5 + \frac{1}{5040}\theta^7 + \cdots .$$

11. First, we rewrite the given equation in the form

$$y'' = -py' - qy + g .$$

On the right-hand side of this equation, the function y' is differentiable (y'' exists) and the functions y, p, q, and g are differentiable (even twice). Thus we conclude that its left-hand side, y'', is differentiable being the product, sum, and difference of differentiable functions. Therefore, $y''' := (y'')'$ exists and is given by

$$y''' = (-py' - qy + g)' = -p'y' - py'' - q'y - qy' + g' .$$

Similarly, we conclude that the right-hand side of the equation above is a differentiable function since all the functions involved are differentiable (notice that we have just proved the differentiability of y''). Hence, y''', its left-hand side is differentiable as well, i.e., $(y''')' =: y^{(4)}$ does exist.

13. With form $k = r = A = 1$ and $\omega = 10$, the Duffing's equation becomes

$$y'' + y + y^3 = \cos 10t \qquad \text{or} \qquad y'' = -y - y^3 + \cos 10t .$$

Substituting the initial conditions, $y(0) = 0$ and $y'(0) = 1$ into the latter equation yields

$$y''(0) = -y(0) - y(0)^3 + \cos(10 \cdot 0) = -0 - (0)^3 + \cos(0) = 1 .$$

Differentiating the given equation, we conclude that

$$y''' = \left(-y - y^3 + \cos 10t\right)' = -y' - 3y'y^2 - 10\sin 10t ,$$

which, at $t = 0$, gives

$$y'''(0) = -y'(0) - 3y'(0)y(0)^2 - 10\sin(10 \cdot 0) = -1 - 3(1)(0)^2 - 10\sin(0) = -1 .$$

Thus, the Taylor polynomial approximations to the solution of the given initial value problem are

$$y(t) = y(0) + \frac{y'(0)}{1!}t + \frac{y''(0)}{2!}t^2 + \frac{y'''(0)}{3!}t^3 + \cdots = t + \frac{1}{2}t^2 - \frac{1}{6}t^3 + \cdots .$$

EXERCISES 8.2: Power Series and Analytic Functions, page 460

3. We will use the ratio test given in Theorem 2 on page 454 of the text to find the radius of convergence for this power series. Since $a_n = \dfrac{n^2}{2^n}$, we see that

$$\frac{a_{n+1}}{a_n} = \frac{\dfrac{(n+1)^2}{2^{n+1}}}{\dfrac{n^2}{2^n}} = \frac{(n+1)^2}{2n^2} .$$

Therefore, we have

$$\lim_{n\to\infty}\left|\frac{a_{n+1}}{a_n}\right| = \lim_{n\to\infty}\left|\frac{(n+1)^2}{2n^2}\right| = \frac{1}{2}\lim_{n\to\infty}\left|\frac{(n+1)^2}{n^2}\right| = \frac{1}{2}\lim_{n\to\infty}\left(1 + \frac{1}{n}\right)^2 = \frac{1}{2} .$$

Thus, the radius of convergence is $\rho = 2$. Hence, this power series converges absolutely for $|x + 2| < 2$. That is, for

$$-2 < x + 2 < 2 \qquad \text{or} \qquad -4 < x < 0 .$$

We must now check the end points of this interval. We first check the end point $x = -4$ or $x + 2 = -2$ which yields the series

$$\sum_{n=0}^{\infty} \frac{n^2(-2)^n}{2^n} = \sum_{n=0}^{\infty} (-1)^n n^2 .$$

This series diverges since the nth term, $a_n = (-1)^n n^2$, does not approach zero as n goes to infinity. (Recall that it is necessary for the nth term of a convergent series to approach zero as n goes to infinity. But this fact in itself does not prove that a series converges.) Next, we check the end point $x=0$ or $x+2=2$ which yields the series

$$\sum_{n=0}^{\infty} \frac{n^2 2^n}{2^n} = \sum_{n=0}^{\infty} n^2 .$$

Again, as above, this series diverges. Therefore, this power series converges in the open interval $(-4,0)$ and diverges outside of this interval.

11. We want to find the product $f(x)g(x)$ of the two series

$$f(x) = \sum_{n=0}^{\infty} \frac{x^n}{n!} = 1 + x + \frac{x^2}{2} + \frac{x^3}{6} + \frac{x^4}{24} + \cdots ,$$

and

$$g(x) = \sin x = \sum_{k=0}^{\infty} \left[\frac{(-1)^k}{(2k+1)!}\right] x^{2k+1} = x - \frac{x^3}{6} + \frac{x^5}{120} - \frac{x^7}{7!} + \cdots .$$

Therefore, we have

$$f(x)g(x) = \left(1 + x + \frac{x^2}{2} + \frac{x^3}{6} + \frac{x^4}{24} + \cdots\right)\left(x - \frac{x^3}{6} + \frac{x^5}{120} - \frac{x^7}{7!} + \cdots\right)$$

$$= x + x^2 + \left(\frac{1}{2} - \frac{1}{6}\right)x^3 + \left(\frac{1}{6} - \frac{1}{6}\right)x^4 + \left(\frac{1}{24} - \frac{1}{12} + \frac{1}{120}\right)x^5 + \cdots$$

$$= x + x^2 + \frac{1}{3}x^3 + \cdots .$$

Note that since the radius of convergence for both of the given series is $\rho = \infty$, the expansion of the product $f(x)g(x)$ also converges for all values of x.

19. Here we will assume that this series has a positive radius of convergence. Thus, since we have

$$f(x) = \sum_{n=0}^{\infty} a_n x^n = a_0 + a_1 x + a_2 x^2 + a_3 x^3 + \cdots + a_n x^n + \cdots ,$$

we can differentiate term by term to obtain

$$f'(x) = 0 + a_1 + a_2 2x + a_3 3x^2 + \cdots + a_n n x^{n-1} + \cdots = \sum_{n=1}^{\infty} a_n n x^{n-1} .$$

Note that the summation for $f(x)$ starts at zero while the summation for $f'(x)$ starts at one.

29. We need to determine the nth derivative of $f(x)$ at the point $x = \pi$. Thus, we observe that

$$f(x) = f^{(0)}(x) = \cos x \quad \Rightarrow \quad f(\pi) = f^{(0)}(\pi) = \cos\pi = -1,$$

$$f'(x) = -\sin x \quad \Rightarrow \quad f'(\pi) = -\sin\pi = 0,$$

$$f''(x) = -\cos x \quad \Rightarrow \quad f''(\pi) = -\cos\pi = 1,$$

$$f'''(x) = \sin x \quad \Rightarrow \quad f'''(\pi) = \sin\pi = 0,$$

$$f^{(4)}(x) = \cos x \quad \Rightarrow \quad f^{(4)}(\pi) = \cos\pi = -1.$$

Since $f^{(4)}(x) = \cos x = f(x)$, the four derivatives given above will be repeated indefinitely. Thus, we see that $f^{(n)}(\pi) = 0$ if n is odd and $f^{(n)}(\pi) = \pm 1$ if n is even (where the signs alternate starting at -1 when $n = 0$). Therefore, the Taylor series for f about the point $x_0 = \pi$ is given by

$$f(x) = -1 + 0 + \frac{1}{2!}(x-\pi)^2 + 0 - \frac{1}{4!}(x-\pi)^4 + \cdots + \frac{(-1)^{n+1}(x-\pi)^{2n}}{(2n)!} + \cdots$$

$$= \sum_{n=0}^{\infty} \frac{(-1)^{n+1}(x-\pi)^{2n}}{(2n)!}.$$

EXERCISES 8.3: Power Series Solutions to Linear Differential Equations, page 471

7. In standard form this equation becomes

$$y'' + \left(\frac{\cos x}{\sin x}\right) y = 0.$$

Thus, $p(x) = 0$ and, hence, is analytic everywhere. We also see that

$$q(x) = \frac{\cos x}{\sin x} = \cot x.$$

Note that $q(x)$ is the quotient of two functions ($\cos x$ and $\sin x$) that each have a power series expansion with a positive radius of convergence about each real number x. Thus, according to page 456 of the text, we see that $q(x)$ will also have a power series expansion with a positive radius of convergence about every real number x as long as the denominator, $\sin x$, is not equal to zero. Since the cotangent function is $\pm\infty$ at integer multiples of π, we see that $q(x)$ is not defined and, therefore, not analytic at $x = n\pi$. Hence, the differential equation is singular only at the points $x = n\pi$, where n is an integer.

15. Zero is an ordinary point for this equation since the functions $p(x) = x - 1$ and $q(x) = 1$ are both analytic everywhere and, hence, at the point $x = 0$. Thus, we can assume that the solution to this linear differential equation has a power series expansion with a positive radius of convergence about the point $x = 0$. That is, we assume that

$$y(x) = a_0 + a_1 x + a_2 x^2 + a_3 x^3 + \cdots = \sum_{n=0}^{\infty} a_n x^n .$$

In order to solve the differential equation we must find the coefficients a_n. To do this, we must substitute $y(x)$ and its derivatives into the given differential equation. Hence, we must find $y'(x)$ and $y''(x)$. Since $y(x)$ has a power series expansion with a positive radius of convergence about the point $x = 0$, we can find its derivative by differentiating term by term. We can similarly differentiate $y'(x)$ to find $y''(x)$. Thus, we have

$$y'(x) = 0 + a_1 + 2a_2 x + 3a_3 x^2 + \cdots = \sum_{n=1}^{\infty} n a_n x^{n-1}$$

$$\Rightarrow \qquad y''(x) = 2a_2 + 6a_3 x + \cdots = \sum_{n=2}^{\infty} n(n-1) a_n x^{n-2} .$$

By substituting these expressions into the differential equation, we obtain

$$y'' + (x-1)y' + y = \sum_{n=2}^{\infty} n(n-1) a_n x^{n-2} + (x-1)\sum_{n=1}^{\infty} n a_n x^{n-1} + \sum_{n=0}^{\infty} a_n x^n = 0 .$$

Simplifying yields

$$\sum_{n=2}^{\infty} n(n-1) a_n x^{n-2} + \sum_{n=1}^{\infty} n a_n x^n - \sum_{n=1}^{\infty} n a_n x^{n-1} + \sum_{n=0}^{\infty} a_n x^n = 0 . \qquad (1)$$

We want to be able to write the left-hand side of this equation as a single power series. This will allow us to find expressions for the coefficient of each power of x. Therefore, we first need to shift the indices in each power series above so that they sum over the same powers of x. Thus, we let $k = n-2$ in the first summation and note that this means that $n = k+2$ and that $k = 0$ when $n = 2$. This yields

$$\sum_{n=2}^{\infty} n(n-1) a_n x^{n-2} = \sum_{k=0}^{\infty} (k+2)(k+1) a_{k+2} x^k .$$

In the third power series, we let $k = n-1$ which implies that $n = k+1$ and $k = 0$ when $n = 1$. Thus, we see that

$$\sum_{n=1}^{\infty} n a_n x^{n-1} = \sum_{k=0}^{\infty} (k+1) a_{k+1} x^k .$$

For the second and last power series we need only to replace *n* with *k*. Substituting all of these expressions into their appropriate places in equation (1) above yields

$$\sum_{k=0}^{\infty} (k+2)(k+1) a_{k+2} x^k + \sum_{k=1}^{\infty} k a_k x^k - \sum_{k=0}^{\infty} (k+1) a_{k+1} x^k + \sum_{k=0}^{\infty} a_k x^k = 0 .$$

Our next step in writing the left-hand side as a single power series is to start all of the summations at the same point. To do this we observe that

$$\sum_{k=0}^{\infty}(k+2)(k+1)a_{k+2}x^k = (2)(1)a_2x^0 + \sum_{k=1}^{\infty}(k+2)(k+1)a_{k+2}x^k,$$

$$\sum_{k=0}^{\infty}(k+1)a_{k+1}x^k = (1)a_1x^0 + \sum_{k=1}^{\infty}(k+1)a_{k+1}x^k,$$

$$\sum_{k=0}^{\infty}a_kx^k = a_0x^0 + \sum_{k=1}^{\infty}a_kx^k.$$

Thus, all of the summations now start at one. Therefore, we have

$$(2)(1)a_2x^0 + \sum_{k=1}^{\infty}(k+2)(k+1)a_{k+2}x^k + \sum_{k=1}^{\infty}ka_kx^k - (1)a_1x^0$$

$$-\sum_{k=1}^{\infty}(k+1)a_{k+1}x^k + a_0x^0 + \sum_{k=1}^{\infty}a_kx^k = 0$$

$$\Rightarrow \quad 2a_2 - a_1 + a_0 + \sum_{k=1}^{\infty}\left((k+2)(k+1)a_{k+2}x^k + ka_kx^k - (k+1)a_{k+1}x^k + a_kx^k\right) = 0$$

$$\Rightarrow \quad 2a_2 - a_1 + a_0 + \sum_{k=1}^{\infty}\left((k+2)(k+1)a_{k+2} + (k+1)a_k - (k+1)a_{k+1}\right)x^k = 0.$$

In order for this power series to equal zero, each coefficient must be zero. Therefore, we obtain

$$2a_2 - a_1 + a_0 = 0 \quad \Rightarrow \quad a_2 = \frac{a_1 - a_0}{2},$$

and

$$(k+2)(k+1)a_{k+2} + (k+1)a_k - (k+1)a_{k+1} = 0, \qquad k \ge 1$$

$$\Rightarrow \quad a_{k+2} = \frac{a_{k+1} - a_k}{k+2}, \qquad k \ge 1,$$

where we have canceled the factor $(k+1)$ from the recurrence relation, the last equation obtained above. *Note that in this recurrence relation we have solved for the coefficient with the largest subscript, namely* a_{k+2}. *Also, note that the first value for* k *in the recurrence relation is the same as the first value for* k *used in the summation notation.* By using the recurrence relation with $k = 1$, we find that

$$a_3 = \frac{a_2 - a_1}{3} = \frac{\frac{a_1 - a_0}{2} - a_1}{3} = \frac{-(a_1 + a_0)}{6},$$

where we have plugged in the expression for a_2 that we found above. By letting $k = 2$ in the recurrence equation, we obtain

$$a_4 = \frac{a_3 - a_2}{4} = \frac{\frac{-(a_1 + a_0)}{6} - \frac{a_1 - a_0}{2}}{4} = \frac{-2a_1 + a_0}{12},$$

where we have plugged in the values for a_2 and a_3 found above. Continuing this process will allow us to find as many coefficients for the power series of the solution to the differential equation as we may want. Notice that the coefficients just found involve only the variables a_0 and a_1. From the recurrence equation, we see that this will be the case for all coefficients of the power series solution. Thus, a_0 and a_1 are arbitrary constants and these variables will be our arbitrary variables in the general solution. Hence, substituting the values for the coefficients that we found above into the solution

$$y(x) = \sum_{n=0}^{\infty} a_n x^n = a_0 + a_1 x + a_2 x^2 + a_3 x^3 + a_4 x^4 + \cdots,$$

yields the solution

$$y(x) = a_0 + a_1 x + \frac{a_1 - a_0}{2} x^2 + \frac{-(a_1 + a_0)}{6} x^3 + \frac{-2a_1 + a_0}{12} x^4 + \cdots$$

$$= a_0\left(1 - \frac{x^2}{2} - \frac{x^3}{6} + \frac{x^4}{12} + \cdots\right) + a_1\left(x + \frac{x^2}{2} - \frac{x^3}{6} - \frac{x^4}{6} + \cdots\right).$$

21. Since $x = 0$ is an ordinary point for this differential equation, we will assume that the solution has a power series expansion with a positive radius of convergence about the point $x = 0$. Thus, we have

$$y(x) = \sum_{n=0}^{\infty} a_n x^n \quad \Rightarrow \quad y'(x) = \sum_{n=1}^{\infty} n a_n x^{n-1} \quad \Rightarrow \quad y''(x) = \sum_{n=2}^{\infty} n(n-1)\, a_n x^{n-2}.$$

By plugging these expressions into the differential equation, we obtain

$$y'' - xy' + 4y = \sum_{n=2}^{\infty} n(n-1)\, a_n x^{n-2} - x\sum_{n=1}^{\infty} n a_n x^{n-1} + 4\sum_{n=0}^{\infty} a_n x^n = 0$$

$$\Rightarrow \quad \sum_{n=2}^{\infty} n(n-1)\, a_n x^{n-2} - \sum_{n=1}^{\infty} n a_n x^n + \sum_{n=0}^{\infty} 4a_n x^n = 0.$$

In order for each power series to sum over the same powers of x, we will shift the index in the first summation by letting $k = n - 2$, and we will let $k = n$ in the other two power series. Thus, we have

$$\sum_{k=0}^{\infty} (k+2)(k+1)a_{k+2} x^k - \sum_{k=1}^{\infty} k a_k x^k + \sum_{k=0}^{\infty} 4a_k x^k = 0.$$

Next we want all of the summations to start at the same point. Therefore, we will take the first term in the first and last power series out of the summation sign. This yields

$$(2)(1)a_2x^0 + \sum_{k=1}^{\infty} (k+2)(k+1)a_{k+2}x^k - \sum_{k=1}^{\infty} ka_kx^k + 4a_0x^0 + \sum_{k=1}^{\infty} 4a_kx^k = 0$$

$$\Rightarrow \qquad 2a_2 + 4a_0 + \sum_{k=1}^{\infty} (k+2)(k+1)a_{k+2}x^k - \sum_{k=1}^{\infty} ka_kx^k + \sum_{k=1}^{\infty} 4a_kx^k = 0$$

$$\Rightarrow \qquad 2a_2 + 4a_0 + \sum_{k=1}^{\infty} \left[(k+2)(k+1)a_{k+2} + (-k+4)a_k\right]x^k = 0 .$$

By setting each coefficient of the power series equal to zero, we see that

$$2a_2 + 4a_0 = 0 \qquad \Rightarrow \qquad a_2 = \frac{-4a_0}{2} = -2a_0 ,$$

$$(k+2)(k+1)a_{k+2} + (-k+4)a_k = 0 \qquad \Rightarrow \qquad a_{k+2} = \frac{(k-4)a_k}{(k+2)(k+1)}, \qquad k \geq 1,$$

where we have solved the recurrence equation, the last equation above, for a_{k+2}, the coefficient with the largest subscript. Thus, we have

$$k=1 \qquad \Rightarrow \qquad a_3 = \frac{-3a_1}{3\cdot 2} = \frac{-a_1}{2} ,$$

$$k=2 \qquad \Rightarrow \qquad a_4 = \frac{-2a_2}{4\cdot 3} = \frac{(-2)(-4)a_0}{4\cdot 3\cdot 2} = \frac{a_0}{3} ,$$

$$k=3 \qquad \Rightarrow \qquad a_5 = \frac{-a_3}{5\cdot 4} = \frac{(-3)(-1)a_1}{5\cdot 4\cdot 3\cdot 2} = \frac{a_1}{40} ,$$

$$k=4 \qquad \Rightarrow \qquad a_6 = 0 ,$$

$$k=5 \qquad \Rightarrow \qquad a_7 = \frac{a_5}{7\cdot 6} = \frac{(-3)(-1)(1)a_1}{7\cdot 6\cdot 5\cdot 4\cdot 3\cdot 2} = \frac{a_1}{560} ,$$

$$k=6 \qquad \Rightarrow \qquad a_8 = \frac{2a_6}{8\cdot 7} = 0 ,$$

$$k=7 \qquad \Rightarrow \qquad a_9 = \frac{3a_7}{9\cdot 8} = \frac{(-3)(-1)(1)(3)a_1}{9!} ,$$

$$k=8 \qquad \Rightarrow \qquad a_{10} = \frac{4a_8}{10\cdot 9} = 0 ,$$

$$k = 9 \quad \Rightarrow \quad a_{11} = \frac{5a_9}{11\cdot 10} = \frac{(-3)(-1)(1)(3)(5)a_1}{11!}.$$

Now we can see a pattern starting to develop. (*Note that it is easier to determine such a pattern if we consider specific coefficients that have not been multiplied out.*) We first note that a_0 and a_1 can be chosen arbitrarily. Next we notice that the coefficients with even subscripts larger than 4 are zero. We also see that the general formula for a coefficient with an odd subscript is given by

$$a_{2n+1} = \frac{(-3)(-1)(1)\cdots(2n-5)a_1}{(2n+1)!}.$$

Notice that this formula is also valid for a_3 and a_5. Substituting these expressions for the coefficients into the solution

$$y(x) = \sum_{n=0}^{\infty} a_n x^n = a_0 + a_1 x + a_2 x^2 + a_3 x^3 + a_4 x^4 + \cdots,$$

yields

$$y(x) = a_0 + a_1 x - 2a_0 x^2 - \frac{a_1}{2}x^3 + \frac{a_0}{3}x^4 + \frac{a_1}{40}x^5 + \cdots$$

$$+ \frac{(-3)(-1)(1)\cdots(2n-5)a_1}{(2n+1)!}x^{2n+1} + \cdots$$

$$= a_0\left(1 - 2x^2 + \frac{x^4}{3}\right) + a_1\left(x - \frac{x^3}{2} + \frac{x^5}{40} + \cdots + \frac{(-3)(-1)(1)\cdots(2n-5)}{(2n+1)!}x^{2n+1} + \cdots\right)$$

$$= a_0\left(1 - 2x^2 + \frac{x^4}{3}\right) + a_1\left[x + \sum_{k=1}^{\infty} \frac{(-3)(-1)(1)\cdots(2k-5)}{(2k+1)!}x^{2k+1}\right].$$

29. Since $x = 0$ is an ordinary point for this differential equation, we can assume that a solution to this problem is given by

$$y(x) = \sum_{n=0}^{\infty} a_n x^n \quad \Rightarrow \quad y'(x) = \sum_{n=1}^{\infty} n a_n x^{n-1} \quad \Rightarrow \quad y''(x) = \sum_{n=2}^{\infty} n(n-1)a_n x^{n-2}.$$

By substituting the initial conditions, $y(0) = 1$ and $y'(0) = -2$, into the first two equations above, we see that

$$y(0) = a_0 = 1, \quad \text{and} \quad y'(0) = a_1 = -2.$$

Next we will substitute the expressions found above for $y(x)$, $y'(x)$, and $y''(x)$ into the differential equation to obtain

$$y'' + y' - xy = \sum_{n=2}^{\infty} n(n-1)a_n x^{n-2} + \sum_{n=1}^{\infty} n a_n x^{n-1} - x\sum_{n=0}^{\infty} a_n x^n = 0$$

$$\Rightarrow \qquad \sum_{n=2}^{\infty} n(n-1)a_n x^{n-2} + \sum_{n=1}^{\infty} na_n x^{n-1} - \sum_{n=0}^{\infty} a_n x^{n+1} = 0 .$$

By setting $k = n-2$ in the first power series above, $k = n-1$ in the second power series above, and $k = n+1$ in the last power series, we can shift the indices so that x is raised to the power k in each power series. Thus, we obtain

$$\sum_{k=0}^{\infty} (k+2)(k+1)a_{k+2}x^k + \sum_{k=0}^{\infty} (k+1)a_{k+1}x^k - \sum_{k=1}^{\infty} a_{k-1}x^k = 0 .$$

We can start all of the summations at the same point if we remove the first term from each of the first two power series above. Therefore, we have

$$(2)(1)a_2 + \sum_{k=1}^{\infty} (k+2)(k+1)a_{k+2}x^k + (1)a_1 + \sum_{k=1}^{\infty} (k+1)a_{k+1}x^k - \sum_{k=1}^{\infty} a_{k-1}x^k = 0$$

$$\Rightarrow \qquad 2a_2 + a_1 + \sum_{k=1}^{\infty} \left[(k+2)(k+1)a_{k+2} + (k+1)a_{k+1} - a_{k-1}\right]x^k = 0 .$$

By equating coefficients, we see that all of the coefficients of the terms in the power series above must be zero. Thus, we have

$$2a_2 + a_1 = 0 \qquad \Rightarrow \qquad a_2 = \frac{-a_1}{2},$$

$$(k+2)(k+1)a_{k+2} + (k+1)a_{k+1} - a_{k-1} = 0$$

$$\Rightarrow \qquad a_{k+2} = \frac{a_{k-1} - (k+1)a_{k+1}}{(k+2)(k+1)}, \qquad k \geq 1 .$$

Thus, we see that

$$k = 1 \qquad \Rightarrow \qquad a_3 = \frac{a_0 - 2a_2}{3 \cdot 2} = \frac{a_0}{6} + \frac{a_1}{6} .$$

Using the fact that $a_0 = 1$ and $a_1 = -2$, which we found from the initial conditions, we calculate

$$a_2 = \frac{-(-2)}{2} = 1 ,$$

$$a_3 = \frac{1}{6} + \frac{(-2)}{6} = -\frac{1}{6} .$$

By substituting these coefficients, we obtain the cubic polynomial approximation

$$y(x) = 1 - 2x + x^2 - \frac{x^3}{6} .$$

The graphs of the linear, quadratic and cubic polynomial approximations are easily generated by using the software supplied with the text.

EXERCISES 8.4: Equations with Analytic Coefficients, page 478

3. For this equation, $p(x)=0$ and $q(x)=\dfrac{-3}{1+x+x^2}$. Therefore, singular points will occur when

$$1+x+x^2=0 \qquad \Rightarrow \qquad x=-\frac{1}{2}\pm\frac{\sqrt{3}}{2}i.$$

Thus, $x=1$ is an ordinary point for this equation, and we can find a power series solution with a radius of convergence of at least the minimum of the distances between 1 and $-\frac{1}{2}\pm\frac{\sqrt{3}}{2}i$, which, in fact, are equal. Recall that the distance between two complex numbers, $z_1=a+bi$ and $z_2=c+di$, is given by

$$\operatorname{dist}(z_1,z_2)=\sqrt{(a-c)^2+(b-d)^2}.$$

Thus, the distance between $(1+0\cdot i)$ and $-\frac{1}{2}+\frac{\sqrt{3}}{2}i$ is

$$\sqrt{\left[1-\left(-\frac{1}{2}\right)\right]^2+\left[0-\frac{\sqrt{3}}{2}\right]^2}=\sqrt{\frac{9}{4}+\frac{3}{4}}=\sqrt{3}.$$

Therefore, the radius of convergence for the power series solution of this differential equation about $x=1$ will be at least $\rho=\sqrt{3}$.

9. We see that $x=0$ and $x=2$ are the only singular points for this differential equation and, thus, $x=1$ is an ordinary point. Therefore, according to Theorem 5 on page 473 of the text, there exists a power series solution of this equation about the point $x = 1$ with a radius of convergence of at least one, the distance from 1 to either 0 or 2. That is, we have a general solution for this differential equation of the form

$$y(x)=\sum_{n=0}^{\infty}a_n(x-1)^n,$$

which is convergent for all x at least in the interval $(0,2)$, the interval on which the inequality $|x-1|<1$ is satisfied. To find this solution we will proceed as in Example 3 on page 475 of the text. Thus, we make the substitution $t=x-1$, which implies that $x=t+1$. (Note that $\frac{dx}{dt}=1$.) We then define a new function

$$Y(t):=y(t+1)=y(x) \qquad \Rightarrow \qquad \frac{dY}{dt}=\left(\frac{dy}{dx}\right)\left(\frac{dx}{dt}\right)=\left(\frac{dy}{dx}\right)\cdot 1=\frac{dy}{dx}$$

$$\Rightarrow \qquad \frac{d^2Y}{dt^2}=\frac{d}{dt}\left(\frac{dY}{dt}\right)=\frac{d}{dt}\left(\frac{dy}{dx}\right)=\left(\frac{d^2y}{dx^2}\right)\left(\frac{dx}{dt}\right)=\frac{d^2y}{dx^2}.$$

Hence, with the substitutions $t=x-1$ and $Y(t)=y(t+1)$, we transform the differential equation,

$$\left(x^2-2x\right)y''(x)+2y(x)=0,$$

into the differential equation

$$\left[(t+1)^2-2(t+1)\right]y''(t+1)+2y(t+1)=0$$

$$\Rightarrow \qquad \left[(t+1)^2-2(t+1)\right]Y''(t)+2Y(t)=0$$

$$\Rightarrow \qquad (t^2-1)Y''(t)+2Y(t)=0. \tag{2}$$

To find a general solution to (2), we first note that zero is an ordinary point of equation (2). Thus, we can assume that we have a power series solution of equation (2) of the form

$$Y(t)=\sum_{n=0}^{\infty}a_nt^n,$$

which converges for all t in $(-1, 1)$. (This means that $x=t+1$ will be in the interval (0, 2) as desired.) Substituting into equation (2) yields

$$\left(t^2-1\right)\sum_{n=2}^{\infty}n(n-1)a_nt^{n-2}+2\sum_{n=0}^{\infty}a_nt^n=0$$

$$\Rightarrow \qquad \sum_{n=2}^{\infty}n(n-1)a_nt^n-\sum_{n=2}^{\infty}n(n-1)a_nt^{n-2}+\sum_{n=0}^{\infty}2a_nt^n=0.$$

Making the shift in the index, $k=n-2$, in the second power series above and replacing n with k in the other two power series allows us to take each summation over the same power of t. This gives us

$$\sum_{k=2}^{\infty}k(k-1)a_kt^k-\sum_{k=0}^{\infty}(k+2)(k+1)a_{k+2}t^k+\sum_{k=0}^{\infty}2a_kt^k=0.$$

In order to start all of these summations at the same point, we must take the first two terms out of the summation sign in the last two power series. Thus we have,

$$\sum_{k=2}^{\infty}k(k-1)a_kt^k-(2)(1)a_2-(3)(2)a_3t-\sum_{k=2}^{\infty}(k+2)(k+1)a_{k+2}t^k+2a_0+2a_1t+\sum_{k=2}^{\infty}2a_kt^k=0$$

$$\Rightarrow \qquad 2a_0-2a_2+\left(2a_1-6a_3\right)t+\sum_{k=2}^{\infty}\left[k(k-1)a_k-(k+2)(k+1)a_{k+2}+2a_k\right]t^k=0.$$

In order for this power series to equal zero, each coefficient must be zero. Thus, we have

$$2a_0-2a_2=0 \quad\Rightarrow\quad a_2=a_0, \qquad 2a_1-6a_3=0 \quad\Rightarrow\quad a_3=\frac{a_1}{3},$$

$$k(k-1)a_k-(k+2)(k+1)a_{k+2}+2a_k=0\,,\qquad k\ge 2$$

$$\Rightarrow \qquad a_{k+2}=\frac{k(k-1)a_k+2a_k}{(k+2)(k+1)}\,,\qquad k\ge 2 \qquad \Rightarrow \qquad a_{k+2}=\frac{(k^2-k+2)a_k}{(k+2)(k+1)}\,,\qquad k\ge 2\,.$$

Therefore, we see that

$$k=2 \qquad \Rightarrow \qquad a_4=\frac{4a_2}{4\cdot 3}=\frac{a_2}{3}=\frac{a_0}{3}\,,$$

$$k=3 \qquad \Rightarrow \qquad a_5=\frac{8a_3}{5\cdot 4}=\frac{2a_1}{15}\,,\qquad \text{etc.}$$

Plugging these values for the coefficients into the power series solution,

$$Y(t)=\sum_{n=0}^{\infty}a_nt^n=a_0+a_1t+a_2t^2+a_3t^3+a_4t^4+\cdots\,,$$

yields

$$Y(t)=a_0+a_1t+a_0t^2+\frac{a_1t^3}{3}+\frac{a_0t^4}{3}+\frac{2\,a_1t^5}{15}+\cdots$$

$$\Rightarrow \qquad Y(t)=a_0\left(1+t^2+\frac{t^4}{3}+\cdots\right)+a_1\left(t+\frac{t^3}{3}+\frac{2t^5}{15}+\cdots\right).$$

Lastly, we want to change back to the independent variable x. To do this, we recall that $Y(t)=y(t+1)$. Thus, if $t=x-1$, then

$$Y(t)=Y(x-1)=y([x-1]+1)=y(x)\,.$$

Thus, we replace t with $x-1$ in the solution just found, and we obtain a power series expansion for a general solution in the independent variable x. Substituting, we have

$$y(x)=a_0\left[1+(x-1)^2+\frac{1}{3}(x-1)^4+\cdots\right]+a_1\left[(x-1)+\frac{1}{3}(x-1)^3+\frac{2}{15}(x-1)^5+\cdots\right].$$

17. Here $p(x)=0$ and $q(x)=-\sin x$ both of which are analytic everywhere. Thus, $x=\pi$ is an ordinary point for this differential equation, and there are no singular points. Therefore, by Theorem 5 on page 473 of the text, we can assume that this equation has a general power series solution about the point $x=\pi$ with an infinite radius of convergence (i.e., $\rho=\infty$). That is, we assume that we have a solution to this differential equation given by

$$y(x)=\sum_{n=0}^{\infty}a_n(x-\pi)^n \qquad \left[\Rightarrow\ y'(x)=\sum_{n=1}^{\infty}na_n(x-\pi)^{n-1}\right],$$

which converges for all x. If we apply the initial conditions, $y(\pi)=1$ and $y'(\pi)=0$, we see that $a_0=1$ and $a_1=0$. To find a general solution of this differential equation, we will combine the methods of Example 3 and Example 4 on page 475 of the text. Thus, we will first define a new function, $Y(t)$, using the transformation $t = x - \pi$. Thus, we define

$$Y(t) := y(t+\pi) = y(x).$$

Hence, by the chain rule (using the fact that $x = t + \pi$ which implies that $\frac{dx}{dt} = 1$), we have $\frac{dY}{dt} = \frac{dy}{dx}\frac{dx}{dt} = \frac{dy}{dx}$, and similarly $\frac{d^2Y}{dt^2} = \frac{d^2y}{dx^2}$. We now solve the transformed differential equation

$$\frac{d^2Y}{dt^2} - [\sin(t+\pi)]Y(t) = 0$$

$$\Rightarrow \qquad \frac{d^2Y}{dt^2} + (\sin t)Y(t) = 0, \qquad (3)$$

where we have used the fact that $\sin(t+\pi) = -\sin t$. When we have found the solution $Y(t)$, we will use the fact that $y(x) = Y(x-\pi)$ to obtain the solution to the original differential equation in terms of the independent variable x. Hence, we seek a power series solution to equation (3) of the form

$$Y(t) = \sum_{n=0}^{\infty} a_n t^n$$

$$\Rightarrow \qquad Y'(t) = \sum_{n=1}^{\infty} n a_n t^{n-1}$$

$$\Rightarrow \qquad Y''(t) = \sum_{n=2}^{\infty} n(n-1) a_n t^{n-2}.$$

Since the initial conditions, $y(\pi) = 1$ and $y'(\pi) = 0$, transform into $Y(0) = 1$ and $Y'(0) = 0$, we must have

$$Y(0) = a_0 = 1 \qquad \text{and} \qquad Y'(0) = a_1 = 0.$$

Next we note that $q(t) = \sin t$ is an analytic function with a Maclaurin series given by

$$\sin t = \sum_{n=0}^{\infty} \frac{(-1)^n t^{2n+1}}{(2n+1)!} = t - \frac{t^3}{6} + \frac{t^5}{120} - \frac{t^7}{5040} + \cdots.$$

By substituting the expressions that we found for $Y(t)$, $Y''(t)$, and $\sin t$ into equation (3), we obtain

$$\sum_{n=2}^{\infty} n(n-1)\, a_n t^{n-2} + \left(t - \frac{t^3}{6} + \frac{t^5}{120} - \frac{t^7}{5040} + \cdots\right)\sum_{n=0}^{\infty} a_n t^n = 0.$$

Therefore, expanding this last equation (and explicitly showing only terms of up to order four), yields

$$\left(2a_2+6a_3t+12a_4t^2+20a_5t^3+30a_6t^4+\cdots\right)+t\left(a_0+a_1t+a_2t^2+a_3t^3+\cdots\right)$$

$$-\frac{t^3}{6}(a_0+a_1t+\cdots)+\cdots=0$$

$$\Rightarrow \quad \left(2a_2+6a_3t+12a_4t^2+20a_5t^3+30a_6t^4+\cdots\right)+t\left(a_0+a_1t+a_2t^2+a_3t^3+\cdots\right)$$

$$+\left(\frac{-a_0t^3}{6}-\frac{a_1t^4}{6}+\cdots\right)+\cdots=0.$$

By grouping these terms according to their powers of t, we obtain

$$2a_2+(6a_3+a_0)t+(12a_4+a_1)t^2+\left(20a_5+a_2-\frac{a_0}{6}\right)t^3+\left(30a_6+a_3-\frac{a_1}{6}\right)t^4+\cdots=0.$$

Setting these coefficients to zero and recalling that $a_0=1$ and $a_1=0$ yields the system of equations

$$2a_2=0 \quad \Rightarrow \quad a_2=0,$$

$$6a_3+a_0=0 \quad \Rightarrow \quad a_3=\frac{-a_0}{6}=\frac{-1}{6},$$

$$12a_4+a_1=0 \quad \Rightarrow \quad a_4=\frac{-a_1}{12}=0,$$

$$20a_5+a_2-\frac{a_0}{6}=0 \quad \Rightarrow \quad a_5=\frac{\frac{a_0}{6}-a_2}{20}=\frac{\frac{1}{6}}{20}=\frac{1}{120},$$

$$30a_6+a_3-\frac{a_1}{6}=0 \quad \Rightarrow \quad a_6=\frac{\frac{a_1}{6}-a_3}{30}=\frac{0+\frac{1}{6}}{30}=\frac{1}{180}.$$

Plugging these coefficients into the power series solution

$$Y(t)=\sum_{n=0}^{\infty}a_nt^n=a_0+a_1t+a_2t^2+\cdots,$$

yields the solution to equation (3):

$$Y(t)=1+0+0-\frac{t^3}{6}+0+\frac{t^5}{120}+\frac{t^6}{180}+\cdots=1-\frac{t^3}{6}+\frac{t^5}{120}+\frac{t^6}{180}+\cdots.$$

Lastly we want to find the solution to the original equation with the independent variable x. In order to do this, we recall that $t=x-\pi$ and $Y(x-\pi)=y(x)$. Therefore, by substituting these values into the equation above, we obtain the solution

$$y(x) = 1 - \frac{1}{6}(x-\pi)^3 + \frac{1}{120}(x-\pi)^5 + \frac{1}{180}(x-\pi)^6 + \cdots .$$

21. We assume that this differential equation has a power series solution with a positive radius of convergence about the point $x = 0$. This is reasonable because all of the coefficients and the forcing function $g(x) = \sin x$ are analytic everywhere. Thus, we assume that

$$y(x) = \sum_{n=0}^{\infty} a_n x^n \qquad \Rightarrow \qquad y'(x) = \sum_{n=1}^{\infty} n a_n x^{n-1} .$$

By substituting these expressions and the Maclaurin expansion for $\sin x$ into the differential equation, $y'(x) - xy(x) = \sin x$, we obtain

$$\sum_{n=1}^{\infty} n a_n x^{n-1} - x\sum_{n=0}^{\infty} a_n x^n = \sum_{n=0}^{\infty} (-1)^n \frac{x^{2n+1}}{(2n+1)!} .$$

In the first power series on the left, we make the shift $k = n-1$. In the second power series on the left, we make the shift $k = n+1$. Thus, we obtain

$$\sum_{k=0}^{\infty} (k+1)a_{k+1}x^k - \sum_{k=1}^{\infty} a_{k-1}x^k = \sum_{n=0}^{\infty} \frac{(-1)^n x^{2n+1}}{(2n+1)!} .$$

Separating out the first term of the first power series on the left yields

$$a_1 + \sum_{k=1}^{\infty} (k+1)a_{k+1}x^k - \sum_{k=1}^{\infty} a_{k-1}x^k = \sum_{n=0}^{\infty} \frac{(-1)^n x^{2n+1}}{(2n+1)!}$$

$$\Rightarrow \qquad a_1 + \sum_{k=1}^{\infty} \left[(k+1)a_{k+1} - a_{k-1}\right]x^k = \sum_{n=0}^{\infty} \frac{(-1)^n x^{2n+1}}{(2n+1)!} .$$

Therefore, by expanding both of the power series, we have

$$a_1 + (2a_2 - a_0)x + (3a_3 - a_1)x^2 + (4a_4 - a_2)x^3 + (5a_5 - a_3)x^4$$

$$+ (6a_6 - a_4)x^5 + (7a_7 - a_5)x^6 + \cdots = x - \frac{x^3}{6} + \frac{x^5}{120} - \frac{x^7}{5040} + \cdots .$$

By equating the coefficients of like powers of x, we obtain

$$a_1 = 0, \qquad 2a_2 - a_0 = 1 \quad \Rightarrow \quad a_2 = \frac{a_0 + 1}{2},$$

$$3a_3 - a_1 = 0 \quad \Rightarrow \quad a_3 = \frac{a_1}{3} = 0,$$

$$4a_4 - a_2 = \frac{-1}{6} \quad \Rightarrow \quad a_4 = \frac{a_2 - 1/6}{4} = \frac{a_0}{8} + \frac{1}{12},$$

$$5a_5 - a_3 = 0 \qquad \Rightarrow \qquad a_5 = \frac{a_3}{5} = 0,$$

$$6a_6 - a_4 = \frac{1}{120} \qquad \Rightarrow \qquad a_6 = \frac{a_4 + 1/120}{6} = \frac{a_0}{48} + \frac{11}{720}.$$

Substituting these coefficients into the power series solution and noting that a_0 is an arbitrary number, yields

$$y(x) = \sum_{n=0}^{\infty} a_n x^n$$

$$= a_0 + 0 + \left(\frac{a_0}{2} + \frac{1}{2}\right)x^2 + 0 + \left(\frac{a_0}{8} + \frac{1}{12}\right)x^4 + 0 + \left(\frac{a_0}{48} + \frac{11}{720}\right)x^6 + \cdots$$

$$= a_0\left[1 + \frac{1}{2}x^2 + \frac{1}{8}x^4 + \frac{1}{48}x^6 + \cdots\right] + \left[\frac{1}{2}x^2 + \frac{1}{12}x^4 + \frac{11}{720}x^6 + \cdots\right].$$

27. Observe that $x = 0$ is an ordinary point for this differential equation. Therefore, we can assume that this equation has a power series solution about the point $x = 0$ with a positive radius of convergence. Thus, we assume that

$$y(x) = \sum_{n=0}^{\infty} a_n x^n$$

$$\Rightarrow \qquad y'(x) = \sum_{n=1}^{\infty} n a_n x^{n-1} \qquad \Rightarrow \qquad y''(x) = \sum_{n=2}^{\infty} n(n-1) a_n x^{n-2}.$$

The Maclaurin series for $\tan x$ is

$$\tan x = x + \frac{x^3}{3} + \frac{2x^5}{15} + \cdots,$$

which is given in the table on the inside front cover of the text. Substituting the expressions for $y(x)$, $y'(x)$, $y''(x)$, and the Maclaurin series for $\tan x$ into the differential equation, $(1 - x^2)y'' - y' + y = \tan x$, yields

$$(1 - x^2)\sum_{n=2}^{\infty} n(n-1) a_n x^{n-2} - \sum_{n=1}^{\infty} n a_n x^{n-1} + \sum_{n=0}^{\infty} a_n x^n = x + \frac{x^3}{3} + \frac{2x^5}{15} + \cdots$$

$$\Rightarrow \qquad \sum_{n=2}^{\infty} n(n-1) a_n x^{n-2} - \sum_{n=2}^{\infty} n(n-1) a_n x^n - \sum_{n=1}^{\infty} n a_n x^{n-1} + \sum_{n=0}^{\infty} a_n x^n$$

$$= x + \frac{x^3}{3} + \frac{2x^5}{15} + \cdots.$$

By shifting the indices of the power series on the left-hand side of the equation above, we obtain

$$\sum_{k=0}^{\infty}(k+2)(k+1)a_{k+2}x^k - \sum_{k=2}^{\infty}k(k-1)a_k x^k - \sum_{k=0}^{\infty}(k+1)a_{k+1}x^k + \sum_{k=0}^{\infty}a_k x^k$$

$$= x + \frac{x^3}{3} + \frac{2x^5}{15} + \cdots .$$

Removing the first two terms from the summation notation in the first, third and fourth power series above yields

$$(2)(1)a_2 + (3)(2)a_3 x + \sum_{k=2}^{\infty}(k+2)(k+1)a_{k+2}x^k - \sum_{k=2}^{\infty}k(k-1)a_k x^k - (1)a_1 - (2)a_2 x$$

$$- \sum_{k=2}^{\infty}(k+1)a_{k+1}x^k + a_0 + a_1 x + \sum_{k=2}^{\infty}a_k x^k = x + \frac{x^3}{3} + \frac{2x^5}{15} + \cdots$$

$$\Rightarrow \quad (2a_2 - a_1 + a_0) + (6a_3 - 2a_2 + a_1)x$$

$$+ \sum_{k=2}^{\infty}\left[(k+2)(k+1)a_{k+2} - k(k-1)a_k - (k+1)a_{k+1} + a_k\right]x = x + \frac{x^3}{3} + \frac{2x^5}{15} + \cdots .$$

By equating the coefficients of the two power series, we see that

$$2a_2 - a_1 + a_0 = 0 \quad \Rightarrow \quad a_2 = \frac{a_1 - a_0}{2},$$

$$6a_3 - 2a_2 + a_1 = 1 \quad \Rightarrow \quad a_3 = \frac{2a_2 - a_1 + 1}{6} = \frac{1 - a_0}{6},$$

$$4 \cdot 3a_4 - 2 \cdot 1a_2 - 3a_3 + a_2 = 0 \quad \Rightarrow \quad a_4 = \frac{a_2 + 3a_3}{12} = \frac{a_1 - 2a_0 + 1}{24}.$$

Therefore, noting that a_0 and a_1 are arbitrary, we can substitute these coefficients into the power series solution

$$y(x) = \sum_{n=0}^{\infty}a_n x^n = a_0 + a_1 x + a_2 x^2 + a_3 x^3 + a_4 x^4 + \cdots ,$$

to obtain

$$y(x) = a_0 + a_1 x + \left(\frac{a_1}{2} - \frac{a_0}{2}\right)x^2 + \left(\frac{1}{6} - \frac{a_0}{6}\right)x^3 + \left(\frac{a_1}{24} - \frac{a_0}{12} + \frac{1}{24}\right)x^4 + \cdots$$

$$= a_0\left[1 - \frac{1}{2}x^2 - \frac{1}{6}x^3 - \frac{1}{12}x^4 + \cdots\right] + a_1\left[x + \frac{1}{2}x^2 + \frac{1}{24}x^4 + \cdots\right] + \left[\frac{1}{6}x^3 + \frac{1}{24}x^4 + \cdots\right].$$

EXERCISES 8.5: Cauchy-Euler (Equidimensional) Equations Revisited, page 482

5. Notice that, since $x > 0$, we can multiply this differential equation by x^2 and rewrite it to obtain

$$x^2 \frac{d^2y}{dx^2} - 5x\frac{dy}{dx} + 13y = 0 .$$

We see that this is a Cauchy-Euler equation. Thus, we will assume that a solution has the form

$$y(x) = x^r \quad \Rightarrow \quad y'(x) = rx^{r-1} \quad \Rightarrow \quad y''(x) = r(r-1)x^{r-2} .$$

Substituting these expressions into the differential equation above yields

$$r(r-1)x^r - 5rx^r + 13x^r = 0$$

$$\Rightarrow \quad (r^2 - 6r + 13)x^r = 0 \quad \Rightarrow \quad r^2 - 6r + 13 = 0 .$$

We obtained this last equation by using the assumption that $x > 0$. (We also could arrive at this equation by using equation (4) on page 480 of the text.) Using the quadratic formula, we see that the roots to this equation are

$$r = \frac{6 \pm \sqrt{36-52}}{2} = 3 \pm 2i .$$

Therefore, using formulas (5) and (6) on page 480 of the text with complex conjugates (or using Euler's formula), we have two linearly independent solutions give by

$$y_1(x) = x^3 \cos(2\ln x), \qquad y_2(x) = x^3 \sin(2\ln x).$$

Hence the general solution to this equation is given by

$$y(x) = c_1 x^3 \cos(2\ln x) + c_2 x^3 \sin(2\ln x).$$

7. This equation is a third order Cauchy-Euler equation, and, thus, we will assume that a solution has the form $y(x) = x^r$. This implies that

$$y'(x) = rx^{r-1} \quad \Rightarrow \quad y''(x) = r(r-1)x^{r-2} \quad \Rightarrow \quad y'''(x) = r(r-1)(r-2)x^{r-3} .$$

By substituting these expressions into the differential equation, we obtain

$$[r(r-1)(r-2) + 4r(r-1) + 10r - 10]x^r = 0$$

$$\Rightarrow \quad [r^3 + r^2 + 8r - 10]x^r = 0 \quad \Rightarrow \quad r^3 + r^2 + 8r - 10 = 0 .$$

By inspection we see that $r = 1$ is a root of this last equation. Thus, one solution to this differential equation will be given by $y_1(x) = x$ and we can factor the indicial equation above as follows:

$$(r-1)(r^2 + 2r + 10) = 0 .$$

Therefore, using the quadratic formula, we see that the roots to this equation are $r = 1, -1 \pm 3i$. Thus, we can find two more linearly independent solutions to this equation by using Euler's formula as was done on page 480 of the text. Thus, three linearly independent solutions to this problem are given by

$$y_1(x) = x, \qquad y_2(x) = x^{-1}\cos(3\ln x), \qquad y_3(x) = x^{-1}\sin(3\ln x).$$

Hence, the general solution to this differential equation is

$$y(x) = c_1 x + c_2 x^{-1}\cos(3\ln x) + c_3 x^{-1}\sin(3\ln x).$$

13. We first must find two linearly independent solutions to the associated homogeneous equation. Since this is a Cauchy-Euler equation, we assume that there are solutions of the form

$$y(x) = x^r \qquad \Rightarrow \qquad y'(x) = rx^{r-1} \qquad \Rightarrow \qquad y''(x) = r(r-1)x^{r-2}.$$

Substituting these expressions into the associated homogeneous equation yields

$$[r(r-1) - 2r + 2]x^r = 0 \qquad \Rightarrow \qquad r^2 - 3r + 2 = 0 \qquad \Rightarrow \qquad (r-1)(r-2) = 0.$$

Thus, the roots to this indicial equation are $r = 1, 2$. Therefore, a general solution to the associated homogeneous equation is

$$y_h(x) = c_1 x + c_2 x^2.$$

For the variation of parameters method, let $y_1(x) = x$ and $y_2(x) = x^2$, and then assume that a particular solution has the form

$$y_p(x) = v_1(x)y_1(x) + v_2(x)y_2(x) = v_1(x)x + v_2(x)x^2.$$

In order to find $v_1(x)$ and $v_2(x)$, we would like to use formula (10) on page 215 of the text. To use equation (10), we must first find the Wronskian of y_1 and y_2. Thus, we compute

$$W[y_1, y_2](x) = y_1(x)y_2'(x) - y_2(x)y_1'(x) = 2x^2 - x^2 = x^2.$$

Next we must write the differential equation given in this problem in standard form. When we do this, we see that $g(x) = x^{-5/2}$. Therefore, by equation (10), we have

$$v_1(x) = \int \frac{-x^{-5/2}x^2}{x^2}\,dx = \int -x^{-5/2}\,dx = \frac{2}{3}x^{-3/2}$$

and

$$v_2(x) = \int \frac{x^{-5/2}x}{x^2}\,dx = \int x^{-7/2}\,dx = \frac{-2}{5}x^{-5/2}.$$

Thus, a particular solution is given by

$$y_p(x) = \left(\frac{2}{3}x^{-3/2}\right) + \left(\frac{-2}{5}x^{-5/2}\right)x^2 = \frac{4}{15}x^{-1/2}.$$

Therefore, a general solution of the nonhomogeneous differential equation is given by

$$y(x) = y_h(x) + y_p(x) = c_1 x + c_2 x^2 + \frac{4}{15} x^{-1/2}.$$

19. **(a)** For this linear differential operator L, we have

$$L[x^r](x) = x^3[r(r-1)(r-2)x^{r-3}] + x[rx^{r-1}] - x^r$$
$$= r(r-1)(r-2)x^r + rx^r - x^r$$
$$= (r^3 - 3r^2 + 3r - 1)x^r = (r-1)^3 x^r.$$

(b) From part (a) above, we see that $r = 1$ is a root of multiplicity three of the indicial equation. Thus, we have one solution given by

$$y_1(x) = x \qquad (4)$$

To find two more linearly independent solutions, we use a method similar to that used in the text. By taking the partial derivative of $L[x^r](x) = (r-1)^3 x^r$ with respect to r, we have

$$\frac{\partial}{\partial r}\{L[x^r](x)\} = \frac{\partial}{\partial r}\{(r-1)^3 x^r\} = 3(r-1)^2 x^r + (r-1)^3 x^r \ln x.$$

$$\Rightarrow \qquad \frac{\partial^2}{\partial r^2}\{L[x^r](x)\} = \frac{\partial}{\partial r}\{3(r-1)^2 x^r + (r-1)^3 x^r \ln x\}$$

$$= 6(r-1)x^r + 6(r-1)^2 x^r \ln x + (r-1)^3 x^r (\ln x)^2.$$

Since $r - 1$ is a factor of every term in $\frac{\partial}{\partial r}\{L[x^r](x)\}$ and $\frac{\partial^2}{\partial r^2}\{L[x^r](x)\}$ above, we see that

$$\left.\frac{\partial}{\partial r}\{L[x^r](x)\}\right|_{r=1} = 0, \qquad (5)$$

and

$$\left.\frac{\partial^2}{\partial r^2}\{L[x^r](x)\}\right|_{r=1} = 0. \qquad (6)$$

We can use these facts to find the two solutions that we seek. In order to find a second solution, we would like an alternative form for $\left.\frac{\partial}{\partial r}\{L[x^r](x)\}\right|_{r=1}$. Using the fact that $L[y](x) = x^3 y'''(x) + xy'(x) - y(x)$ and proceeding as in equation (9) on page 480 of the text with $w(r,x) = x^r$, we have

$$\frac{\partial}{\partial r}\{L[x^r](x)\}=\frac{\partial}{\partial r}\{L[w](x)\}=\frac{\partial}{\partial r}\left\{x^3\frac{\partial^3 w}{\partial x^3}+x\frac{\partial w}{\partial x}-w\right\}$$

$$=x^3\frac{\partial^4 w}{\partial r\partial x^3}+x\frac{\partial^2 w}{\partial r\partial x}-\frac{\partial w}{\partial r}=x^3\frac{\partial^4 w}{\partial x^3\partial r}+x\frac{\partial^2 w}{\partial x\partial r}-\frac{\partial w}{\partial r}$$

$$=x^3\frac{\partial^3}{\partial x^3}\left(\frac{\partial w}{\partial r}\right)+x\frac{\partial}{\partial x}\left(\frac{\partial w}{\partial r}\right)-\frac{\partial}{\partial x}\left(\frac{\partial w}{\partial r}\right)=L\left[\frac{\partial w}{\partial r}\right](x),$$

where we are using the fact that mixed partials of $w(r,x)$ are equal. Therefore, combining this with equation (5) above yields

$$\frac{\partial}{\partial r}\{L[x^r](x)\}\bigg|_{r=1}=L\left[\frac{\partial x^r}{\partial r}\bigg|_{r=1}\right]=L\left[x^r\ln x\big|_{r=1}\right]=L[x\ln x]=0.$$

Thus, a second linearly independent solution is given by

$y_2(x)=x\ln x$.

To find a third solution, we will use equation (6) above. Hence, we would like to find an alternative form for $\frac{\partial^2}{\partial r^2}\{L[x^r](x)\}$. To do this, we use the fact that

$$\frac{\partial}{\partial r}\{L[x^r](x)\}=x^3\frac{\partial^4 w}{\partial r\partial x^3}+x\frac{\partial^2 w}{\partial r\partial x}-\frac{\partial w}{\partial r},$$

which we found above and the fact that mixed partial derivatives of $w(r, x)$ are equal. Thus, we have

$$\frac{\partial^2}{\partial r^2}\{L[x^r](x)\}=\frac{\partial}{\partial r}\left[\frac{\partial}{\partial r}\{L[x^r](x)\}\right]=\frac{\partial}{\partial r}\left\{x^3\frac{\partial^4 w}{\partial r\partial x^3}+x\frac{\partial^2 w}{\partial r\partial x}-\frac{\partial w}{\partial r}\right\}$$

$$=x^3\frac{\partial^5 w}{\partial r^2\partial x^3}+x\frac{\partial^3 w}{\partial r^2\partial x}-\frac{\partial^2 w}{\partial r^2}=x^3\frac{\partial^5 w}{\partial x^3\partial r^2}+x\frac{\partial^3 w}{\partial x\partial r^2}-\frac{\partial^2 w}{\partial r^2}$$

$$=x^3\frac{\partial}{\partial x}\left(\frac{\partial^2 w}{\partial r^2}\right)+x\frac{\partial}{\partial x}\left(\frac{\partial^2 w}{\partial r^2}\right)-\frac{\partial^2 w}{\partial r^2}=L\left[\frac{\partial^2 w}{\partial r^2}\right](x).$$

Therefore, combining this with equation (6) above yields

$$\frac{\partial^2}{\partial r^2}\{L[x^r](x)\}\bigg|_{r=1}=L\left[\frac{\partial^2(x^r)}{\partial r^2}\bigg|_{r=1}\right]=L\left[x(\ln x)^2\right]=0,$$

where we have used the fact that $\frac{\partial^2 x^r}{\partial r^2}=x^r(\ln x)^2$. Thus we see that another solution is

$y_3(x)=x(\ln x)^2$,

which, by inspection, is linearly independent from y_1 and y_2. Thus, a general solution to the differential equation is $y(x) = C_1 x + C_2 x \ln x + C_3 x(\ln x)^2$.

EXERCISES 8.6: Method of Frobenius, page 494

5. By putting this equation in standard form, we see that

$$p(x) = -\frac{x-1}{(x^2-1)^2} = -\frac{x-1}{(x-1)^2(x+1)^2} = -\frac{1}{(x-1)(x+1)^2},$$

and

$$q(x) = \frac{3}{(x^2-1)^2} = \frac{3}{(x-1)^2(x+1)^2}.$$

Thus, $x = 1, -1$ are singular points of this equation. To check if $x = 1$ is regular, we note that

$$(x-1)p(x) = -\frac{1}{(x+1)^2} \quad \text{and} \quad (x-1)^2 q(x) = \frac{3}{(x+1)^2}.$$

These functions are analytic at $x = 1$. Therefore, $x = 1$ is a regular singular point for this differential equation. Next we check the singular point $x = -1$. Here

$$(x+1)p(x) = \frac{-1}{(x-1)(x+1)}$$

is not analytic at $x = -1$. Therefore, $x = -1$ is an irregular singular point for this differential equation.

13. By putting this equation in standard form, we see that

$$p(x) = \frac{x^2-4}{(x^2-x-2)^2} = \frac{(x-2)(x+2)}{(x-2)^2(x+1)^2} = \frac{x+2}{(x-2)(x+1)^2},$$

$$q(x) = \frac{-6x}{(x-2)^2(x+1)^2}.$$

Thus, we have

$$(x-2)p(x) = \frac{x+2}{(x+1)^2} \quad \text{and} \quad (x-2)^2 q(x) = \frac{-6x}{(x+1)^2}.$$

Therefore, $x = 2$ is a regular singular point of this differential equation. We also observe that

$$\lim_{x\to 2}(x-2)p(x) = \lim_{x\to 2}\frac{x+2}{(x+1)^2} = \frac{4}{9} = p_0,$$

$$\lim_{x\to 2}(x-2)^2 q(x) = -\lim_{x\to 2}\frac{x+2}{(x+1)^2} = \frac{-12}{9} = \frac{-4}{3} = q_0 .$$

Thus, we can use equation (16) on page 485 of the text to obtain the indicial equation

$$r(r-1)+\frac{4r}{9}-\frac{4}{3}=0 \qquad \Rightarrow \qquad r^2-\frac{5r}{9}-\frac{4}{3}=0 .$$

By the quadratic formula, we see that the roots to this equation and, therefore, the exponents of the singularity $x=2$, are given by

$$r_1 = \frac{5+\sqrt{25+432}}{18} = \frac{5+\sqrt{457}}{18}$$

$$r_2 = \frac{5-\sqrt{457}}{18} .$$

21. Here $p(x)=x^{-1}$ and $q(x)=1$. This implies that $xp(x)=1$ and $x^2q(x)=x^2$. Therefore, we see that $x=0$ is a regular singular point for this differential equation, and so we can use the method of Frobenius to find a solution to this problem. (Note also that $x=0$ is the only singular point for this equation.) Thus, we will assume that this solution has the form

$$w(r,x)=x^r\sum_{n=0}^{\infty} a_n x^n = \sum_{n=0}^{\infty} a_n x^{n+r} .$$

We also notice that

$$p_0 = \lim_{x\to 0} xp(x) = \lim_{x\to 0} 1 = 1,$$

$$q_0 = \lim_{x\to 0} x^2 q(x) = \lim_{x\to 0} x^2 = 0 .$$

Hence, we see that the indicial equation is given by

$$r(r-1)+r=r^2=0 .$$

This means that $r_1=r_2=0$. Since $x=0$ is the only singular point for this differential equation, we observe that the series solution $w(0,x)$ which we will find by the method of Frobenius converges for all $x>0$. To find the solution, we note that

$$w(r,x)=\sum_{n=0}^{\infty} a_n x^{n+r}$$

$$\Rightarrow \qquad w'(r,x)=\sum_{n=0}^{\infty}(n+r)\,a_n x^{n+r-1}$$

$$\Rightarrow \qquad w''(r,x)=\sum_{n=0}^{\infty}(n+r-1)(n+r)\,a_n x^{n+r-2} .$$

Notice that the power series for w' and w'' start at $n=0$. Substituting these expressions into the differential equation and simplifying yields

$$\sum_{n=0}^{\infty}(n+r-1)(n+r)a_n x^{n+r}+\sum_{n=0}^{\infty}(n+r)a_n x^{n+r}+\sum_{n=0}^{\infty}a_n x^{n+r+2}=0.$$

Next we want each power series to sum over x^{k+r}. Thus, we shift the index in the first and second power series by letting $k=n$ and in the last power series by letting $k=n+2$. Therefore, we have

$$\sum_{k=0}^{\infty}(k+r-1)(k+r)a_k x^{k+r}+\sum_{k=0}^{\infty}(k+r)a_k x^{k+r}+\sum_{k=2}^{\infty}a_{k-2}x^{k+r}=0 .$$

We will separate out the first two terms from the first two power series above so that we can start all of our power series at the same place. Thus, we have

$$(r-1)ra_0x^r+r(1+r)a_1x^{1+r}+\sum_{k=2}^{\infty}(k+r-1)(k+r)\,a_k x^{k+r}+ra_0x^r$$

$$+(1+r)a_1x^{1+r}+\sum_{k=2}^{\infty}(k+r)a_k x^{k+r}+\sum_{k=2}^{\infty}a_{k-2}x^{k+r}=0$$

$$\Rightarrow\qquad [r(r-1)+r]a_0x^r+[r(r+1)+(r+1)]a_1x^{1+r}$$

$$+\sum_{k=2}^{\infty}[(k+r-1)(k+r)a_k+(k+r)a_k+a_{k-2}]x^{k+r}=0.$$

By equating coefficients and assuming that $a_0\neq 0$, we obtain

$r(r-1)+r=0$, (the indicial equation),

$[r(r+1)+(r+1)]a_1=0 \quad\Rightarrow\quad (r+1)^2a_1=0$,

and, for $k\geq 2$, the recurrence relation

$$(k+r-1)(k+r)a_k+(k+r)a_k+a_{k-2}=0 \qquad\Rightarrow\qquad a_k=\frac{-a_{k-2}}{(k+r)^2},\qquad k\geq 2.$$

Using the fact (which we found from the indicial equation above) that $r_1=0$, we observe that $a_1=0$. Next, using the recurrence relation (and the fact that $r_1=0$), we see that

$$a_k=\frac{-a_{k-2}}{k^2},\qquad k\geq 2.$$

Hence,

$$k=2\qquad\Rightarrow\qquad a_2=\frac{-a_0}{4},$$

$$k=3\qquad\Rightarrow\qquad a_3=\frac{-a_1}{9}=0,$$

$$k=4 \quad \Rightarrow \quad a_4=\frac{-a_2}{16}=\frac{-\frac{-a_0}{4}}{16}=\frac{a_0}{64},$$

$$k=5 \quad \Rightarrow \quad a_5=\frac{-a_3}{25}=0,$$

$$k=6 \quad \Rightarrow \quad a_6=\frac{-a_4}{36}=\frac{-\frac{a_0}{64}}{36}=\frac{-a_0}{2304}.$$

Substituting these coefficients into the solution

$$w(0,x)=\sum_{n=0}^{\infty} a_n x^n = a_0+a_1x+a_2x^2+a_3x^3+a_4x^4+a_5x^5+a_6x^6+\cdots,$$

we obtain the series solution for $x>0$ given by

$$w(0,x)=a_0\left[1-\frac{1}{4}x^2+\frac{1}{64}x^4-\frac{1}{2304}x^6+\cdots\right].$$

25. For this equation, we see that $xp(x)=\frac{x}{2}$ and $x^2q(x)=\frac{-(x+3)}{4}$. Thus, $x=0$ is a regular singular point for this equation and we can use the method of Frobenius to find a solution. To this end, we compute

$$\lim_{x\to 0} xp(x)=\lim_{x\to 0}\frac{x}{2}=0, \qquad \text{and} \qquad \lim_{x\to 0} x^2q(x)=\lim_{x\to 0}\frac{-(x+3)}{4}=\frac{-3}{4}.$$

Therefore, by equation (16) on page 485 of the text, the indicial equation is

$$r(r-1)-\frac{3}{4}=0 \quad \Rightarrow \quad 4r^2-4r-3=0 \quad \Rightarrow \quad (2r+1)(2r-3)=0.$$

This indicial equation has roots $r_1=\frac{3}{2}$ and $r_2=\frac{-1}{2}$. By the method of Frobenius, we can assume that a solution to this differential equation will have the form

$$w(r,x)=\sum_{n=0}^{\infty} a_n x^{n+r}$$

$$\Rightarrow \quad w'(r,x)=\sum_{n=0}^{\infty}(n+r)a_n x^{n+r-1}$$

$$\Rightarrow \quad w''(r,x)=\sum_{n=0}^{\infty}(n+r-1)(n+r)\,a_n x^{n+r-2},$$

where $r = r_1 = \frac{3}{2}$. Since $x = 0$ is the only singular point for this equation, we see that the solution, $w\left(\frac{3}{2}, x\right)$, converges for all $x > 0$. The first step in finding this solution is to plug $w(r,x)$ and its first and second derivatives (which we have found above by term by term differentiation) into the differential equation. Thus, we obtain

$$\sum_{n=0}^{\infty} 4(n+r-1)(n+r)\,a_n x^{n+r} + \sum_{n=0}^{\infty} 2(n+r)\,a_n x^{n+r+1} - \sum_{n=0}^{\infty} a_n x^{n+r+1} - \sum_{n=0}^{\infty} 3a_n x^{n+r} = 0.$$

By shifting indices, we can sum each power series over the same power of x, namely x^{k+r}. Thus, with the substitution $k = n$ in the first and last power series and the substitution $k = n+1$ in the two remaining power series, we obtain

$$\sum_{k=0}^{\infty} 4(k+r-1)(k+r)\,a_k x^{k+r} + \sum_{k=1}^{\infty} 2(k+r-1)a_{k-1} x^{k+r} - \sum_{k=1}^{\infty} a_{k-1} x^{k+r} - \sum_{k=0}^{\infty} 3a_k x^{k+r} = 0 .$$

Next removing the first term (the $k = 0$ term) from the first and last power series above and writing the result as a single power series yields

$$4(r-1)ra_0 x^r + \sum_{k=1}^{\infty} 4(k+r-1)(k+r)\,a_k x^{k+r} + \sum_{k=1}^{\infty} 2(k+r-1)\,a_{k-1} x^{k+r}$$

$$- \sum_{k=1}^{\infty} a_{k-1} x^{k+r} - 3a_0 x^r - \sum_{k=1}^{\infty} 3a_k x^{k+r} = 0 .$$

$$\Rightarrow \qquad [4(r-1)r - 3]a_0 x^r + \sum_{k=1}^{\infty} [4(k+r-1)(k+r)a_k + 2(k+r-1)a_{k-1} - a_{k-1} - 3a_k]x^{k+r} = 0.$$

By equating coefficients we see that each coefficient in the power series must be zero. Also we are assuming that $a_0 \neq 0$. Therefore, we have

$$4(r-1)r - 3 = 0, \qquad \text{(the indicial equation)},$$

$$4(k+r-1)(k+r)a_k + 2(k+r-1)a_{k-1} - a_{k-1} - 3a_k = 0, \qquad k \geq 1.$$

Thus, the recurrence equation is given by

$$a_k = \frac{(3-2k-2r)a_{k-1}}{4(k+r-1)(k+r)-3}, \qquad k \geq 1.$$

Therefore, for $r = r_1 = \frac{3}{2}$, we have

$$a_k = \frac{-2k\,a_{k-1}}{4\left(k+\frac{1}{2}\right)\left(k+\frac{3}{2}\right)-3}, \qquad k \geq 1 \qquad \Rightarrow \qquad a_k = \frac{-a_{k-1}}{2(k+2)}, \qquad k \geq 1.$$

Thus, we see that

$$k = 1 \quad \Rightarrow \quad a_1 = \frac{-a_0}{2\cdot 3},$$

$$k = 2 \quad \Rightarrow \quad a_2 = \frac{-a_1}{2\cdot 4} = \frac{a_0}{2\cdot 2\cdot 3\cdot 4} = \frac{a_0}{2\cdot 4!},$$

$$k = 3 \quad \Rightarrow \quad a_3 = \frac{-a_2}{2\cdot 5} = \frac{-a_0}{2^2\cdot 5!},$$

$$k = 4 \quad \Rightarrow \quad a_4 = \frac{-a_3}{2\cdot 6} = \frac{a_0}{2^3\cdot 6!}.$$

Inspection of this sequence shows that we can write the *n*th coefficient, a_n, for $n \ge 1$ as

$$a_n = \frac{(-1)^n a_0}{2^{n-1}(n+2)!}.$$

Substituting these coefficients into the solution given by

$$w\left(\frac{3}{2}, x\right) = \sum_{n=0}^{\infty} a_n x^{n+(3/2)},$$

yields a power series solution for $x > 0$ given by

$$w\left(\frac{3}{2}, x\right) = a_0 x^{3/2} + a_0 \sum_{n=1}^{\infty} \frac{(-1)^n x^{n+(3/2)}}{2^{n-1}(n+2)!}.$$

But since substituting $n = 0$ into the general coefficient, $a_n = \dfrac{(-1)^n a_0}{2^{n-1}(n+2)!}$, yields $\dfrac{(-1)^0 a_0}{2^{-1}(2)!} = a_0$, the solution that we found above can be written as

$$w\left(\frac{3}{2}, x\right) = a_0 \sum_{n=0}^{\infty} \frac{(-1)^n x^{n+(3/2)}}{2^{n-1}(n+2)!}.$$

27. In this equation, we see that $p(x) = \dfrac{-1}{x}$ and $q(x) = -1$. Thus, the only singular point is $x = 0$. Since $xp(x) = -1$ and $x^2 q(x) = -x^2$, we see that $x = 0$ is a regular singular point for this equation and so we can use the method of Frobenius to find a solution to this equation. We also note that the solution that we find by this method will converge for all $x > 0$. To find this solution we observe that

$$p_0 = \lim_{x\to 0} xp(x) = \lim_{x\to 0}(-1) = -1, \qquad \text{and} \qquad q_0 = \lim_{x\to 0} x^2 q(x) = \lim_{x\to 0}\left(-x^2\right) = 0.$$

Thus, according to equation (16) on page 485 of the text, the indicial equation for the point $x = 0$ is

$$r(r-1) - r = 0 \quad \Rightarrow \quad r(r-2) = 0.$$

Therefore, the roots to the indicial equation are $r_1 = 2$, $r_2 = 0$. Hence, we will use the method of Frobenius to find the solution $w(2,x)$. If we let

$$w(r,x) = \sum_{n=0}^{\infty} a_n x^{n+r},$$

then

$$w'(r,x) = \sum_{n=0}^{\infty} (n+r)\, a_n x^{n+r-1} \qquad \text{and} \qquad w''(r,x) = \sum_{n=0}^{\infty} (n+r-1)(n+r)\, a_n x^{n+r-2}.$$

By substituting these expressions into the differential equation and simplifying, we obtain

$$\sum_{n=0}^{\infty} (n+r-1)(n+r)\, a_n x^{n+r-1} - \sum_{n=0}^{\infty} (n+r)\, a_n x^{n+r-1} - \sum_{n=0}^{\infty} a_n x^{n+r+1} = 0.$$

Next we shift the indices by letting $k = n-1$ in the first two power series above and $k = n+1$ in the last power series above. Therefore, we have

$$\sum_{k=-1}^{\infty} (k+r)(k+r+1)\, a_{k+1} x^{k+r} - \sum_{k=-1}^{\infty} (k+r+1)\, a_{k+1} x^{k+r} - \sum_{k=1}^{\infty} a_{k-1} x^{k+r} = 0.$$

We can start all three of these summations at the same term, the $k = 1$ term, if we separate out the first two terms (the $k = -1$ and $k = 0$ terms) from the first two power series. Thus, we have

$$(r-1)\, r a_0 x^{r-1} + r(r+1)\, a_1 x^r + \sum_{k=1}^{\infty} (k+r)(k+r+1)\, a_{k+1} x^{k+r} - r a_0 x^{r-1}$$
$$- (r+1) a_1 x^r - \sum_{k=1}^{\infty} (k+r+1)\, a_{k+1} x^{k+r} - \sum_{k=1}^{\infty} a_{k-1} x^{k+r} = 0$$

$$\Rightarrow \qquad \left[(r-1)r - r\right] a_0 x^{r-1} + \left[r(r+1) - (r+1)\right] a_1 x^r$$
$$+ \sum_{k=1}^{\infty} \left[(k+r)(k+r+1)\, a_{k+1} - (k+r+1)\, a_{k+1} - a_{k-1}\right] x^{k+r} = 0.$$

By equating coefficients and assuming that $a_0 \neq 0$, we obtain

$$r(r-1) - r = 0, \qquad \text{(the indicial equation)},$$
$$(r+1)(r-1)\, a_1 = 0, \tag{7}$$
$$(k+r)(k+r+1)\, a_{k+1} - (k+r+1)\, a_{k+1} - a_{k-1} = 0, \qquad k \geq 1,$$

where the last equation above is the recurrence relation. Simplifying this recurrence relation yields

$$a_{k+1} = \frac{a_{k-1}}{(k+r+1)(k+r-1)}, \qquad k \ge 1. \tag{8}$$

Next we let $r = r_1 = 2$ in equation (7) and in the recurrence relation, equation (8), to obtain

$$3a_1 = 0 \qquad \Rightarrow \qquad a_1 = 0,$$

$$a_{k+1} = \frac{a_{k-1}}{(k+3)(k+1)}, \qquad k \ge 1.$$

Thus, we have

$$k = 1 \qquad \Rightarrow \qquad a_2 = \frac{a_0}{4 \cdot 2},$$

$$k = 2 \qquad \Rightarrow \qquad a_3 = \frac{a_1}{5 \cdot 3} = 0,$$

$$k = 3 \qquad \Rightarrow \qquad a_4 = \frac{a_2}{6 \cdot 4} = \frac{a_0}{6 \cdot 4 \cdot 4 \cdot 2} = \frac{a_0}{2^4 \cdot 3 \cdot 2 \cdot 2 \cdot 1 \cdot 1} = \frac{a_0}{2^4 \cdot 3! \cdot 2!},$$

$$k = 4 \qquad \Rightarrow \qquad a_5 = \frac{a_3}{7 \cdot 5} = 0,$$

$$k = 5 \qquad \Rightarrow \qquad a_6 = \frac{a_4}{8 \cdot 6} = \frac{a_0}{8 \cdot 6 \cdot 2^4 \cdot 3! \cdot 2!} = \frac{a_0}{2^6 \cdot 4! \cdot 3!}.$$

By inspection we can now see that the coefficients of the power series solution $w(2, x)$ are given by

$$a_{2n-1} = 0, \qquad \text{and} \qquad a_{2n} = \frac{a_0}{2^{2n} n!(n+1)!}, \qquad n \ge 1.$$

Thus, substituting these coefficients into the power series solution yields the solution

$$w(2, x) = a_0 \sum_{n=0}^{\infty} \frac{x^{2n+2}}{2^{2n} n!(n+1)!}.$$

35. In applying the method of Frobenius to this third order linear differential equation, we will seek a solution of the form

$$w(r, x) = \sum_{n=0}^{\infty} a_n x^{n+r}$$

$$\Rightarrow \qquad w'(r, x) = \sum_{n=0}^{\infty} (n+r) a_n x^{n+r-1}$$

$$\Rightarrow \qquad w''(r,x) = \sum_{n=0}^{\infty} (n+r-1)(n+r)\,a_n x^{n+r-2}$$

$$\Rightarrow \qquad w'''(r,x) = \sum_{n=0}^{\infty} (n+r-2)(n+r-1)(n+r)\,a_n x^{n+r-3}\,,$$

where we have differentiated term by term. Substituting these expressions into the differential equation and simplifying yields

$$\sum_{n=0}^{\infty} 6(n+r-2)(n+r-1)(n+r)a_n x^{n+r} + \sum_{n=0}^{\infty} 13(n+r-1)(n+r)a_n x^{n+r}$$
$$+\sum_{n=0}^{\infty} (n+r)a_n x^{n+r} + \sum_{n=0}^{\infty} (n+r)a_n x^{n+r+1} + \sum_{n=0}^{\infty} a_n x^{n+r+1} = 0\,.$$

By the shift of index $k = n+1$ in the last two power series above and the shift $k = n$ in all of the other power series, we obtain

$$\sum_{k=0}^{\infty} 6(k+r-2)(k+r-1)(k+r)\,a_k x^{k+r} + \sum_{k=0}^{\infty} 13(k+r-1)(k+r)\,a_k x^{k+r}$$
$$+\sum_{k=0}^{\infty} (k+r)\,a_k x^{k+r} + \sum_{k=1}^{\infty} (k-1+r)\,a_{k-1} x^{k+r} + \sum_{k=1}^{\infty} a_{k-1} x^{k+r} = 0\,.$$

Next we remove the first term from each of the first three power series above so that all of these series start at $k = 1$. Thus, we have

$$6(r-2)(r-1)\,ra_0 x^r + \sum_{k=1}^{\infty} 6(k+r-2)(k+r-1)(k+r)\,a_k x^{k+r} + 13(r-1)\,ra_0 x^r$$
$$+\sum_{k=1}^{\infty} 13(k+r-1)(k+r)\,a_k x^{k+r} + ra_0 x^r + \sum_{k=1}^{\infty} (k+r)\,a_k x^{k+r}$$
$$+\sum_{k=1}^{\infty} (k-1+r)\,a_{k-1} x^{k+r} + \sum_{k=1}^{\infty} a_{k-1} x^{k+r} = 0$$

$$\Rightarrow \qquad [6(r-2)(r-1)r + 13(r-1)r + r]a_0 x^r$$
$$+\sum_{k=1}^{\infty} \big[6(k+r-2)(k+r-1)(k+r)\,a_k + 13(k+r-1)(k+r)\,a_k + (k+r)\,a_k$$
$$+(k-1+r)\,a_{k-1} + a_{k-1}\big]x^{k+r} = 0\,. \qquad (9)$$

If we assume that $a_0 \neq 0$ and set the coefficient of x^r equal to zero, we find that the *indicial equation* is

$$6(r-2)(r-1)r + 13(r-1)r + r = 0 \qquad \Rightarrow \qquad r^2(6r-5) = 0\,.$$

Hence, the roots to the indicial equation are $0, 0, 5/6$. We will find the solution associated with the largest of these roots. That is, we will find $w(5/6, x)$. Also, from equation (9), we see that we have the recurrence relation

$$6(k+r-2)(k+r-1)(k+r)a_k + 13(k+r-1)(k+r)a_k + (k+r)a_k$$
$$+ (k-1+r)a_{k-1} + a_{k-1} = 0, \qquad k \geq 1$$

$$\Rightarrow \qquad a_k = \frac{-a_{k-1}}{6(k+r-2)(k+r-1)+13(k+r-1)+1}, \qquad k \geq 1.$$

If we assume that $r = 5/6$, then this recurrence relation simplifies to

$$a_k = \frac{-a_{k-1}}{k(6k+5)}, \qquad k \geq 1.$$

Therefore, we have

$$k=1 \qquad \Rightarrow \qquad a_1 = \frac{-a_0}{11},$$

$$k=2 \qquad \Rightarrow \qquad a_2 = \frac{-a_1}{34} = \frac{a_0}{374},$$

$$k=3 \qquad \Rightarrow \qquad a_3 = \frac{-a_2}{69} = \frac{-a_0}{25{,}806}.$$

By substituting these coefficients into the solution $w(5/6, x) = \sum_{n=0}^{\infty} a_n x^{n+(5/6)}$, we obtain

$$w\left(\frac{5}{6}, x\right) = a_0\left(x^{5/6} - \frac{x^{11/6}}{11} + \frac{x^{17/6}}{374} - \frac{x^{23/6}}{25{,}806} + \cdots\right).$$

41. If we let $z = \frac{1}{x}\left(\Rightarrow \frac{dz}{dx} = \frac{-1}{x^2}\right)$, then we can define a new function $Y(z)$ as

$$Y(z) := y\left(\frac{1}{z}\right) = y(x).$$

Thus, by the chain rule, we have

$$\frac{dy}{dx} = \frac{dY}{dx} = \left(\frac{dY}{dz}\right)\left(\frac{dz}{dx}\right) = \left(\frac{-1}{x^2}\right)\left(\frac{dY}{dz}\right) \tag{10}$$

$$\Rightarrow \qquad -x^2\frac{dy}{dx} = \frac{dY}{dz}. \tag{11}$$

Therefore, using the product rule and chain rule, we see that

$$\frac{d^2y}{dx^2} = \frac{d^2Y}{dx^2} = \frac{d}{dx}\left(\frac{dY}{dx}\right) = \frac{d}{dx}\left[\left(\frac{-1}{x^2}\right)\left(\frac{dY}{dz}\right)\right] \qquad \text{(by (10) above)}$$

$$= \frac{d}{dx}\left(\frac{-1}{x^2}\right)\times\left(\frac{dY}{dz}\right) + \left(\frac{-1}{x^2}\right)\times\frac{d}{dx}\left(\frac{dY}{dz}\right) \qquad \text{(by product rule)}$$

$$= \left(\frac{2}{x^3}\right)\times\left(\frac{dY}{dz}\right) + \left(\frac{-1}{x^2}\right)\times\left[\left(\frac{d^2Y}{dz^2}\right)\left(\frac{dz}{dx}\right)\right] \qquad \text{(by chain rule)}$$

$$= \left(\frac{2}{x^3}\right)\times\left(\frac{dY}{dz}\right) + \left(\frac{-1}{x^2}\right)^2\times\left(\frac{d^2Y}{dz^2}\right) \qquad \left(\text{since } \frac{dz}{dx} = \frac{-1}{x^2}\right)$$

$$= \frac{2}{x^3}\frac{dY}{dz} + \frac{1}{x^4}\frac{d^2Y}{dz^2}.$$

Hence, we have

$$x^3\frac{d^2y}{dx^2} = 2\frac{dY}{dz} + \frac{1}{x}\frac{d^2Y}{dz^2} = 2\frac{dY}{dz} + z\frac{d^2Y}{dz^2}. \qquad (12)$$

By using the fact that $Y(z) = y(x)$ and equations (11) and (12) above, we can now transform the original differential equation into the differential equation

$$2\frac{dY}{dz} + z\frac{d^2Y}{dz^2} + \frac{dY}{dz} - Y = 0$$

$$\Rightarrow \qquad zY'' + 3Y' - Y = 0. \qquad (13)$$

We will now solve this transformed differential equation. To this end, we first note that

$$p(z) = \frac{3}{z} \qquad \Rightarrow \qquad zp(z) = 3,$$

and

$$q(z) = \frac{-1}{z} \qquad \Rightarrow \qquad z^2q(z) = -z.$$

Therefore, $z = 0$ is a regular singular point of this equation and so infinity is a regular singular point of the original equation.

To find a power series solution for equation (13), we first compute

$$p_0 = \lim_{z\to 0} zp(z) = 3, \qquad \text{and} \qquad q_0 = \lim_{z\to 0} z^2q(z) = 0.$$

Thus, the indicial equation for equation (13) is

$$r(r-1) + 3r = 0 \qquad \Rightarrow \qquad r(r+2) = 0.$$

Hence, this indicial equation has roots $r_1 = 0$ and $r_2 = -2$. We seek a solution of the form

$$w(r,z)=\sum_{n=0}^{\infty} a_n z^{n+r}.$$

Substituting this expression into equation (13) above yields

$$z\sum_{n=0}^{\infty}(n+r-1)(n+r)\,a_n z^{n+r-2}+3\sum_{n=0}^{\infty}(n+r)\,a_n z^{n+r-1}-\sum_{n=0}^{\infty}a_n z^{n+r}=0.$$

By simplifying, this equation becomes

$$\sum_{n=0}^{\infty}(n+r-1)(n+r)\,a_n z^{n+r-1}+\sum_{n=0}^{\infty}3(n+r)\,a_n z^{n+r-1}-\sum_{n=0}^{\infty}a_n z^{n+r}=0.$$

Making the shift of index $k=n-1$ in the first two power series and $k=n$ in the last power series allows us to sum each power series over the same powers of z, namely z^{k+r}. Thus, we have

$$\sum_{k=-1}^{\infty}(k+r)(k+r+1)\,a_{k+1}z^{k+r}+\sum_{k=-1}^{\infty}3(k+r+1)\,a_{k+1}z^{k+r}-\sum_{k=0}^{\infty}a_k z^{k+r}=0.$$

By removing the first term from the first two power series above, we can write these three summations as a single power series. Therefore, we have

$$(r-1)\,ra_0 z^{r-1}+\sum_{k=0}^{\infty}(k+r)(k+r+1)\,a_{k+1}z^{k+r}+3ra_0 z^{r-1}$$

$$+\sum_{k=0}^{\infty}3(k+r+1)\,a_{k+1}z^{k+r}-\sum_{k=0}^{\infty}a_k z^{k+r}=0$$

$$\Rightarrow\qquad\big[(r-1)r+3r\big]a_0 z^{r-1}+\sum_{k=0}^{\infty}\big[(k+r)(k+r+1)\,a_{k+1}+3(k+r+1)\,a_{k+1}-a_k\big]z^{k+r}=0.$$

Equating coefficients and assuming that $a_0\neq 0$ yields the indicial equation, $(r-1)r+3r=0$, and the recurrence relation

$$(k+r)(k+r+1)\,a_{k+1}+3(k+r+1)\,a_{k+1}-a_k=0,\qquad k\geq 0$$

$$\Rightarrow\qquad a_{k+1}=\frac{a_k}{(k+r+1)(k+r+3)},\qquad k\geq 0.$$

Thus, with $r=r_1=0$, we obtain the recurrence relation

$$a_{k+1}=\frac{a_k}{(k+1)(k+3)},\qquad k\geq 0.$$

Since a_0 is an arbitrary number, we see from this recurrence equation that the next three coefficients are given by

$$k=0\qquad\Rightarrow\qquad a_1=\frac{a_0}{3},$$

$$k = 1 \qquad \Rightarrow \qquad a_2 = \frac{a_1}{8} = \frac{a_0}{24},$$

$$k = 2 \qquad \Rightarrow \qquad a_3 = \frac{a_2}{15} = \frac{a_0}{360}.$$

Thus, from the method of Frobenius, we obtain a power series solution for equation (13) given by

$$Y(z) = w(0,z) = \sum_{n=0}^{\infty} a_n z^n = a_0\left(1 + \frac{1}{3}z + \frac{1}{24}z^2 + \frac{1}{360}z^3 + \cdots\right).$$

In order to find the solution of the original differential equation, we again make the substitution $z = \frac{1}{x}$ and $Y(z) = Y\left(x^{-1}\right) = y(x)$. Therefore, in the solution found above, we replace the z's with $\frac{1}{x}$ to obtain the solution given by

$$y(x) = Y\left(x^{-1}\right) = a_0\left(1 + \frac{1}{3}x^{-1} + \frac{1}{24}x^{-2} + \frac{1}{360}x^{-3} + \cdots\right).$$

EXERCISES 8.7: Finding a Second Linearly Independent Solution, page 506

3. In Problem 21 of Exercises 8.6, we found one power series solution for this differential equation about the point $x = 0$ given by

$$y_1(x) = 1 - \frac{1}{4}x^2 + \frac{1}{64}x^4 - \frac{1}{2304}x^6 + \cdots,$$

where we let $a_0 = 1$. We also found that the roots to the indicial equation are $r_1 = r_2 = 0$. Thus, to find a second linearly independent solution about the regular singular point $x = 0$, we will use part (b) of Theorem 7 on page 499 of the text. Therefore, we see that this second linearly independent solution will have the form given by

$$y_2(x) = y_1(x)\ln x + \sum_{n=1}^{\infty} b_n x^n$$

$$\Rightarrow \qquad y_2'(x) = y_1'(x)\ln x + x^{-1}y_1(x) + \sum_{n=1}^{\infty} nb_n x^{n-1}$$

$$\Rightarrow \qquad y_2''(x) = y_1''(x)\ln x - x^{-2}y_1(x) + 2x^{-1}y_1'(x) + \sum_{n=1}^{\infty} n(n-1)b_n x^{n-2}.$$

Substituting these expressions into the differential equation yields

$$x^2\left\{y_1''(x)\ln x - x^{-2}y_1(x) + 2x^{-1}y_1'(x) + \sum_{n=1}^{\infty} n(n-1)\,b_n x^{n-2}\right\}$$

$$+x\left\{y_1'(x)\ln x + x^{-1}y_1(x) + \sum_{n=1}^{\infty} nb_n x^{n-1}\right\} + x^2\left\{y_1(x)\ln x + \sum_{n=1}^{\infty} b_n x^n\right\} = 0\,,$$

which simplifies to

$$x^2 y_1''(x)\ln x - y_1(x) + 2xy_1'(x) + \sum_{n=1}^{\infty} n(n-1)\,b_n x^n + xy_1'(x)\ln x + y_1(x)$$

$$+\sum_{n=1}^{\infty} n\,b_n x^n + x^2 y_1(x)\ln x + \sum_{n=1}^{\infty} b_n x^{n+2} = 0$$

$$\Rightarrow \qquad \left(x^2 y_1''(x) + xy_1'(x) + x^2 y_1(x)\right)\ln x + 2xy_1'(x)$$

$$+\sum_{n=1}^{\infty} n(n-1)\,b_n x^n + \sum_{n=1}^{\infty} nb_n x^n + \sum_{n=1}^{\infty} b_n x^{n+2} = 0\,.$$

Therefore, since $y_1(x)$ is a solution to the differential equation, the term in braces is zero and the above equation reduces to

$$2xy_1'(x) + \sum_{n=1}^{\infty} n(n-1)\,b_n x^n + \sum_{n=1}^{\infty} nb_n x^n + \sum_{n=1}^{\infty} b_n x^{n+2} = 0\,.$$

Next we make the substitution $k = n+2$ in the last power series above and the substitution $k = n$ in the other two power series so that we can sum all three of the power series over the same power of x, namely x^k. Thus, we have

$$2xy_1'(x) + \sum_{k=1}^{\infty} k(k-1)b_k x^k + \sum_{k=1}^{\infty} k\,b_k x^k + \sum_{k=3}^{\infty} b_{k-2} x^k = 0\,.$$

By separating out the first two terms in the first two summations above and simplifying, we obtain

$$2xy_1'(x) + 0 + 2b_2 x^2 + \sum_{k=3}^{\infty} k(k-1)\,b_k x^k + b_1 x + 2b_2 x^2 + \sum_{k=3}^{\infty} kb_k x^k + \sum_{k=3}^{\infty} b_{k-2} x^k = 0$$

$$\Rightarrow \qquad 2xy_1'(x) + b_1 x + 4b_2 x^2 + \sum_{k=3}^{\infty}\left(k^2 b_k + b_{k-2}\right)x^k = 0\,. \qquad (14)$$

By differentiating the series for $y_1(x)$ term by term, we obtain

$$y_1'(x) = -\frac{1}{2}x + \frac{1}{16}x^3 - \frac{1}{384}x^5 + \cdots\,.$$

Thus, substituting this expression for $y_1'(x)$ into equation (14) above and simplifying yields

$$\left\{-x^2 + \frac{1}{8}x^4 - \frac{1}{192}x^6 + \cdots\right\} + b_1 x + 4b_2 x^2 + \sum_{k=3}^{\infty}\left(k^2 b_k + b_{k-2}\right)x^k = 0\,.$$

Therefore, by equating coefficients, we see that

$b_1 = 0$;

$$4b_2 - 1 = 0 \quad \Rightarrow \quad b_2 = \frac{1}{4};$$

$$9b_3 + b_1 = 0 \quad \Rightarrow \quad b_3 = 0;$$

$$\frac{1}{8} + 16b_4 + b_2 = 0 \quad \Rightarrow \quad b_4 = \frac{-3}{128};$$

$$25b_5 + b_3 = 0 \quad \Rightarrow \quad b_5 = 0;$$

$$\frac{-1}{192} + 36b_6 + b_4 = 0 \quad \Rightarrow \quad b_6 = \frac{11}{13{,}824}.$$

Substituting these coefficients into the solution $y_2(x) = y_1(x)\ln x + \sum_{n=1}^{\infty} b_n x^n$, yields

$$y_2(x) = y_1(x)\ln x + \frac{1}{4}x^2 - \frac{3}{128}x^4 + \frac{11}{13{,}824}x^6 + \cdots.$$

Thus, a general solution of this differential equation is given by

$$y(x) = c_1 y_1(x) + c_2 y_2(x),$$

where

$$y_1(x) = 1 - \frac{1}{4}x^2 + \frac{1}{64}x^4 - \frac{1}{2304}x^6 + \cdots,$$

and

$$y_2(x) = y_1(x)\ln x + \frac{1}{4}x^2 - \frac{3}{128}x^4 + \frac{11}{13{,}824}x^6 + \cdots.$$

7. In Problem 25 of Section 8.6, we found a solution to this differential equation about the regular singular point $x = 0$ given by

$$y_1(x) = \sum_{n=0}^{\infty} \frac{(-1)^n x^{n+(3/2)}}{2^{n-1}(n+2)!} = x^{3/2} - \frac{1}{6}x^{5/2} + \frac{1}{48}x^{7/2} + \cdots,$$

where we let $a_0 = 1$. We also found that the roots to the indicial equation for this problem are $r_1 = \frac{3}{2}$ and $r_2 = \frac{-1}{2}$, and so $r_1 - r_2 = 2$. Thus, in order to find a second linearly independent solution about $x = 0$, we will use part (c) of Theorem 7 on page 499 of the text. Therefore, we will assume that this second solution has the form

$$y_2(x) = C_1 y_1(x)\ln x + \sum_{n=0}^{\infty} b_n x^{n-(1/2)}, \qquad b_0 \neq 0$$

$$\Rightarrow \qquad y_2'(x) = Cy_1'(x)\ln x + C\frac{1}{x}y_1(x) + \sum_{n=0}^{\infty}\left(n-\frac{1}{2}\right)b_n x^{n-(3/2)}$$

$$\Rightarrow \qquad y_2''(x) = Cy_1''(x)\ln x + 2C\frac{1}{x}y_1'(x) - C\frac{1}{x^2}y_1(x) + \sum_{n=0}^{\infty}\left(n-\frac{3}{2}\right)\left(n-\frac{1}{2}\right)b_n x^{n-(5/2)}.$$

Substituting these expressions into the differential equation yields

$$4x^2\left\{Cy_1''(x)\ln x + 2C\frac{1}{x}y_1'(x) - C\frac{1}{x^2}y_1(x) + \sum_{n=0}^{\infty}\left(n-\frac{3}{2}\right)\left(n-\frac{1}{2}\right)b_n x^{n-(5/2)}\right\}$$

$$+2x^2\left\{Cy_1'(x)\ln x + C\frac{1}{x}y_1(x) + \sum_{n=0}^{\infty}\left(n-\frac{1}{2}\right)b_n x^{n-(3/2)}\right\}$$

$$-(x+3)\left\{Cy_1(x)\ln x + \sum_{n=0}^{\infty} b_n x^{n-(1/2)}\right\} = 0$$

$$\Rightarrow \qquad 4x^2Cy_1''(x)\ln x + 8Cxy_1'(x) - 4Cy_1(x) + \sum_{n=0}^{\infty}4\left(n-\frac{3}{2}\right)\left(n-\frac{1}{2}\right)b_n x^{n-(1/2)}$$

$$+2x^2Cy_1'(x)\ln x + 2Cxy_1(x) + \sum_{n=0}^{\infty}2\left(n-\frac{1}{2}\right)b_n x^{n+(1/2)}$$

$$-Cxy_1(x)\ln x - \sum_{n=0}^{\infty} b_n x^{n+(1/2)} - 3Cy_1(x)\ln x - \sum_{n=0}^{\infty}3b_n x^{n-(1/2)} = 0$$

$$\Rightarrow \qquad C\left(4x^2y_1''(x) + 2x^2y_1'(x) - xy_1(x) - 3y_1(x)\right)\ln x + 8Cxy_1'(x) + \left(2Cx - 4C\right)y_1(x)$$

$$+\sum_{n=0}^{\infty}(2n-3)(2n-1)\,b_n x^{n-(1/2)} + \sum_{n=0}^{\infty}(2n-1)\,b_n x^{n+(1/2)} - \sum_{n=0}^{\infty} b_n x^{n+(1/2)} - \sum_{n=0}^{\infty}3b_n x^{n-(1/2)} = 0.$$

Since $y_1(x)$ is a solution to the differential equation, the term in braces is zero. By shifting indices so that each power series is summed over the same power of x, we have

$$8Cxy_1'(x) + \left(2Cx - 4C\right)y_1(x) + \sum_{k=0}^{\infty}(2k-3)(2k-1)\,b_k x^{k-(1/2)}$$

$$+\sum_{k=1}^{\infty}(2k-3)\,b_{k-1}x^{k-(1/2)} - \sum_{k=1}^{\infty} b_{k-1}x^{k-(1/2)} - \sum_{k=0}^{\infty}3b_k x^{k-(1/2)} = 0.$$

By writing all of these summations as a single power series (noting that the $k=0$ term of the first and last summations add to zero), we obtain

$$8Cxy_1'(x)+(2Cx-4C)y_1(x)+\sum_{k=1}^{\infty}[(2k-3)(2k-1)b_k+(2k-3)b_{k-1}-b_{k-1}-3b_k]x^{k-(1/2)}=0$$

Substituting into this equation the expressions for $y_1(x)$ and $y_1'(x)$ given by

$$y_1(x)=\sum_{n=0}^{\infty}\frac{(-1)^n x^{n+(3/2)}}{2^{n-1}(n+2)!}, \qquad y_1'(x)=\sum_{n=0}^{\infty}\frac{(-1)^n\left(n+\frac{3}{2}\right)x^{n+(1/2)}}{2^{n-1}(n+2)!},$$

yields

$$\sum_{n=0}^{\infty}\frac{8C(-1)^n(n+3/2)x^{n+(3/2)}}{2^{n-1}(n+2)!}+\sum_{n=0}^{\infty}\frac{2C(-1)^n x^{n+(5/2)}}{2^{n-1}(n+2)!}$$

$$-\sum_{n=0}^{\infty}\frac{4C(-1)^n x^{n+(3/2)}}{2^{n-1}(n+2)!}+\sum_{k=1}^{\infty}[4k(k-2)b_k+2(k-2)b_{k-1}]x^{k-(1/2)}=0,$$

where we have simplified the expression inside the last summation. Combining the first and third power series yields

$$\sum_{n=0}^{\infty}\frac{8C(-1)^n(n+1)x^{n+(3/2)}}{2^{n-1}(n+2)!}+\sum_{n=0}^{\infty}\frac{2C(-1)^n x^{n+(5/2)}}{2^{n-1}(n+2)!}$$

$$+\sum_{k=1}^{\infty}[4k(k-2)b_k+2(k-2)b_{k-1}]x^{k-(1/2)}=0. \qquad (15)$$

By writing out the terms up to order $x^{7/2}$, we obtain

$$8C\left\{x^{3/2}-\frac{1}{3}x^{5/2}+\frac{3}{16}x^{7/2}+\cdots\right\}+2C\left\{x^{5/2}-\frac{1}{6}x^{7/2}+\cdots\right\}$$

$$+\left[(-4b_1-2b_0)x^{1/2}+(12b_3+2b_2)x^{5/2}+(32b_4+4b_3)x^{7/2}+\cdots\right]=0.$$

Setting the coefficients equal to zero, yields

$$-4b_1-2b_0=0 \qquad \Rightarrow \qquad b_1=\frac{-b_0}{2};$$

$$8C=0 \qquad \Rightarrow \qquad C=0;$$

$$-\frac{8}{3}C+2C+12b_3+2b_2=0 \qquad \Rightarrow \qquad b_3=\frac{-b_2}{6};$$

$$\frac{2}{3}C-\frac{1}{3}C+32b_4+4b_3=0 \qquad \Rightarrow \qquad b_4=\frac{b_2}{48}.$$

From this we see that b_0 and b_2 are arbitrary constants and that $C = 0$. Also, since $C = 0$, we can use the last power series in equation (15) to obtain the recurrence equation $b_k = \frac{b_{k-1}}{2k}$. Thus, every coefficient after b_4 will depend only on b_2 (not on b_0). Substituting these coefficients into the solution

$$y_2(x) = Cy_1(x)\ln x + \sum_{n=0}^{\infty} b_n x^{n-(1/2)},$$

yields

$$y_2(x) = b_0\left[x^{-1/2} - \frac{1}{2}x^{1/2}\right] + b_2\left[x^{3/2} - \frac{1}{6}x^{5/2} + \frac{1}{48}x^{7/2} + \cdots\right].$$

The expression in the brackets following b_2 is just the series expansion for $y_1(x)$. Hence, in order to obtain a second linearly independent solution, we must choose b_0 to be nonzero. Taking $b_0 = 1$ and $b_2 = 0$ gives

$$y_2(x) = x^{-1/2} - \frac{1}{2}x^{1/2}.$$

Therefore, a general solution is

$$y(x) = c_1 y_1(x) + c_2 y_2(x),$$

where

$$y_1(x) = x^{3/2} - \frac{1}{6}x^{5/2} + \frac{1}{48}x^{7/2} + \cdots \qquad \text{and} \qquad y_2(x) = x^{-1/2} - \frac{1}{2}x^{1/2}.$$

17. In Problem 35 of Section 8.6, we assumed that there exists a series solution to this problem of the form $w(r,x) = \sum_{n=0}^{\infty} a_n x^{n+r}$. This led to the equation (cf. equation (9), of the solution to Problem 35, Exercises 8.6)

$$r^2(6r-5)a_0x^r + \sum_{k=1}^{\infty}\left\{(k+r)^2[6(k+r)-5]\,a_k + (k+r)\,a_{k-1}\right\}x^{k+r} = 0. \tag{16}$$

From this we found the indicial equation $r^2(6r-5) = 0$, which has roots $r = 0, 0, \frac{5}{6}$. By using the root $r = \frac{5}{6}$, we found the solution $w\left(\frac{5}{6}, x\right)$. Hence one solution is

$$y_1(x) = x^{5/6} - \frac{x^{11/6}}{11} + \frac{x^{17/6}}{374} - \frac{x^{23/6}}{25{,}806} + \cdots,$$ where we have chosen $a_0 = 1$ in $w\left(\frac{5}{6}, x\right)$.

We now seek two more linearly independent solutions to this differential equation. To find a second linearly independent solution, we will use the root $r = 0$ and set the coefficients in equation (16) to zero to obtain the recurrence relation

$$k^2(6k-5)a_k + ka_{k-1} = 0, \qquad k \geq 1.$$

Solving for a_k in terms of a_{k-1} gives

$$a_k = \frac{-a_{k-1}}{k(6k-5)}, \qquad k \geq 1.$$

Thus, we have

$$k = 1 \qquad \Rightarrow \qquad a_1 = -a_0,$$

$$k = 2 \qquad \Rightarrow \qquad a_2 = \frac{-a_1}{14} = \frac{a_0}{14},$$

$$k = 3 \qquad \Rightarrow \qquad a_3 = \frac{-a_2}{39} = \frac{-a_0}{546},$$

$$k = 4 \qquad \Rightarrow \qquad a_4 = \frac{-a_3}{76} = \frac{a_0}{41{,}496},$$

$$k = 5 \qquad \Rightarrow \qquad a_5 = \frac{-a_4}{125} = \frac{-a_0}{5{,}187{,}000}.$$

Plugging these coefficients into the solution $w(0, x)$ and setting $a_0 = 1$ yields a second linearly independent solution

$$y_2(x) = 1 - x + \frac{1}{14}x^2 - \frac{1}{546}x^3 + \frac{1}{41{,}496}x^4 - \frac{1}{5{,}187{,}000}x^5 + \cdots.$$

To find a third linearly independent solution, we will use the repeated root $r = 0$ and assume that, as in the case of second order equations with repeated roots, the solution that we seek will have the form

$$y_3(x) = y_2(x)\ln x + \sum_{n=1}^{\infty} c_n x^n.$$

Since the first three derivatives of $y_3(x)$ are given by

$$y_3'(x) = y_2'(x)\ln x + x^{-1}y_2(x) + \sum_{n=1}^{\infty} nc_n x^{n-1}$$

$$\Rightarrow \qquad y_3''(x) = y_2''(x)\ln x + 2x^{-1}y_2'(x) - x^{-2}y_2(x) + \sum_{n=1}^{\infty}(n-1)n\,c_n x^{n-2}$$

$$\Rightarrow \qquad y_3'''(x) = y_2'''(x)\ln x + 3x^{-1}y_2''(x) - 3x^{-2}y_2'(x) + 2x^{-3}y_2(x) + \sum_{n=1}^{\infty}(n-2)(n-1)nc_n x^{n-3},$$

substituting $y_3(x)$ into the differential equation and recalling that $y_2(x)$ is a solution to this equation yields

$$6x^3y''' + 13x^2y'' + (x+x^2)y' + xy$$

$$= 6x^3\left\{y_2'''(x)\ln x + 3x^{-1}y_2''(x) - 3x^{-2}y_2'(x) + 2x^{-3}y_2(x) + \sum_{n=1}^{\infty}(n-2)(n-1)\,nc_n x^{n-3}\right\}$$

$$+13x^2\left\{y_2''(x)\ln x + 2x^{-1}y_2'(x) - x^{-2}y_2(x) + \sum_{n=1}^{\infty}(n-1)\,nc_n x^{n-2}\right\}$$

$$+(x+x^2)\left\{y_2'(x)\ln x + x^{-1}y_2(x) + \sum_{n=1}^{\infty}nc_n x^{n-1}\right\} + x\left\{y_2(x)\ln x + \sum_{n=1}^{\infty}c_n x^n\right\} = 0$$

$$\Rightarrow \qquad 18x^2y_2''(x) + 8xy_2'(x) + xy_2(x) + \sum_{n=1}^{\infty}6(n-2)(n-1)\,nc_n x^n$$

$$+\sum_{n=1}^{\infty}13(n-1)\,nc_n x^n + \sum_{n=1}^{\infty}nc_n x^n + \sum_{n=1}^{\infty}nc_n x^{n+1} + \sum_{n=1}^{\infty}c_n x^{n+1} = 0.$$

By shifting indices and then starting all of the resulting power series at the same point, we can combine all of the summations above into a single power series. Thus, we have

$$18x^2y_2''(x) + 8xy_2'(x) + xy_2(x) + c_1x$$

$$+\sum_{k=2}^{\infty}\left[6(k-2)(k-1)\,kc_k + 13(k-1)\,kc_k + kc_k + kc_{k-1}\right]x^k = 0$$

$$\Rightarrow \qquad 18x^2y_2''(x) + 8xy_2'(x) + xy_2(x) + c_1x + \sum_{k=2}^{\infty}\left[(6k^3-5k^2)\,c_k + kc_{k-1}\right]x^k = 0. \qquad (17)$$

By computing $y_2'(x)$ and $y_2''(x)$, we obtain

$$y_2'(x) = -1 + \frac{1}{7}x - \frac{1}{182}x^2 + \frac{1}{10374}x^3 + \cdots,$$

$$y_2''(x) = \frac{1}{7} - \frac{1}{91}x + \frac{1}{3458}x^2 + \cdots.$$

By substituting these expressions into equation (17), we have

$$18x^2\left\{\frac{1}{7}-\frac{1}{91}x+\frac{1}{3458}x^2+\cdots\right\}+8x\left\{-1+\frac{1}{7}x-\frac{1}{182}x^2+\frac{1}{10374}x^3+\cdots\right\}$$

$$+x\left\{1-x+\frac{1}{14}x^2-\frac{1}{546}x^3+\frac{1}{41{,}496}x^4+\cdots\right\}+c_1x+\sum_{k=2}^{\infty}\left[(6k^3-5k^2)c_k+kc_{k-1}\right]x^k=0.$$

Writing out the terms up to order x^3, we find

$$(-7+c_1)x+\left(\frac{19}{7}+28c_2+2c_1\right)x^2+\left(-\frac{31}{182}+117c_3+3c_2\right)x^3+\cdots=0.$$

By equating coefficients to zero, we obtain

$$-7+c_1=0 \qquad \Rightarrow \qquad c_1=7;$$

$$\frac{19}{7}+28c_2+2c_1=0 \qquad \Rightarrow \qquad c_2=\frac{-117}{196};$$

$$\frac{-31}{182}+117c_3+3c_2=0 \qquad \Rightarrow \qquad c_3=\frac{4997}{298116}.$$

Therefore, plugging these coefficients into the expansion

$$y_3(x)=y_2(x)\ln x+\sum_{n=1}^{\infty}c_nx^n,$$

yields a third linearly independent solution is given by

$$y_3(x)=y_2(x)\ln x+7x-\frac{117}{196}x^2+\frac{4997}{298116}x^3+\cdots.$$

Thus, a general solution is

$$y(x)=c_1y_1(x)+c_2y_2(x)+c_3y_3(x),$$

where

$$y_1(x)=x^{5/6}-\frac{x^{11/6}}{11}+\frac{x^{17/6}}{374}-\frac{x^{23/6}}{25{,}806}+\cdots,$$

$$y_2(x)=1-x+\frac{1}{14}x^2-\frac{1}{546}x^3+\frac{1}{41{,}496}x^4-\frac{1}{5{,}187{,}000}x^5+\cdots,$$

and

$$y_3(x)=y_2(x)\ln x+7x-\frac{117}{196}x^2+\frac{4997}{298116}x^3+\cdots.$$

23. We will try to find a solution of the form

$$y(x)=\sum_{n=0}^{\infty}a_n x^{n+r}$$

$$\Rightarrow \qquad y'(x)=\sum_{n=0}^{\infty}(n+r)\,a_n x^{n+r-1}$$

$$\Rightarrow \qquad y''(x)=\sum_{n=0}^{\infty}(n+r)(n+r-1)\,a_n x^{n+r-2}\,.$$

Therefore, we substitute these expressions into the differential equation to obtain

$$x^2y''+y'-2y=x^2\sum_{n=0}^{\infty}(n+r)(n+r-1)\,a_n x^{n+r-2}+\sum_{n=0}^{\infty}(n+r)\,a_n x^{n+r-1}-2\sum_{n=0}^{\infty}a_n x^{n+r}=0$$

$$\Rightarrow \qquad \sum_{k=0}^{\infty}(k+r)(k+r-1)\,a_k x^{k+r}+\sum_{k=-1}^{\infty}(k+r+1)\,a_{k+1}x^{k+r}\;-\sum_{k=0}^{\infty}2a_k x^{k+r}=0$$

$$\Rightarrow \qquad ra_0x^{r-1}+\sum_{k=0}^{\infty}\left[(k+r)(k+r-1)\,a_k+(k+r+1)\,a_{k+1}-2a_k\right]x^{k+r}=0\,,$$

where we have changed all of the indices and the starting point for the second summation so that we could write these three power series as a single power series. By assuming that $a_0\neq 0$ and $ra_0x^{r-1}=0$, we see that $r=0$. Plugging $r=0$ into the coefficients in the summation and noting that each of these coefficients must be zero yields the recurrence relation

$$k(k-1)a_k+(k+1)a_{k+1}-2a_k=0\,, \qquad k\geq 0$$

$$\Rightarrow \qquad a_{k+1}=(2-k)a_k\,, \quad k\geq 0\,.$$

Thus, we see that the coefficients to the solution are given by

$$k=0 \quad\Rightarrow\quad a_1=2a_0\,; \qquad k=1 \quad\Rightarrow\quad a_2=a_1=2a_0\,;$$

$$k=2 \quad\Rightarrow\quad a_3=0\,; \qquad k=3 \quad\Rightarrow\quad a_4=-a_3=0\,.$$

Since each coefficient is a multiple of the previous coefficient, we see that $a_n=0$ for $n\geq 3$. If we take $a_0=1$, then one solution is

$$y_1(x)=1+2x+2x^2\,.$$

We will now use the reduction of order formula (see equation (11) on page 497 of the text) to find a second linearly independent solution. We first note that $p(x)=x^{-2}$. Therefore, we have

$$e^{-\int p(x)\,dx}=\exp\left[\int\left(-x^{-2}\right)dx\right]=\exp\left[x^{-1}\right]=1+x^{-1}+\frac{x^{-2}}{2}+\frac{x^{-3}}{6}+\frac{x^{-4}}{24}+\cdots,$$

where we have used the Maclaurin expansion for e^z. We next note that

$[y_1(x)]^2 = 4x^4 + 8x^3 + 8x^2 + 4x + 1$.

Using long division (with descending powers of x in each polynomial), we see that

$$\frac{\exp\left[\int(-x^{-2})dx\right]}{[y_1(x)]^2} = \frac{1 + x^{-1} + \frac{x^{-2}}{2} + \frac{x^{-3}}{6} + \frac{x^{-4}}{24} + \cdots}{4x^4 + 8x^3 + 8x^2 + 4x + 1} = \frac{1}{4}x^{-4} - \frac{1}{4}x^{-5} + \frac{1}{8}x^{-6} + \cdots .$$

Therefore, by equation (11) on page 497 of the text, we have

$$y_2(x) = y_1(x)\int \frac{\exp\left[\int p(x)\,dx\right]}{[y_1(x)]^2}dx$$

$$= \left[1 + 2x + 2x^2\right]\int\left(\frac{1}{4}x^{-4} - \frac{1}{4}x^{-5} + \frac{1}{8}x^{-6} + \cdots\right)dx$$

$$= \left[1 + 2x + 2x^2\right]\left[-\frac{1}{12}x^{-3} + \frac{1}{16}x^{-4} - \frac{1}{40}x^{-5} + \cdots\right]$$

$$= -\frac{1}{6}x^{-1} - \frac{1}{24}x^{-2} - \frac{1}{120}x^{-3} + \cdots .$$

Thus, a general solution to this differential equation is given by

$$y(x) = c_1 y_1(x) + c_2 y_2(x),$$

where

$$y_1(x) = 1 + 2x + 2x^2 \qquad \text{and} \qquad y_2(x) = -\frac{1}{6}x^{-1} - \frac{1}{24}x^{-2} - \frac{1}{120}x^{-3} + \cdots .$$

EXERCISES 8.8: Special Functions, page 517

1. For this problem, we see that $\gamma = \frac{1}{2}$, $\alpha + \beta + 1 = 4$, and $\alpha \times \beta = 2$. First we note that γ is not an integer. Next, by solving in the last two equations above simultaneously for α and β, we see that either $\alpha = 1$ and $\beta = 2$ or $\alpha = 2$ and $\beta = 1$. Therefore, by assuming that $\alpha = 1$ and $\beta = 2$, equations (10) and (17) on pages 509 and 510 of the text yield the two solutions

$$y_1(x) = F\left(1, 2; \frac{1}{2}; x\right) \qquad \text{and} \qquad y_2(x) = x^{1/2}F\left(\frac{3}{2}, \frac{5}{2}; \frac{3}{2}; x\right).$$

Therefore, a general solution for this differential equation is given by

$$y(x) = c_1 F\left(1, 2; \frac{1}{2}; x\right) + c_2 x^{1/2} F\left(\frac{3}{2}, \frac{5}{2}; \frac{3}{2}; x\right).$$

Notice that

$$F(\alpha,\beta;\gamma;x)=1+\sum_{n=0}^{\infty}\left\{\frac{(\alpha)_n(\beta)_n}{n!(\gamma)_n}\right\}x^n=1+\sum_{n=0}^{\infty}\left\{\frac{(\beta)_n(\alpha)_n}{n!(\gamma)_n}\right\}x^n=F(\beta,\alpha;\gamma;x)$$

Therefore, letting $\alpha=2$ and $\beta=1$ yields an equivalent form of the same solution given by

$$y(x)=c_1F\left(2,1;\frac{1}{2};x\right)+c_2x^{1/2}F\left(\frac{5}{2},\frac{3}{2};\frac{3}{2};x\right).$$

13. This equation can be written as

$$x^2y''+xy'+\left(x^2-\frac{1}{4}\right)y=0.$$

Thus, $\nu^2=\frac{1}{4}$ which implies that $\nu=\frac{1}{2}$. Since this is not an integer (even though 2ν is an integer), by the discussion on page 511 of the text, two linearly independent solutions to this problem are given by equations (25) and (26) also on page 511, that is

$$y_1(x)=J_{1/2}(x)=\sum_{n=0}^{\infty}\left(\frac{(-1)^n}{\left[n!\Gamma(\frac{3}{2}+n)\right]}\right)\left(\frac{x}{2}\right)^{2n+1/2},$$

$$y_2(x)=J_{-1/2}(x)=\sum_{n=0}^{\infty}\left(\frac{(-1)^n}{\left[n!\Gamma(\frac{1}{2}+n)\right]}\right)\left(\frac{x}{2}\right)^{2n-1/2}.$$

Therefore, a general solution to this differential equation is given by

$$y(x)=c_1J_{1/2}(x)+c_2J_{-1/2}(x).$$

15. In this problem $\nu=1$. Thus, one solution to this differential equation is given by

$$y_1(x)=J_1(x)=\sum_{n=0}^{\infty}\left(\frac{(-1)^n}{\left[n!\Gamma(2+n)\right]}\right)\left(\frac{x}{2}\right)^{2n+1}.$$

By the discussion on page 511 of the text, $J_{-1}(x)$ and $J_1(x)$ are linearly dependent. Thus, $J_{-1}(x)$ will not be a second linearly independent solution for this problem. But, a second linearly independent solution will be given by equation (30) on page 512 of the text with $m=1$. That is we have

$$y_2(x)=Y_1(x)=\lim_{\nu\to1}\frac{\cos(\nu\pi)J_\nu(x)-J_{-\nu}(\nu)}{\sin(\nu\pi)}.$$

Therefore, a general solution to this differential equation is given by

$$y(x)=c_1J_1(x)+c_2Y_1(x).$$

21. Let $y(x)=x^{\nu}J_{\nu}(x)$. Then, by equation (31) on page 512 of the text, we have

$$y'(x)=x^{\nu}J_{\nu-1}(x).$$

Therefore, we see that

$$y''(x)=D_x[y'(x)]=D_x[x^{\nu}J_{\nu-1}(x)]=D_x\{x\cdot[x^{\nu-1}J_{\nu-1}(x)]\}$$

$$=x^{\nu-1}J_{\nu-1}(x)+x\cdot D_x[x^{\nu-1}J_{\nu-1}(x)]=x^{\nu-1}J_{\nu-1}(x)+x^{\nu}J_{\nu-2}(x).$$

Notice that in order to take the last derivative above, we have again used equation (31) on page 512 of the text. By substituting these expressions into the left-hand side of the first differential equation given in the problem, we obtain

$$xy''+(1-2\nu)y'+xy=x[x^{\nu-1}J_{\nu-1}(x)+x^{\nu}J_{\nu-2}(x)]+(1-2\nu)x^{\nu}J_{\nu-1}(x)+x[x^{\nu}J_{\nu}(x)]$$

$$=x^{\nu}J_{\nu-1}(x)+x^{\nu+1}J_{\nu-2}(x)+x^{\nu}J_{\nu-1}(x)-2\nu x^{\nu}J_{\nu-1}(x)+x^{\nu+1}J_{\nu}(x). \tag{18}$$

Notice that by equation (33) on page 512 of the text, we have

$$J_{\nu}(x)=\frac{2(\nu-1)}{x}J_{\nu-1}(x)-J_{\nu-2}(x)$$

$$\Rightarrow \qquad x^{\nu+1}J_{\nu}(x)=2(\nu-1)x^{\nu}J_{\nu-1}(x)-x^{\nu+1}J_{\nu-2}(x).$$

Replacing $x^{\nu+1}J_{\nu}(x)$ in equation (18) with the above expression and simplifying yields

$$xy''+(1-2\nu)y'+xy=x^{\nu}J_{\nu-1}(x)+x^{\nu+1}J_{\nu-2}(x)+x^{\nu}J_{\nu-1}(x)$$

$$-2\nu x^{\nu}J_{\nu-1}(x)+2(\nu-1)x^{\nu}J_{\nu-1}(x)-x^{\nu+1}J_{\nu-2}(x)=0.$$

Therefore, $y(x)=x^{\nu}J_{\nu}(x)$ is a solution to this type of differential equation.

In order to find a solution to the differential equation $xy''-2y'+xy=0$, we observe that this equation is of the same type as the equation given above with

$$1-2\nu=-2 \qquad \Rightarrow \qquad \nu=\frac{3}{2}.$$

Thus, a solution to this equation will be

$$y(x)=x^{3/2}J_{3/2}(x)=x^{3/2}\sum_{n=0}^{\infty}\left[\frac{(-1)^n}{n!\Gamma(\frac{5}{2}+n)}\right]\left(\frac{x}{2}\right)^{2n+3/2}.$$

29. In Legendre polynomials, n is a fixed nonnegative integer. Thus, in the first such polynomial, n equals zero. Therefore, we see that $\left[\frac{n}{2}\right]=\left[\frac{1}{2}\right]=0$ and, by equation (43) on page 515 of the text, we have

$$P_0(x) = 2^{-0}\left(\frac{(-1)^0 0!}{0!\cdot 0!\cdot 0!}\right)x^0 = 1.$$

Similarly, we have

$$n = 1 \quad \Rightarrow \quad \left[\frac{1}{2}\right] = 0$$

$$\Rightarrow \quad P_1(x) = 2^{-1}\left(\frac{(-1)^0 2!}{1!\cdot 0!\cdot 1!}\right)x^1 = x,$$

$$n = 2 \quad \Rightarrow \quad \left[\frac{2}{2}\right] = 1$$

$$\Rightarrow \quad P_2(x) = 2^{-2}\left\{\left(\frac{(-1)^0 4!}{2!\cdot 0!\cdot 2!}\right)x^2 + \left(\frac{(-1)^1 2!}{1!\cdot 1!\cdot 0!}\right)x^0\right\} = \frac{3x^2 - 1}{2},$$

$$n = 3 \quad \Rightarrow \quad \left[\frac{3}{2}\right] = 1$$

$$\Rightarrow \quad P_3(x) = 2^{-3}\left\{\left(\frac{(-1)^0 6!}{3!\cdot 0!\cdot 3!}\right)x^3 + \left(\frac{(-1)^1 4!}{2!\cdot 1!\cdot 1!}\right)x^1\right\} = \frac{5x^3 - 3x}{2},$$

$$n = 4 \quad \Rightarrow \quad \left[\frac{4}{2}\right] = 2$$

$$\Rightarrow \quad P_4(x) = 2^{-4}\left\{\left(\frac{(-1)^0 8!}{4!\cdot 0!\cdot 4!}\right)x^4 + \left(\frac{(-1)^1 6!}{3!\cdot 1!\cdot 2!}\right)x^2 + \left(\frac{(-1)^2 4!}{2!\cdot 2!\cdot 0!}\right)x^0\right\} = \frac{35x^4 - 30x^2 + 3}{8}.$$

37. Since the Taylor series expansion of an analytic function $f(t)$ about $t = 0$ is given by

$$f(t) = \sum_{n=0}^{\infty} \frac{f^{(n)}(0)}{n!} t^n,$$

we see that $H_n(x)$ is just the nth derivative of $y(t) = \mathrm{e}^{2tx - t^2}$ with respect to t evaluated at the point $t = 0$ (treating x as a fixed parameter). Therefore, we have

$$y^{(0)}(t) = \mathrm{e}^{2tx - t^2}$$

$$\Rightarrow \quad H_0(x) = y(0) = \mathrm{e}^0 = 1,$$

$$y'(t) = (2x - 2t)e^{2tx-t^2}$$

$$\Rightarrow \qquad H_1(x) = y'(0) = 2xe^0 = 2x,$$

$$y''(t) = -2e^{2tx-t^2} + (2x - 2t)^2 e^{2tx-t^2}$$

$$\Rightarrow \qquad H_2(x) = y''(0) = 4x^2 - 2,$$

$$y'''(t) = \left[-2(2x-2t) + 2(2x-2t)(-2) + (2x-2t)^3\right]e^{2tx-t^2}$$

$$\Rightarrow \qquad H_3(x) = y'''(0) = 8x^3 - 12x.$$

39. To find the first four Laguerre polynomials, we need to find the first four derivatives of the function $y(x) = x^n e^{-x}$. Therefore, we have

$$y^{(0)}(x) = x^n e^{-x},$$

$$y'(x) = \left[nx^{n-1} - x^n\right]e^{-x},$$

$$y''(x) = \left[n(n-1)x^{n-2} - 2nx^{n-1} + x^n\right]e^{-x},$$

$$y'''(x) = \left[n(n-1)(n-2)x^{n-3} - 3n(n-1)x^{n-2} + 3nx^{n-1} - x^n\right]e^{-x}.$$

Substituting these expressions into Rodrigues's formula and plugging in the appropriate values of n yields

$$L_0(x) = \frac{e^x}{0!}x^0 e^{-x} = 1,$$

$$L_1(x) = \frac{e^x}{1!}\left[1x^{1-1} - x^1\right]e^{-x} = 1 - x,$$

$$L_2(x) = \frac{e^x}{2!}\left[2(2-1)x^{2-2} - 2\cdot 2x^{2-1} + x^2\right]e^{-x} = \frac{2 - 4x + x^2}{2},$$

$$L_3(x) = \frac{e^x}{3!}\left[3(3-1)(3-2)x^{3-3} - 3(3)(3-1)x^{3-2} + 3\cdot 3x^{3-1} - x^3\right]e^{-x} = \frac{6 - 18x + 9x^2 - x^3}{6}.$$

CHAPTER 9: Matrix Methods for Linear Systems

EXERCISES 9.1: Introduction, page 531

3. We start by expressing right-hand sides of all equations as dot products.

$$x+y+z=[1,1,1]\cdot[x,y,z],\quad 2z-x=[-1,0,2]\cdot[x,y,z],\quad 4y=[0,4,0]\cdot[x,y,z].$$

Thus, by definition of the product of a matrix and column vector, the matrix form is given by

$$\begin{bmatrix} x \\ y \\ z \end{bmatrix}' = \begin{bmatrix} 1 & 1 & 1 \\ -1 & 0 & 2 \\ 0 & 4 & 0 \end{bmatrix}\begin{bmatrix} x \\ y \\ z \end{bmatrix}.$$

7. First we have to express the second derivative, y'', as a first derivative in order to rewrite the equation as a first order system. Denoting y' by v we get

$$\begin{aligned} y' &= v, \\ mv' + bv + ky &= 0 \end{aligned} \qquad \text{or} \qquad \begin{aligned} y' &= v, \\ v' &= -\frac{k}{m}y - \frac{b}{m}v. \end{aligned}$$

Expressing the right-hand side of each equation as a dot product, we obtain

$$v=[0,1]\cdot[y,v],\qquad -\frac{k}{m}y-\frac{b}{m}v=\left[-\frac{k}{m},-\frac{b}{m}\right]\cdot[y,v].$$

Thus, the matrix form of the system is

$$\begin{bmatrix} y \\ v \end{bmatrix}' = \begin{bmatrix} 0 & 1 \\ -k/m & -b/m \end{bmatrix}\begin{bmatrix} y \\ v \end{bmatrix}.$$

11. Introducing the auxiliary variables

$$x_1 = x,\qquad x_2 = x',\qquad x_3 = y,\qquad x_4 = y',$$

we can rewrite the given system in normal form:

$$\begin{aligned} x_1' &= x_2 \\ x_3' &= x_4 \\ x_2' + 3x_1 + 2x_3 &= 0 \\ x_4' - 2x_1 &= 0 \end{aligned} \qquad \text{or} \qquad \begin{aligned} x_1' &= x_2 \\ x_2' &= -3x_1 - 2x_3 \\ x_3' &= x_4 \\ x_4' &= 2x_1. \end{aligned}$$

Since

$$x_2=[0,1,0,0]\cdot[x_1,x_2,x_3,x_4],\qquad -3x_1-2x_3=[-3,0,-2,0]\cdot[x_1,x_2,x_3,x_4],$$

$$x_4 = [0,0,0,1]\cdot[x_1, x_2, x_3, x_4], \qquad 2x_1 = [2,0,0,0]\cdot[x_1, x_2, x_3, x_4],$$

the matrix is given by

$$\begin{bmatrix} x_1 \\ x_2 \\ x_3 \\ x_4 \end{bmatrix}' = \begin{bmatrix} 0 & 1 & 0 & 0 \\ -3 & 0 & -2 & 0 \\ 0 & 0 & 0 & 1 \\ 2 & 0 & 0 & 0 \end{bmatrix} \begin{bmatrix} x_1 \\ x_2 \\ x_3 \\ x_4 \end{bmatrix}.$$

EXERCISES 9.2: Review 1: Linear Algebraic Equations, page 536

3. By subtracting 2 times the first equation from the second, we eliminate x_1 from the latter. Similarly, x_1 is eliminated from the third equation by subtracting the first equation from it. So we get

$$\begin{aligned} x_1 + 2x_2 + x_3 &= -3, \\ -3x_3 &= 6, \\ x_2 - 3x_3 &= 6 \end{aligned} \qquad \text{or (interchanging last two equations)} \qquad \begin{aligned} x_1 + 2x_2 + x_3 &= -3, \\ x_2 - 3x_3 &= 6, \\ x_3 &= -2. \end{aligned}$$

The second unknown, x_2, can be eliminated from the first equation by subtracting 2 times the first one from it:

$$\begin{aligned} x_1 \quad + 7x_3 &= -15, \\ x_2 - 3x_3 &= 6, \\ x_3 &= -2. \end{aligned}$$

Finally, we eliminate x_3 from the first two equations by adding (-7) times and 3 times, respectively, the third equation. This gives

$$\begin{aligned} x_1 &= -1, \\ x_2 &= 0, \\ x_3 &= -2. \end{aligned}$$

7. Subtracting 3 times the first equation from the second equation yields

$$\begin{aligned} -x_1 + 3x_2 &= 0, \\ 0 &= 0. \end{aligned}$$

The last equation is trivially satisfied, so we ignore it. Thus, just one equation remains:

$$-x_1 + 3x_2 = 0 \qquad \Rightarrow \qquad x_1 = 3x_2.$$

Choosing x_2 as a free variable, we get $x_1 = 3s$, $x_2 = s$, where s is any number.

9. We eliminate x_1 from the first equation by adding $(1-i)$ times the second equation to it:

$$\begin{aligned}\left[2-(1+i)(1-i)\right]x_2 &= 0, \\ -x_1-(1+i)\,x_2 &= 0.\end{aligned}$$

Since $(1-i)(1+i)=1^2-i^2=1-(-1)=2$, we obtain

$$\begin{aligned}0&=0, \\ -x_1-(1+i)\,x_2&=0\end{aligned} \qquad \Rightarrow \qquad x_2=-\frac{1}{1+i}x_1=\frac{-1+i}{2}x_1.$$

Assigning an arbitrary complex value to x_1, say $2s$, we see that the system has infinitely many solutions given by

$$x_1=2s, \qquad x_2=(-1+i)s, \qquad \text{where } s \text{ is any complex number.}$$

11. It is slightly more convenient to put the last equation at the top:

$$\begin{aligned}-x_1+x_2+5x_3&=0, \\ 2x_1+\phantom{x_2+{}}x_3&=-1, \\ -3x_1+x_2+4x_3&=1.\end{aligned}$$

We then eliminate x_1 from the second equation by adding 2 times the first one to it; and by subtracting 3 times the first equation from the third, we eliminate x_1 in the latter.

$$\begin{aligned}-x_1+x_2+5x_3&=0, \\ -2x_2-11x_3&=1, \\ 2x_2+11x_3&=-1.\end{aligned}$$

To make the computations more convenient, we multiply the first equation by 2.

$$\begin{aligned}-2x_1+2x_2+10x_3&=0, \\ -2x_2-11x_3&=1, \\ 2x_2+11x_3&=-1.\end{aligned}$$

Now we add the second equation to each of the remaining, and obtain

$$\begin{aligned}-2x_1-x_3&=1, \\ -2x_2-11x_3&=1, \\ 0&=0\end{aligned} \qquad \text{or} \qquad \begin{aligned}-2x_1-x_3&=1, \\ -2x_2-11x_3&=1.\end{aligned}$$

Choosing x_3 as free variable, i.e., $x_3=s$, yields $x_1=-(s+1)/2$, $x_2=-(11s+1)/2$, $-\infty<s<+\infty$.

13. The given system can be written in the equivalent form

$$(2-r)x_1 - 3x_2 = 0,$$
$$x_1 - (2+r)x_2 = 0.$$

The variable x_1 can be eliminated from the first equation by subtracting $(2-r)$ times the second equation:

$$[-3+(2-r)(2+r)]x_2 = 0, \qquad \qquad (1-r^2)x_2 = 0,$$
$$x_1 - (2+r)x_2 = 0 \qquad \text{or} \qquad x_1 - (2+r)x_2 = 0.$$

If $1-r^2 \neq 0$, i.e., $r \neq \pm 1$, then the first equation implies $x_2 = 0$. Substituting this into the second equation, we get $x_1 = 0$. Thus, the given system has a unique (zero) solution for any $r \neq \pm 1$, in particular, for $r = 2$.
If $r = 1$ or $r = -1$, then the first equation in the latter system becomes trivial $0 = 0$, and the system degenerates to

$$x_1 - (2+r)x_2 = 0 \qquad \Rightarrow \qquad x_1 = (2+r)x_2.$$

Therefore, there are infinitely many solutions to the given system of the form

$$x_1 = (2+r)s, \qquad x_2 = s, \qquad s \in (-\infty, +\infty), \quad r = \pm 1.$$

In particular, for $r = 1$ we obtain

$$x_1 = 3s, \qquad x_2 = s, \qquad s \in (-\infty, +\infty).$$

EXERCISES 9.3: Review 2: Matrices and Vectors, page 545

5. **(a)** $\mathbf{AB} = \begin{bmatrix} 1 & -2 \\ 2 & -3 \end{bmatrix} \begin{bmatrix} 1 & 0 \\ 1 & 1 \end{bmatrix} = \begin{bmatrix} 1-2 & 0-2 \\ 2-3 & 0-3 \end{bmatrix} = \begin{bmatrix} -1 & -2 \\ -1 & -3 \end{bmatrix}.$

(b) $\mathbf{AC} = \begin{bmatrix} 1 & -2 \\ 2 & -3 \end{bmatrix} \begin{bmatrix} -1 & 1 \\ 2 & 1 \end{bmatrix} = \begin{bmatrix} -1-4 & 1-2 \\ -2-6 & 2-3 \end{bmatrix} = \begin{bmatrix} -5 & -1 \\ -8 & -1 \end{bmatrix}.$

(c) By the Distributive Property of matrix multiplication given on page 539 of the text, we have

$$\mathbf{A}(\mathbf{B}+\mathbf{C}) = \mathbf{AB} + \mathbf{AC} = \begin{bmatrix} -1 & -2 \\ -1 & -3 \end{bmatrix} + \begin{bmatrix} -5 & -1 \\ -8 & -1 \end{bmatrix} = \begin{bmatrix} -6 & -3 \\ -9 & -4 \end{bmatrix}.$$

13. Authors note: We will use $R_i + cR_j \to R_k$ to denote the row operation "*add row i to c times row j and place the result into row k.*" We will use $cR_j \to R_k$ to denote the row operation "*multiply row j by c and place the result into row k.*"

As in Example 1 on page 541 of the text, we will perform row-reduction on the matrix $[\mathbf{A}\,|\,\mathbf{I}]$. Thus, we have

$$[\mathbf{A}\,|\,\mathbf{I}] = \left[\begin{array}{ccc|ccc} -2 & -1 & 1 & 1 & 0 & 0 \\ 2 & 1 & 0 & 0 & 1 & 0 \\ 3 & 1 & -1 & 0 & 0 & 1 \end{array}\right]$$

$$\begin{array}{l} \\ R_2 + R_1 \to R_2 \\ 2R_3 + 3R_1 \to R_3 \end{array} \quad \left[\begin{array}{ccc|ccc} -2 & -1 & 1 & 1 & 0 & 0 \\ 0 & 0 & 1 & 1 & 1 & 0 \\ 0 & -1 & 1 & 3 & 0 & 2 \end{array}\right]$$

$$R_1 - R_3 \to R_1 \quad \left[\begin{array}{ccc|ccc} -2 & 0 & 0 & -2 & 0 & -2 \\ 0 & 0 & 1 & 1 & 1 & 0 \\ 0 & -1 & 1 & 3 & 0 & 2 \end{array}\right]$$

$$\begin{array}{l} -\frac{1}{2}R_1 \to R_1 \\ R_3 \to R_2 \\ R_2 \to R_3 \end{array} \quad \left[\begin{array}{ccc|ccc} 1 & 0 & 0 & 1 & 0 & 1 \\ 0 & -1 & 1 & 3 & 0 & 2 \\ 0 & 0 & 1 & 1 & 1 & 0 \end{array}\right]$$

$$-R_2 + R_3 \to R_2 \quad \left[\begin{array}{ccc|ccc} 1 & 0 & 0 & 1 & 0 & 1 \\ 0 & 1 & 0 & -2 & 1 & -2 \\ 0 & 0 & 1 & 1 & 1 & 0 \end{array}\right]$$

Therefore, the inverse matrix $\mathbf{A}^{-1}$ is $\begin{bmatrix} 1 & 0 & 1 \\ -2 & 1 & -2 \\ 1 & 1 & 0 \end{bmatrix}$.

To check the algebra, it's a good idea to multiply $\mathbf{A}$ by $\mathbf{A}^{-1}$ to verify that the product is the identity matrix.

19. Authors note: We will use $R_i + cR_j \to R_k$ to denote the row operation "*add row i to c times row j and place the result into row k.*" We will use $cR_j \to R_k$ to denote the row operation "*multiply row j by c and place the result into row k.*"

To find the inverse matrix $\mathbf{X}^{-1}(t)$, we will again use the method of Example 1 on page 541 of the text. Thus, we start with

$$[\mathbf{X}(t)\,|\,\mathbf{I}] = \left[\begin{array}{ccc|ccc} e^t & e^{-t} & e^{2t} & 1 & 0 & 0 \\ e^t & -e^{-t} & 2e^{2t} & 0 & 1 & 0 \\ e^t & e^{-t} & 4e^{2t} & 0 & 0 & 1 \end{array}\right]$$

$$\begin{array}{l} R_2 - R_1 \to R_2 \\ R_3 - R_1 \to R_3 \end{array} \quad \left[\begin{array}{ccc|ccc} e^t & e^{-t} & e^{2t} & 1 & 0 & 0 \\ 0 & -2e^{-t} & e^{2t} & -1 & 1 & 0 \\ 0 & 0 & 3e^{2t} & -1 & 0 & 1 \end{array}\right]$$

$$\begin{array}{l} -\frac{1}{2}R_2 \to R_2 \\ \frac{1}{3}R_3 \to R_3 \end{array} \quad \left[\begin{array}{ccc|ccc} e^t & e^{-t} & e^{2t} & 1 & 0 & 0 \\ 0 & e^{-t} & -\frac{1}{2}e^{2t} & \frac{1}{2} & -\frac{1}{2} & 0 \\ 0 & 0 & e^{2t} & -\frac{1}{3} & 0 & \frac{1}{3} \end{array}\right]$$

$$\begin{array}{l} R_1 - R_3 \to R_1 \\ R_2 - \frac{1}{2}R_3 \to R_2 \end{array} \quad \left[\begin{array}{ccc|ccc} e^t & e^{-t} & 0 & \frac{4}{3} & 0 & -\frac{1}{3} \\ 0 & c^{-t} & 0 & \frac{1}{3} & -\frac{1}{2} & \frac{1}{6} \\ 0 & 0 & e^{2t} & -\frac{1}{3} & 0 & \frac{1}{3} \end{array}\right]$$

$$R_1 - R_2 \to R_1 \quad \left[\begin{array}{ccc|ccc} e^t & 0 & 0 & 1 & \frac{1}{2} & -\frac{1}{2} \\ 0 & e^{-t} & 0 & \frac{1}{3} & -\frac{1}{2} & \frac{1}{6} \\ 0 & 0 & e^{2t} & -\frac{1}{3} & 0 & \frac{1}{3} \end{array}\right]$$

$$\begin{array}{l} e^{-t}R_1 \to R_1 \\ e^{t}R_2 \to R_2 \\ e^{-2t}R_3 \to R_3 \end{array} \quad \left[\begin{array}{ccc|ccc} 1 & 0 & 0 & e^{-t} & \frac{1}{2}e^{-t} & -\frac{1}{2}e^{-t} \\ 0 & 1 & 0 & \frac{1}{3}e^{t} & -\frac{1}{2}e^{t} & \frac{1}{6}e^{t} \\ 0 & 0 & 1 & -\frac{1}{3}e^{-2t} & 0 & \frac{1}{3}e^{-2t} \end{array}\right].$$

Thus, the inverse matrix $\mathbf{X}^{-1}(t)$ is given by the matrix

$$\begin{bmatrix} e^{-t} & \frac{1}{2}e^{-t} & -\frac{1}{2}e^{-t} \\ \frac{1}{3}e^{t} & -\frac{1}{2}e^{t} & \frac{1}{6}e^{t} \\ -\frac{1}{3}e^{-2t} & 0 & \frac{1}{3}e^{-2t} \end{bmatrix}.$$

23. We will calculate this determinant by first finding its cofactor expansion about row 1. Therefore, we have

$$\begin{vmatrix} 1 & 0 & 0 \\ 3 & 1 & 2 \\ 1 & 5 & -2 \end{vmatrix} = 1\begin{vmatrix} 1 & 2 \\ 5 & -2 \end{vmatrix} - 0 + 0 = -2 - 10 = -12.$$

37. We first calculate $\mathbf{X}'(t)$ by differentiating each entry of $\mathbf{X}(t)$. Therefore, we have

$$\mathbf{X}'(t) = \begin{bmatrix} 2e^{2t} & 3e^{3t} \\ -2e^{2t} & -6e^{3t} \end{bmatrix}.$$

Thus, substituting the matrix $\mathbf{X}(t)$ into the differential equation and performing matrix multiplication yields

$$\begin{bmatrix} 2e^{2t} & 3e^{3t} \\ -2e^{2t} & -6e^{3t} \end{bmatrix} = \begin{bmatrix} 1 & -1 \\ 2 & 4 \end{bmatrix}\begin{bmatrix} e^{2t} & e^{3t} \\ -e^{2t} & -2e^{3t} \end{bmatrix} = \begin{bmatrix} e^{2t} + e^{2t} & e^{3t} + 2e^{3t} \\ 2e^{2t} - 4e^{2t} & 2e^{3t} - 8e^{3t} \end{bmatrix}.$$

Since this equation is true, we see that $\mathbf{X}(t)$ does satisfy the given differential equation.

39. **(a)** To calculate $\int \mathbf{A}(t)\,dt$, we integrate each entry of $\mathbf{A}(t)$ to obtain

$$\int \mathbf{A}(t)dt = \begin{bmatrix} \int t\,dt & \int e^t\,dt \\ \int 1\,dt & \int e^t\,dt \end{bmatrix} = \begin{bmatrix} \frac{1}{2}t^2 + c_1 & e^t + c_2 \\ t + c_3 & e^t + c_4 \end{bmatrix}.$$

(b) Taking the definite integral of each entry of $\mathbf{B}(t)$ yields

$$\int_0^1 \mathbf{B}(t)dt = \begin{bmatrix} \int_0^1 \cos t\,dt & -\int_0^1 \sin t\,dt \\ \int_0^1 \sin t\,dt & \int_0^1 \cos t\,dt \end{bmatrix} = \begin{bmatrix} \sin t\big|_0^1 & \cos t\big|_0^1 \\ -\cos t\big|_0^1 & \sin t\big|_0^1 \end{bmatrix} = \begin{bmatrix} \sin 1 & \cos 1 - 1 \\ 1 - \cos 1 & \sin 1 \end{bmatrix}.$$

(c) By the product rule on page 545 of the text, we see that

$$\frac{d}{dt}(\mathbf{A}(t)\,\mathbf{B}(t)) = \mathbf{A}(t)\,\mathbf{B}'(t) + \mathbf{A}'(t)\,\mathbf{B}(t).$$

Therefore, we first calculate $\mathbf{A}'(t)$ and $\mathbf{B}'(t)$ by differentiating each entry of $\mathbf{A}(t)$ and $\mathbf{B}(t)$, respectively to obtain

$$\mathbf{A}'(t)=\begin{bmatrix}1 & e^t\\ 0 & e^t\end{bmatrix} \quad \text{and} \quad \mathbf{B}'(t)=\begin{bmatrix}-\sin t & -\cos t\\ \cos t & -\sin t\end{bmatrix}.$$

Hence, by matrix multiplication we have

$$\frac{d}{dt}\left(\mathbf{A}(t)\,\mathbf{B}(t)\right) = \mathbf{A}(t)\,\mathbf{B}'(t)+\mathbf{A}'(t)\,\mathbf{B}(t)$$

$$=\begin{bmatrix}t & e^t\\ 1 & e^t\end{bmatrix}\begin{bmatrix}-\sin t & -\cos t\\ \cos t & -\sin t\end{bmatrix}+\begin{bmatrix}1 & e^t\\ 0 & e^t\end{bmatrix}\begin{bmatrix}\cos t & -\sin t\\ \sin t & \cos t\end{bmatrix}$$

$$=\begin{bmatrix}e^t\cos t-t\sin t & -t\cos t-e^t\sin t\\ e^t\cos t-\sin t & -\cos t-e^t\sin t\end{bmatrix}+\begin{bmatrix}\cos t+e^t\sin t & e^t\cos t-\sin t\\ e^t\sin t & e^t\cos t\end{bmatrix}$$

$$=\begin{bmatrix}(1+e^t)\cos t+(e^t-t)\sin t & (e^t-t)\cos t-(e^t+1)\sin t\\ e^t\cos t+(e^t-1)\sin t & (e^t-1)\cos t-e^t\sin t\end{bmatrix}.$$

EXERCISES 9.4: Linear Systems in Normal Form, page 554

1. To write this system in matrix form, we will define the vectors $\mathbf{x}(t)=\mathrm{col}[x(t),y(t)]$ (which means that $\mathbf{x}'(t)=\mathrm{col}[x'(t),y'(t)]$) and $f(t)=\mathrm{col}[t^2,e^t]$, and the matrix

$$\mathbf{A}(t)=\begin{bmatrix}3 & -1\\ -1 & 2\end{bmatrix}.$$

Thus, this system becomes the equation in matrix form given by

$$\begin{bmatrix}x'(t)\\ y'(t)\end{bmatrix}=\begin{bmatrix}3 & -1\\ -1 & 2\end{bmatrix}\begin{bmatrix}x(t)\\ y(t)\end{bmatrix}+\begin{bmatrix}t^2\\ e^t\end{bmatrix}.$$

We can see that this equation is equivalent to the original system by performing matrix multiplication and addition to obtain the vector equation

$$\begin{bmatrix}x'(t)\\ y'(t)\end{bmatrix}=\begin{bmatrix}3x(t)-y(t)\\ -x(t)+2y(t)\end{bmatrix}+\begin{bmatrix}t^2\\ e^t\end{bmatrix}$$

$$\Rightarrow \quad \begin{bmatrix}x'(t)\\ y'(t)\end{bmatrix}=\begin{bmatrix}3x(t)-y(t)+t^2\\ -x(t)+2y(t)+e^t\end{bmatrix}.$$

Since two vectors are equal only when their corresponding components are equal, we see that this vector equation implies that

$$x'(t) = 3x(t) - y(t) + t^2,$$
$$y'(t) = -x(t) + 2y(t) + e^t,$$

which is the original system.

5. This equation can be written as a first order system in normal form by using the substitutions $x_1(t) = y(t)$ and $x_2(t) = y'(t)$. With these substitutions this differential equation becomes the system

$$x_1'(t) = 0 \cdot x_1(t) + x_2(t),$$

$$x_2'(t) = 10x_1(t) + 3x_2(t) + \sin t .$$

We can then write this system as a matrix differential equation by defining the vectors $\mathbf{x}(t) = \text{col}[x_1(t), x_2(t)]$ (which means that $\mathbf{x}'(t) = \text{col}[x_1'(t), x_2'(t)]$), $f(t) = \text{col}[0, \sin t]$, and the matrix

$$\mathbf{A} = \begin{bmatrix} 0 & 1 \\ 10 & 3 \end{bmatrix}.$$

Hence, the system above in normal form becomes the differential equation given in matrix form by

$$\begin{bmatrix} x_1'(t) \\ x_2'(t) \end{bmatrix} = \begin{bmatrix} 0 & 1 \\ 10 & 3 \end{bmatrix} \begin{bmatrix} x_1(t) \\ x_2(t) \end{bmatrix} + \begin{bmatrix} 0 \\ \sin t \end{bmatrix}.$$

(As in Problem 1 above, we can see that this equation in matrix form is equivalent to the system by performing matrix multiplication and addition and then noting that corresponding components of equal vectors are equal.)

17. Notice that by scalar multiplication these vector functions can be written as

$$\begin{bmatrix} e^{2t} \\ 0 \\ 5e^{2t} \end{bmatrix}, \quad \begin{bmatrix} e^{2t} \\ e^{2t} \\ -e^{2t} \end{bmatrix}, \quad \begin{bmatrix} 0 \\ e^{3t} \\ 0 \end{bmatrix}.$$

Thus, as in Example 2 on page 550 of the text, we will prove that these vectors are linearly independent by showing that the only way that we can have

$$c_1 \begin{bmatrix} e^{2t} \\ 0 \\ 5e^{2t} \end{bmatrix} + c_2 \begin{bmatrix} e^{2t} \\ e^{2t} \\ -e^{2t} \end{bmatrix} + c_3 \begin{bmatrix} 0 \\ e^{3t} \\ 0 \end{bmatrix} = 0$$

for all t in $(-\infty, \infty)$ is for $c_1 = c_2 = c_3 = 0$. Since the equation above must be true for all t, it must be true for $t = 0$. Thus, c_1, c_2, and c_3 must satisfy

$$c_1 \begin{bmatrix} 1 \\ 0 \\ 5 \end{bmatrix} + c_2 \begin{bmatrix} 1 \\ 1 \\ -1 \end{bmatrix} + c_3 \begin{bmatrix} 0 \\ 1 \\ 0 \end{bmatrix} = 0 ,$$

which is equivalent to the system

$$\begin{aligned} c_1 + c_2 &= 0, \\ c_2 + c_3 &= 0, \\ 5c_1 - c_2 &= 0. \end{aligned}$$

By solving the first and last of these equations simultaneously, we see that $c_1 = c_2 = 0$. Substituting these values into the second equation above yields $c_3 = 0$. Therefore, the original set of vectors must be linearly independent on the interval $(-\infty, \infty)$.

21. Since it is given that these vectors are solutions to the system $\mathbf{x}'(t) = \mathbf{A}\mathbf{x}(t)$, in order to determine whether they are linearly independent, we need only calculate their Wronskian. If their Wronskian is never zero, then these vectors are linearly independent and so form a fundamental solution set. If the Wronskian is identically zero, then the vectors are linearly dependent, and they do not form a fundamental solution set. Thus, we observe

$$W[\mathbf{x}_1, \mathbf{x}_2, \mathbf{x}_3](t) = \begin{vmatrix} e^{-t} & e^{t} & e^{3t} \\ 2e^{-t} & 0 & -e^{3t} \\ e^{-t} & e^{t} & 2e^{3t} \end{vmatrix}$$

$$= e^{-t} \begin{vmatrix} 0 & -e^{3t} \\ e^{t} & 2e^{3t} \end{vmatrix} - e^{t} \begin{vmatrix} 2e^{-t} & -e^{3t} \\ e^{-t} & 2e^{3t} \end{vmatrix} + e^{3t} \begin{vmatrix} 2e^{-t} & 0 \\ e^{-t} & e^{t} \end{vmatrix}$$

$$= e^{-t}\left(0 + e^{4t}\right) - e^{t}\left(4e^{2t} + e^{2t}\right) + e^{3t}(2 - 0) = -2e^{3t} \neq 0,$$

where we have used cofactors to calculate the determinant. Therefore, this set of vectors is linearly independent and so forms a fundamental solution set for the system. Thus, a fundamental matrix is given by

$$\mathbf{X}(t) = \begin{bmatrix} e^{-t} & e^{t} & e^{3t} \\ 2e^{-t} & 0 & -e^{3t} \\ e^{-t} & e^{t} & 2e^{3t} \end{bmatrix},$$

and a general solution of the system will be

$$\mathbf{x}(t)=\mathbf{X}(t)\mathbf{c}=c_1\begin{bmatrix} e^{-t} \\ 2e^{-t} \\ e^{-t} \end{bmatrix}+c_2\begin{bmatrix} e^{t} \\ 0 \\ e^{t} \end{bmatrix}+c_3\begin{bmatrix} e^{3t} \\ -e^{3t} \\ 2e^{3t} \end{bmatrix}.$$

27. In order to show that $\mathbf{X}(t)$ is a fundamental matrix for the system, we must first show that each of its column vectors is a solution. Thus, we substitute each of the vectors

$$\mathbf{x}_1(t)=\begin{bmatrix} 6e^{-t} \\ -e^{-t} \\ -5e^{-t} \end{bmatrix},\quad \mathbf{x}_2(t)=\begin{bmatrix} -3e^{-2t} \\ e^{-2t} \\ e^{-2t} \end{bmatrix},\quad \mathbf{x}_3(t)=\begin{bmatrix} 2e^{3t} \\ e^{3t} \\ e^{3t} \end{bmatrix},$$

into the given system to obtain

$$\mathbf{A}\mathbf{x}_1=\begin{bmatrix} 0 & 6 & 0 \\ 1 & 0 & 1 \\ 1 & 1 & 0 \end{bmatrix}\begin{bmatrix} 6e^{-t} \\ -e^{-t} \\ -5e^{-t} \end{bmatrix}=\begin{bmatrix} -6e^{-t} \\ e^{-t} \\ 5e^{-t} \end{bmatrix}=\mathbf{x}_1'(t),$$

$$\mathbf{A}\mathbf{x}_2=\begin{bmatrix} 0 & 6 & 0 \\ 1 & 0 & 1 \\ 1 & 1 & 0 \end{bmatrix}\begin{bmatrix} -3e^{-2t} \\ e^{-2t} \\ e^{-2t} \end{bmatrix}=\begin{bmatrix} 6e^{-2t} \\ -2e^{-2t} \\ -2e^{-2t} \end{bmatrix}=\mathbf{x}_2'(t),$$

$$\mathbf{A}\mathbf{x}_3=\begin{bmatrix} 0 & 6 & 0 \\ 1 & 0 & 1 \\ 1 & 1 & 0 \end{bmatrix}\begin{bmatrix} 2e^{3t} \\ e^{3t} \\ e^{3t} \end{bmatrix}=\begin{bmatrix} 6e^{3t} \\ 3e^{3t} \\ 3e^{3t} \end{bmatrix}=\mathbf{x}_3'(t).$$

Therefore, each column vector of $\mathbf{X}(t)$ is a solution to the system on $(-\infty, \infty)$.

Next we must show that these vectors are linearly independent. Since they are solutions to a differential equation in matrix form, it is enough to show that their Wronskian is never zero. Thus, we find

$$W(t)=\begin{vmatrix} 6e^{-t} & -3e^{-2t} & 2e^{3t} \\ -e^{-t} & e^{-2t} & e^{3t} \\ -5e^{-t} & e^{-2t} & e^{3t} \end{vmatrix}$$

$$=6e^{-t}\begin{vmatrix} e^{-2t} & e^{3t} \\ e^{-2t} & e^{3t} \end{vmatrix}+3e^{-2t}\begin{vmatrix} -e^{-t} & e^{3t} \\ -5e^{-t} & e^{3t} \end{vmatrix}+2e^{3t}\begin{vmatrix} -e^{-t} & e^{-2t} \\ -5e^{-t} & e^{-2t} \end{vmatrix}$$

$$=6e^{-t}(e^{t}-e^{t})+3e^{-2t}(-e^{2t}+5e^{2t})+2e^{3t}(-e^{-3t}+5e^{-3t})=20\neq 0,$$

where we have used cofactors to calculate the determinant. Hence, these three vectors are linearly independent. Therefore, $\mathbf{X}(t)$ is a fundamental matrix for this system.

We will now find the inverse of the matrix $\mathbf{X}(t)$ by performing row-reduction on the matrix $[\mathbf{X}(t)|\mathbf{I}]$. Thus, we have

$$[\mathbf{X}(t)\,|\,\mathbf{I}] = \left[\begin{array}{ccc|ccc} 6e^{-t} & -3e^{-2t} & 2e^{3t} & 1 & 0 & 0 \\ -e^{-t} & e^{-2t} & e^{3t} & 0 & 1 & 0 \\ -5e^{-t} & e^{-2t} & e^{3t} & 0 & 0 & 1 \end{array}\right]$$

$$\begin{array}{l} -R_2 \to R_1 \\ R_1 \to R_2 \end{array} \quad \left[\begin{array}{ccc|ccc} e^{-t} & -e^{-2t} & -e^{3t} & 0 & -1 & 0 \\ 6e^{-t} & -3e^{-2t} & 2e^{3t} & 1 & 0 & 0 \\ -5e^{-t} & e^{-2t} & e^{3t} & 0 & 0 & 1 \end{array}\right]$$

$$\begin{array}{l} R_2 - 6R_1 \to R_2 \\ R_3 + 5R_1 \to R_3 \end{array} \quad \left[\begin{array}{ccc|ccc} e^{-t} & -e^{-2t} & -e^{3t} & 0 & -1 & 0 \\ 0 & 3e^{-2t} & 8e^{3t} & 1 & 6 & 0 \\ 0 & -4e^{-2t} & -4e^{3t} & 0 & -5 & 1 \end{array}\right]$$

$$\begin{array}{l} -\frac{1}{4}R_3 \to R_2 \\ R_2 \to R_3 \end{array} \quad \left[\begin{array}{ccc|ccc} e^{-t} & -e^{-2t} & -e^{3t} & 0 & -1 & 0 \\ 0 & e^{-2t} & e^{3t} & 0 & \frac{5}{4} & -\frac{1}{4} \\ 0 & 3e^{-2t} & 8e^{3t} & 1 & 6 & 0 \end{array}\right]$$

$$\begin{array}{l} R_1 + R_2 \to R_1 \\ R_3 - 3R_2 \to R_3 \end{array} \quad \left[\begin{array}{ccc|ccc} e^{-t} & 0 & 0 & 0 & \frac{1}{4} & -\frac{1}{4} \\ 0 & e^{-2t} & e^{3t} & 0 & \frac{5}{4} & -\frac{1}{4} \\ 0 & 0 & 5e^{3t} & 1 & \frac{9}{4} & \frac{3}{4} \end{array}\right]$$

$$\frac{1}{5}R_3 \to R_3 \quad \left[\begin{array}{ccc|ccc} e^{-t} & 0 & 0 & 0 & \frac{1}{4} & -\frac{1}{4} \\ 0 & e^{-2t} & e^{3t} & 0 & \frac{5}{4} & -\frac{1}{4} \\ 0 & 0 & e^{3t} & \frac{1}{5} & \frac{9}{20} & \frac{3}{20} \end{array}\right]$$

$$R_2 - R_3 \to R_2 \quad \left[\begin{array}{ccc|ccc} e^{-t} & 0 & 0 & 0 & \frac{1}{4} & -\frac{1}{4} \\ 0 & e^{-2t} & 0 & -\frac{1}{5} & \frac{4}{5} & -\frac{2}{5} \\ 0 & 0 & e^{3t} & \frac{1}{5} & \frac{9}{20} & \frac{3}{20} \end{array}\right]$$

$$\begin{matrix} e^{t} R_1 \to R_1 \\ e^{2t} R_2 \to R_2 \\ e^{-3t} R_3 \to R_3 \end{matrix} \left[\begin{array}{ccc|ccc} 1 & 0 & 0 & 0 & \frac{1}{4}e^{t} & -\frac{1}{4}e^{t} \\ 0 & 1 & 0 & -\frac{1}{5}e^{2t} & \frac{4}{5}e^{2t} & -\frac{2}{5}e^{2t} \\ 0 & 0 & 1 & \frac{1}{5}e^{-3t} & \frac{9}{20}e^{-3t} & \frac{3}{20}e^{-3t} \end{array}\right].$$

Therefore, we see that

$$\mathbf{X}^{-1}(t) = \begin{bmatrix} 0 & \frac{1}{4}e^{t} & -\frac{1}{4}e^{t} \\ -\frac{1}{5}e^{2t} & \frac{4}{5}e^{2t} & -\frac{2}{5}e^{2t} \\ \frac{1}{5}e^{-3t} & \frac{9}{20}e^{-3t} & \frac{3}{20}e^{-3t} \end{bmatrix}.$$

We now can use Problem 26 to find the solution to this differential equation for *any* initial value. For the initial value given here we note that $t_0 = 0$. Thus, substituting $t_0 = 0$ into the matrix $\mathbf{X}^{-1}(t)$ above yields

$$\mathbf{X}^{-1}(0) = \begin{bmatrix} 0 & \frac{1}{4} & -\frac{1}{4} \\ -\frac{1}{5} & \frac{4}{5} & -\frac{2}{5} \\ \frac{1}{5} & \frac{9}{20} & \frac{3}{20} \end{bmatrix}.$$

Hence, we see that the solution to this problem is given by

$$\mathbf{x}(t) = \mathbf{X}(t)\mathbf{X}^{-1}(t)\mathbf{x}(0)$$

$$= \begin{bmatrix} 6e^{-t} & -3e^{-2t} & 2e^{3t} \\ -e^{-t} & e^{-2t} & e^{3t} \\ -5e^{-t} & e^{-2t} & e^{3t} \end{bmatrix} \begin{bmatrix} 0 & \frac{1}{4} & -\frac{1}{4} \\ -\frac{1}{5} & \frac{4}{5} & -\frac{2}{5} \\ \frac{1}{5} & \frac{9}{20} & \frac{3}{20} \end{bmatrix} \begin{bmatrix} -1 \\ 0 \\ 1 \end{bmatrix}$$

$$= \begin{bmatrix} 6e^{-t} & -3e^{-2t} & 2e^{3t} \\ -e^{-t} & e^{-2t} & e^{3t} \\ -5e^{-t} & e^{-2t} & e^{3t} \end{bmatrix} \begin{bmatrix} -\frac{1}{4} \\ -\frac{1}{5} \\ -\frac{1}{20} \end{bmatrix}$$

$$= \begin{bmatrix} -\frac{3}{2}e^{-t} + \frac{3}{5}e^{-2t} - \frac{1}{10}e^{3t} \\ \frac{1}{4}e^{-t} - \frac{1}{5}e^{-2t} - \frac{1}{20}e^{3t} \\ \frac{5}{4}e^{-t} - \frac{1}{5}e^{-2t} - \frac{1}{20}e^{3t} \end{bmatrix}.$$

There are two short cuts that can be taken to solve the given problem. First, since we only need $\mathbf{X}^{-1}(0)$, it suffices to compute the inverse of $\mathbf{X}(0)$, not $\mathbf{X}(t)$. Second, by producing $\mathbf{X}^{-1}(t)$ we automatically know that $\det \mathbf{X}(0) \neq 0$ and hence $\mathbf{X}(t)$ is a fundamental matrix. Thus, it was not really necessary to compute the Wronskian.

33. Let $\phi(t)$ be an arbitrary solution to the system $\mathbf{x}'(t) = \mathbf{A}(t)\mathbf{x}(t)$ on the interval I. We want to find $\mathbf{c} = \mathrm{col}(c_1, c_2, \ldots, c_n)$ so that

$$\phi(t) = c_1\mathbf{x}_1(t) + c_2\mathbf{x}_2(t) + \cdots + c_n\mathbf{x}_n(t),$$

where $\mathbf{x}_1, \mathbf{x}_2, \ldots, \mathbf{x}_n$ are n linearly independent solutions for this system. Since $c_1\mathbf{x}_1(t) + c_2\mathbf{x}_2(t) + \ldots + c_n\mathbf{x}_n(t) = \mathbf{X}(t)\,\mathbf{c}$, where $\mathbf{X}(t)$ is the fundamental matrix whose columns are the vectors $\mathbf{x}_1, \mathbf{x}_2, \ldots, \mathbf{x}_n$, this equation can be written as

$$\phi(t) = \mathbf{X}(t)\,\mathbf{c}. \qquad (1)$$

Since $\mathbf{x}_1, \mathbf{x}_2, \ldots, \mathbf{x}_n$ are linearly independent solutions of the system $\mathbf{x}'(t) = \mathbf{A}(t)\mathbf{x}(t)$, their Wronskian is never zero. Therefore, as was discussed on page 551 of the text, $\mathbf{X}(t)$ has an inverse at each point in I. Thus, at t_0, a point in I, $\mathbf{X}^{-1}(t_0)$ exists and equation (1) becomes

$$\phi(t_0) = \mathbf{X}(t_0)\mathbf{c}$$

$$\Rightarrow \quad \mathbf{X}^{-1}(t_0)\phi(t_0) = \mathbf{X}^{-1}(t_0)\mathbf{X}(t_0)\mathbf{c} = \mathbf{c}.$$

Hence, if we define $\mathbf{c}_0$ to be the vector $\mathbf{c}_0 = \mathbf{X}^{-1}(t_0)\phi(t_0)$, then equation (1) is true at the point t_0 (i. e. $\phi(t_0) = \mathbf{X}(t_0)\mathbf{X}^{-1}(t_0)\phi(t_0)$). To see that, for this definition of $\mathbf{c}_0$, equation (1) is true for all t in I (and so this is the vector that we seek), notice that $\phi(t)$ and $\mathbf{X}(t)\mathbf{c}_0$ are both solutions to same initial value problem (with the initial value given at the point t_0). Therefore, by the uniqueness of solutions, Theorem 2 on page 549 of the text, these solutions must be equal on I, which means that $\phi(t) = \mathbf{X}(t)\mathbf{c}_0$ for all t in I.

EXERCISES 9.5: Homogeneous Linear Systems with Constant Coefficients, page 565

5. The characteristic equation for this matrix is given by

$$|\mathbf{A}-r\mathbf{I}|=\begin{vmatrix}1-r & 0 & 0\\ 0 & -r & 2\\ 0 & 2 & -r\end{vmatrix}=(1-r)\begin{vmatrix}-r & 2\\ 2 & -r\end{vmatrix}$$

$$=(1-r)(r^2-4)=(1-r)(r-2)(r+2)=0.$$

Thus, the eigenvalues of this matrix are r = 1, 2, –2. Substituting the eigenvalue $r=1$, into equation $(\mathbf{A}-r\mathbf{I})\mathbf{u}=\mathbf{0}$ yields

$$(\mathbf{A}-\mathbf{I})\mathbf{u}=\begin{bmatrix}0 & 0 & 0\\ 0 & -1 & 2\\ 0 & 2 & -1\end{bmatrix}\begin{bmatrix}u_1\\ u_2\\ u_3\end{bmatrix}=\begin{bmatrix}0\\ 0\\ 0\end{bmatrix}, \qquad (2)$$

which is equivalent to the system

$$-u_2+2u_3=0,$$

$$2u_2-u_3=0.$$

This system reduces to the system $u_2=0$, $u_3=0$, which does not assign any value to u_1. Thus, we can let u_1 be any value, say $u_1=s$, and $u_2=0$, $u_3=0$ and the system given by (2) will be satisfied. From this we see that the eigenvectors associated with the eigenvalue $r=1$ are given by

$$\mathbf{u}_1=\operatorname{col}(u_1,u_2,u_3)=\operatorname{col}(s,0,0)=s\operatorname{col}(1,0,0).$$

For $r=2$ we observe that the equation $(\mathbf{A}-r\mathbf{I})\mathbf{u}=\mathbf{0}$ becomes

$$(\mathbf{A}-2\mathbf{I})\mathbf{u}=\begin{bmatrix}-1 & 0 & 0\\ 0 & -2 & 2\\ 0 & 2 & -2\end{bmatrix}\begin{bmatrix}u_1\\ u_2\\ u_3\end{bmatrix}=\begin{bmatrix}0\\ 0\\ 0\end{bmatrix},$$

whose corresponding system of equations reduces to $u_1=0$, $u_2=u_3$. Therefore, we can pick u_2 to be any value, say $u_2=s$ (which means that $u_3=s$), and we find that the eigenvectors for this matrix associated with the eigenvalue $r=2$ are given by

$$\mathbf{u}_2=\operatorname{col}(u_1,u_2,u_3)=\operatorname{col}(0,s,s)=s\operatorname{col}(0,1,1).$$

For $r=-2$, we solve the equation

$$(\mathbf{A}+2\mathbf{I})\mathbf{u} = \begin{bmatrix} 3 & 0 & 0 \\ 0 & 2 & 2 \\ 0 & 2 & 2 \end{bmatrix} \begin{bmatrix} u_1 \\ u_2 \\ u_3 \end{bmatrix} = \begin{bmatrix} 0 \\ 0 \\ 0 \end{bmatrix},$$

which reduces to the system $u_1 = 0$, $u_2 = -u_3$. Hence, u_3 is arbitrary, and so we will let $u_3 = s$ (which means that $u_2 = -s$). Thus, solutions to this system and, therefore, eigenvectors for this matrix associated with the eigenvalue $r = -2$ are given by the vectors

$$\mathbf{u_2} = \text{col}(u_1, u_2, u_3) = \text{col}(0, -s, s) = s\,\text{col}(0, -1, 1).$$

13. We must first find the eigenvalues and eigenvectors associated with the given matrix **A**. Thus, we note that the characteristic equation for this matrix is given by

$$|\mathbf{A} - r\mathbf{I}| = \begin{vmatrix} 1-r & 2 & 2 \\ 2 & -r & 3 \\ 2 & 3 & -r \end{vmatrix} = 0$$

$$\Rightarrow \quad (1-r)\begin{vmatrix} -r & 3 \\ 3 & -r \end{vmatrix} - 2\begin{vmatrix} 2 & 3 \\ 2 & -r \end{vmatrix} + 2\begin{vmatrix} 2 & -r \\ 2 & 3 \end{vmatrix} = 0$$

$$\Rightarrow \quad (1-r)(r^2 - 9) - 2(-2r - 6) + 2(6 + 2r) = 0$$

$$\Rightarrow \quad (r+3)[(1-r)(r-3) + 8] = 0 \qquad \Rightarrow \qquad (r+3)(r-5)(r+1) = 0.$$

Therefore, the eigenvalues are r = –3, –1, 5. To find an eigenvector associated with the eigenvalue
$r = -3$, we must find a vector $\mathbf{u} = \text{col}(u_1, u_2, u_3)$ which satisfies the equation $(\mathbf{A} + 3\mathbf{I})\mathbf{u} = \mathbf{0}$. Thus, we have

$$(\mathbf{A}+3\mathbf{I})\mathbf{u} = \begin{bmatrix} 4 & 2 & 2 \\ 2 & 3 & 3 \\ 2 & 3 & 3 \end{bmatrix} \begin{bmatrix} u_1 \\ u_2 \\ u_3 \end{bmatrix} = \begin{bmatrix} 0 \\ 0 \\ 0 \end{bmatrix}$$

$$\Rightarrow \quad \begin{bmatrix} 2 & 0 & 0 \\ 0 & 1 & 1 \\ 0 & 0 & 0 \end{bmatrix} \begin{bmatrix} u_1 \\ u_2 \\ u_3 \end{bmatrix} = \begin{bmatrix} 0 \\ 0 \\ 0 \end{bmatrix},$$

where we have obtained the last equation above by using elementary row operations. This equation is equivalent to the system $u_1 = 0$, $u_2 = -u_3$. Hence, if we let u_3 have the arbitrary value s_1, then we see that, for the matrix **A**, the eigenvectors associated with the eigenvalue $r = -3$ are given by

$$\mathbf{u} = \text{col}(u_1, u_2, u_3) = \text{col}(0, -s_1, s_1) = s_1\,\text{col}(0, -1, 1).$$

Thus, if we choose $s_1 = 1$, then vector $\mathbf{u_1} = \text{col}(0,-1,1)$ is one eigenvector associated with this eigenvalue. For the eigenvalue $r = -1$, we must find a vector $\mathbf{u}$ which satisfies the equation $(\mathbf{A} + \mathbf{I})\mathbf{u} = \mathbf{0}$. Thus, we see that

$$(\mathbf{A} + \mathbf{I})\mathbf{u} = \begin{bmatrix} 2 & 2 & 2 \\ 2 & 1 & 3 \\ 2 & 3 & 1 \end{bmatrix} \begin{bmatrix} u_1 \\ u_2 \\ u_3 \end{bmatrix} = \begin{bmatrix} 0 \\ 0 \\ 0 \end{bmatrix}$$

$$\Rightarrow \qquad \begin{bmatrix} 1 & 2 & 0 \\ 0 & 1 & -1 \\ 0 & 0 & 0 \end{bmatrix} \begin{bmatrix} u_1 \\ u_2 \\ u_3 \end{bmatrix} = \begin{bmatrix} 0 \\ 0 \\ 0 \end{bmatrix},$$

which is equivalent to the system $u_1 = -2u_2$, $u_3 = u_2$. Therefore, if we let $u_2 = s_2$, then we see that vectors which satisfy the equation $(\mathbf{A} + \mathbf{I})\mathbf{u} = \mathbf{0}$ and, hence, eigenvectors for the matrix $\mathbf{A}$ associated with the eigenvalue $r = -1$ are given by

$$\mathbf{u} = \text{col}(u_1, u_2, u_3) = \text{col}(-2s_2, s_2, s_2) = s_2\,\text{col}(-2,1,1).$$

By letting $s_2 = 1$, we find that one such vector will be the vector $\mathbf{u_2} = \text{col}(-2,1,1)$. In order to find an eigenvector associated with the eigenvalue $r = 5$, we will solve the equation $(\mathbf{A} - 5\mathbf{I})\mathbf{u} = \mathbf{0}$. Thus, we have

$$(\mathbf{A} - 5\mathbf{I})\mathbf{u} = \begin{bmatrix} -4 & 2 & 2 \\ 2 & -5 & 3 \\ 2 & 3 & -5 \end{bmatrix} \begin{bmatrix} u_1 \\ u_2 \\ u_3 \end{bmatrix} = \begin{bmatrix} 0 \\ 0 \\ 0 \end{bmatrix}$$

$$\Rightarrow \qquad \begin{bmatrix} 1 & 0 & -1 \\ 0 & 1 & -1 \\ 0 & 0 & 0 \end{bmatrix} \begin{bmatrix} u_1 \\ u_2 \\ u_3 \end{bmatrix} = \begin{bmatrix} 0 \\ 0 \\ 0 \end{bmatrix},$$

which is equivalent to the system $u_1 = u_3$, $u_2 = u_3$. Thus, if we let $u_3 = s_3$, then, for the matrix $\mathbf{A}$, the eigenvectors associated with the eigenvalue $r = 5$ are given by

$$\mathbf{u} = \text{col}(u_1, u_2, u_3) = \text{col}(s_3, s_3, s_3) = s_3\,\text{col}(1,1,1).$$

Hence, by letting $s_3 = 1$, we see that one such vector will be the vector $\mathbf{u_3} = \text{col}(1,1,1)$. Therefore, by Corollary 1 on page 562 of the text, we see that a fundamental solution set for this equation is given by

$$\{e^{-3t}\mathbf{u_1}, e^{-t}\mathbf{u_2}, e^{5t}\mathbf{u_3}\}.$$

Thus, a general solution for this system is

$$\mathbf{x}(t) = c_1 e^{-3t}\mathbf{u_1} + c_2 e^{-t}\mathbf{u_2} + c_3 e^{5t}\mathbf{u_3}$$

$$= c_1 e^{-3t}\begin{bmatrix} 0 \\ -1 \\ 1 \end{bmatrix} + c_2 e^{-t}\begin{bmatrix} -2 \\ 1 \\ 1 \end{bmatrix} + c_3 e^{5t}\begin{bmatrix} 1 \\ 1 \\ 1 \end{bmatrix}.$$

21. A fundamental matrix for this system has three columns which are linearly independent solutions. Therefore, we will first find three such solutions. To this end, we will first find the eigenvalues for the matrix **A** by solving the characteristic equation given by

$$|\mathbf{A} - r\mathbf{I}| = \begin{vmatrix} -r & 1 & 0 \\ 0 & -r & 1 \\ 8 & -14 & 7-r \end{vmatrix} = 0$$

$$\Rightarrow \quad -r\begin{vmatrix} -r & 1 \\ -14 & 7-r \end{vmatrix} - \begin{vmatrix} 0 & 1 \\ 8 & 7-r \end{vmatrix} = 0$$

$$\Rightarrow \quad r^3 - 7r^2 + 14r - 8 = 0$$

$$\Rightarrow \quad (r-1)(r-2)(r-4) = 0.$$

Hence, this matrix has three distinct eigenvalues, $r = 1, 2, 4$, and, according to Theorem 6 on page 562 of the text, the eigenvectors associated with these eigenvalues will be linearly independent. Thus, these eigenvectors will be used in finding the three linearly independent solutions which we seek. To find an eigenvector, $\mathbf{u} = \text{col}(u_1, u_2, u_3)$, associated with the eigenvalue $r = 1$, we will solve the equation $(\mathbf{A} - \mathbf{I})\mathbf{u} = \mathbf{0}$. Therefore, we have

$$(\mathbf{A} - \mathbf{I})\mathbf{u} = \begin{bmatrix} -1 & 1 & 0 \\ 0 & -1 & 1 \\ 8 & -14 & 6 \end{bmatrix}\begin{bmatrix} u_1 \\ u_2 \\ u_3 \end{bmatrix} = \begin{bmatrix} 0 \\ 0 \\ 0 \end{bmatrix}$$

$$\Rightarrow \quad \begin{bmatrix} -1 & 0 & 1 \\ 0 & -1 & 1 \\ 0 & 0 & 0 \end{bmatrix}\begin{bmatrix} u_1 \\ u_2 \\ u_3 \end{bmatrix} = \begin{bmatrix} 0 \\ 0 \\ 0 \end{bmatrix},$$

which is equivalent to the system $u_1 = u_3$, $u_2 = u_3$. Thus, by letting $u_3 = 1$ (which implies that $u_1 = u_2 = 1$), we find that one eigenvector associated with the eigenvalue $r = 1$ is given by the vector $\mathbf{u}_1 = \text{col}(1,1,1)$. To find an eigenvector associated with the eigenvalue $r = 2$, we solve the equation

$$(\mathbf{A} - 2\mathbf{I})\mathbf{u} = \begin{bmatrix} -2 & 1 & 0 \\ 0 & -2 & 1 \\ 8 & -14 & 5 \end{bmatrix}\begin{bmatrix} u_1 \\ u_2 \\ u_3 \end{bmatrix} = \begin{bmatrix} 0 \\ 0 \\ 0 \end{bmatrix}$$

$$\Rightarrow \qquad \begin{bmatrix} 4 & 0 & -1 \\ 0 & 2 & -1 \\ 0 & 0 & 0 \end{bmatrix} \begin{bmatrix} u_1 \\ u_2 \\ u_3 \end{bmatrix} = \begin{bmatrix} 0 \\ 0 \\ 0 \end{bmatrix},$$

which is equivalent to the system $4u_1 = u_3, 2u_2 = u_3$. Hence, letting $u_3 = 4$ implies that $u_1 = 1$ and $u_2 = 2$. Therefore, one eigenvector associated with the eigenvalue $r = 2$ is the vector $\mathbf{u_2} = \mathrm{col}(1,2,4)$. In order to find an eigenvector associated with the eigenvalue $r = 4$, we will solve the equation

$$(\mathbf{A} - 4\mathbf{I})\mathbf{u} = \begin{bmatrix} -4 & 1 & 0 \\ 0 & -4 & 1 \\ 8 & -14 & 3 \end{bmatrix} \begin{bmatrix} u_1 \\ u_2 \\ u_3 \end{bmatrix} = \begin{bmatrix} 0 \\ 0 \\ 0 \end{bmatrix}$$

$$\Rightarrow \qquad \begin{bmatrix} 16 & 0 & -1 \\ 0 & 4 & -1 \\ 0 & 0 & 0 \end{bmatrix} \begin{bmatrix} u_1 \\ u_2 \\ u_3 \end{bmatrix} = \begin{bmatrix} 0 \\ 0 \\ 0 \end{bmatrix},$$

which is equivalent to the system $16u_1 = u_3$, $4u_2 = u_3$. Therefore, letting $u_3 = 16$ implies that $u_1 = 1$ and $u_2 = 4$. Thus, one eigenvector associated with the eigenvalue $r = 4$ is the vector $\mathbf{u_3} = \mathrm{col}(1,4,16)$. Therefore, by Theorem 5 on page 560 of the text (or Corollary 1), we see that three linearly independent solutions of this system are given by $e^t\mathbf{u_1}$, $e^{2t}\mathbf{u_2}$, and $e^{4t}\mathbf{u_3}$. Thus, a fundamental matrix for this system will be the matrix

$$\mathbf{X}(t) = \begin{bmatrix} e^t & e^{2t} & e^{4t} \\ e^t & 2e^{2t} & 4e^{4t} \\ e^t & 4e^{2t} & 16e^{4t} \end{bmatrix}.$$

33. Since the coefficient matrix for this system is a 3 × 3 real symmetric matrix, by the discussion on page 564 of the text, we know that we can find three linearly independent eigenvectors for this matrix. Therefore, to find the solution to this initial value problem, we must first find three such eigenvectors. To do this we first find eigenvalues for this matrix. Therefore, we solve the characteristic equation given by

$$|\mathbf{A} - r\mathbf{I}| = \begin{vmatrix} 1-r & -2 & 2 \\ -2 & 1-r & -2 \\ 2 & -2 & 1-r \end{vmatrix} = 0$$

$$\Rightarrow \qquad (1-r)\begin{vmatrix} 1-r & -2 \\ -2 & 1-r \end{vmatrix} + 2\begin{vmatrix} -2 & -2 \\ 2 & 1-r \end{vmatrix} + 2\begin{vmatrix} -2 & 1-r \\ 2 & -2 \end{vmatrix} = 0$$

$$\Rightarrow \qquad (1-r)\left[(1-r)^2 - 4\right] + 2[-2(1-r) + 4] + 2[4 - 2(1-r)] = 0$$

$$\Rightarrow \quad (1-r)(r-3)(r+1)+8(r+1)=-(r+1)(r-5)(r+1)=0 .$$

Thus, the eigenvalues are $r=-1$ and $r=5$, with $r=-1$ an eigenvalue of multiplicity two. In order to find an eigenvector associated with the eigenvalue $r=5$, we solve the equation

$$(\mathbf{A}-5\mathbf{I})\mathbf{u}=\begin{bmatrix}-4 & -2 & 2\\ -2 & -4 & -2\\ 2 & -2 & -4\end{bmatrix}\begin{bmatrix}u_1\\ u_2\\ u_3\end{bmatrix}=\begin{bmatrix}0\\ 0\\ 0\end{bmatrix}$$

$$\Rightarrow \quad \begin{bmatrix}1 & 0 & -1\\ 0 & 1 & 1\\ 0 & 0 & 0\end{bmatrix}\begin{bmatrix}u_1\\ u_2\\ u_3\end{bmatrix}=\begin{bmatrix}0\\ 0\\ 0\end{bmatrix}.$$

This equation is equivalent to the system $u_1=u_3$, $u_2=-u_3$. Thus, if we let $u_3=1$, we see that for this coefficient matrix an eigenvector associated with the eigenvalue $r=5$ is given by the vector $\mathbf{u}_1=\operatorname{col}(u_1,u_2,u_3)=\operatorname{col}(1,-1,1)$. We must now find two more linearly independent eigenvectors for this coefficient matrix. By the discussion above, these eigenvectors will be associated with the eigenvalue $r=-1$. Thus, we solve the equation

$$(\mathbf{A}+\mathbf{I})\mathbf{u}=\begin{bmatrix}2 & -2 & 2\\ -2 & 2 & -2\\ 2 & -2 & 2\end{bmatrix}\begin{bmatrix}u_1\\ u_2\\ u_3\end{bmatrix}=\begin{bmatrix}0\\ 0\\ 0\end{bmatrix} \qquad (3)$$

$$\Rightarrow \quad \begin{bmatrix}1 & -1 & 1\\ 0 & 0 & 0\\ 0 & 0 & 0\end{bmatrix}\begin{bmatrix}u_1\\ u_2\\ u_3\end{bmatrix}=\begin{bmatrix}0\\ 0\\ 0\end{bmatrix},$$

which is equivalent to the equation $u_1-u_2+u_3=0$. Therefore, if we arbitrarily assign the value s to u_2 and v to u_3, we see that $u_1=s-v$, and solutions to equation (3) above will be given by

$$\mathbf{u}=\begin{bmatrix}s-v\\ s\\ v\end{bmatrix}=s\begin{bmatrix}1\\ 1\\ 0\end{bmatrix}+v\begin{bmatrix}-1\\ 0\\ 1\end{bmatrix}.$$

By taking $s=1$ and $v=0$, we see that one solution to equation (3) will be the vector $\mathbf{u}_2=\operatorname{col}(1,1,0)$. Hence, this is one eigenvector for the coefficient matrix. Similarly, by letting $s=0$ and $v=1$, we find a second eigenvector will be the vector $\mathbf{u}_3=\operatorname{col}(-1,0,1)$. Since the eigenvectors $\mathbf{u}_1$, $\mathbf{u}_2$, and $\mathbf{u}_3$ are linearly independent, by Theorem 5 on page 560 of the text, we see that a general solution for this system will be given by

$$\mathbf{x}(t)=c_1e^{5t}\begin{bmatrix}1\\-1\\1\end{bmatrix}+c_2e^{-t}\begin{bmatrix}1\\1\\0\end{bmatrix}+c_3e^{-t}\begin{bmatrix}-1\\0\\1\end{bmatrix}.$$

To find a solution which satisfies the initial condition, we must solve the equation

$$\mathbf{x}(0)=c_1\begin{bmatrix}1\\-1\\1\end{bmatrix}+c_2\begin{bmatrix}1\\1\\0\end{bmatrix}+c_3\begin{bmatrix}-1\\0\\1\end{bmatrix}$$

$$\Rightarrow \quad \begin{bmatrix}1&1&-1\\-1&1&0\\1&0&1\end{bmatrix}\begin{bmatrix}c_1\\c_2\\c_3\end{bmatrix}=\begin{bmatrix}-2\\-3\\2\end{bmatrix}.$$

This equation can be solved by either using elementary row operations on the augmented matrix associated with this equation or by solving the system

$$c_1+c_2-c_3=-2,$$
$$-c_1+c_2=-3,$$
$$c_1+c_3=2.$$

By either method we find that $c_1=1$, $c_2=-2$, and $c_3=1$. Therefore, the solution to this initial value problem is given by

$$\mathbf{x}(t)=e^{5t}\begin{bmatrix}1\\-1\\1\end{bmatrix}-2e^{-t}\begin{bmatrix}1\\1\\0\end{bmatrix}+e^{-t}\begin{bmatrix}-1\\0\\1\end{bmatrix}$$

$$=\begin{bmatrix}e^{5t}-2e^{-t}-e^{-t}\\-e^{5t}-2e^{-t}+0\\e^{5t}+0+e^{-t}\end{bmatrix}=\begin{bmatrix}-3e^{-t}+e^{5t}\\-2e^{-t}-e^{5t}\\e^{-t}+e^{5t}\end{bmatrix}.$$

37. **(a)** In order to find the eigenvalues for the matrix **A**, we will solve the characteristic equation

$$|\mathbf{A}-r\mathbf{I}|=\begin{vmatrix}2-r&1&6\\0&2-r&5\\0&0&2-r\end{vmatrix}=0$$

$$\Rightarrow \quad (2-r)\cdot\begin{vmatrix}2-r&5\\0&2-r\end{vmatrix}-\begin{vmatrix}0&5\\0&2-r\end{vmatrix}+6\begin{vmatrix}0&2-r\\0&0\end{vmatrix}=0$$

$$\Rightarrow \quad (2-r)^3=0.$$

Thus, $r = 2$ is an eigenvalue of multiplicity three. To find the eigenvectors for the matrix **A** associated with this eigenvalue, we solve the equation

$$(\mathbf{A}-2\mathbf{I})\mathbf{u} = \begin{bmatrix} 0 & 1 & 6 \\ 0 & 0 & 5 \\ 0 & 0 & 0 \end{bmatrix}\begin{bmatrix} u_1 \\ u_2 \\ u_3 \end{bmatrix} = \begin{bmatrix} 0 \\ 0 \\ 0 \end{bmatrix}.$$

This equation is equivalent to the system $u_2 = 0,\ u_3 = 0$. Therefore, we can assign u_1 to be any arbitrary value, say $u_1 = s$, and we find that the vector $\mathbf{u} = \operatorname{col}(u_1, u_2, u_3) = \operatorname{col}(s,0,0) = s\operatorname{col}(1,0,0)$ will solve this equation and will, thus, be an eigenvector for the matrix **A**. We also notice that the vectors $\mathbf{u} = s\operatorname{col}(1,0,0)$ are the only vectors that will solve this equation, and, hence, they will be the only eigenvectors for the matrix **A**.

(b) By taking $s = 1$, we find that, for the matrix **A**, one eigenvector associated with the eigenvalue $r = 2$ will be the vector $\mathbf{u}_1 = \operatorname{col}(1,0,0)$. Therefore, by the way eigenvalues and eigenvectors were defined (as was discussed in the text on page 557), we see that one solution to the system $\mathbf{x}' = \mathbf{A}\mathbf{x}$ will be given by the vector

$$\mathbf{x}_1(t) = e^{2t}\mathbf{u}_1 = e^{2t}\begin{bmatrix} 1 \\ 0 \\ 0 \end{bmatrix}.$$

(c) We know that $\mathbf{u}_1 = \operatorname{col}(1,0,0)$ is an eigenvector for the matrix **A** associated with the eigenvalue $r = 2$. Thus, $\mathbf{u}_1$ satisfies the equation

$$(\mathbf{A}-2\mathbf{I})\mathbf{u}_1 = \mathbf{0} \qquad \Rightarrow \qquad \mathbf{A}\mathbf{u}_1 = 2\mathbf{u}_1. \tag{4}$$

We want to find a constant vector $\mathbf{u}_2 = \operatorname{col}(v_1, v_2, v_3)$ such that $\mathbf{x}_2(t) = te^{2t}\mathbf{u}_1 + e^{2t}\mathbf{u}_2$ will be a second solution to the system $\mathbf{x}' = \mathbf{A}\mathbf{x}$. To do this, we will first show that $\mathbf{x}_2$ will satisfy the equation $\mathbf{x}' = \mathbf{A}\mathbf{x}$ if and only if the vector $\mathbf{u}_2$ satisfies the equation $(\mathbf{A}-2\mathbf{I})\mathbf{u}_2 = \mathbf{u}_1$. To this end, we find that

$$\mathbf{x}_2'(t) = 2te^{2t}\mathbf{u}_1 + (\mathbf{u}_1 + 2\mathbf{u}_2)e^{2t},$$

where we have used the fact that $\mathbf{u}_1$ and $\mathbf{u}_2$ are constant vectors. We also have

$$\mathbf{A}\mathbf{x}_2(t) = \mathbf{A}\left(te^{2t}\mathbf{u}_1 + e^{2t}\mathbf{u}_2\right)$$

$= \mathbf{A}\left(te^{2t}\mathbf{u}_1\right) + \mathbf{A}\left(e^{2t}\mathbf{u}_2\right),$ by the distributive property of matrix multiplication (page 539 of the text)

$= te^{2t}(\mathbf{A}\mathbf{u}_1) + e^{2t}(\mathbf{A}\mathbf{u}_2),$ by the associative property of matrix multiplication (page 539 of the text)

$= 2te^{2t}\mathbf{u}_1 + e^{2t}\mathbf{A}\mathbf{u}_2,$ by equation (4) above.

Thus, if $\mathbf{x}_2$ is to be a solution to the given system we, must have

$$\mathbf{x}_2'(t) = \mathbf{A}\mathbf{x}_2(t)$$

$$\Rightarrow \qquad 2te^{2t}\mathbf{u}_1 + (\mathbf{u}_1 + 2\mathbf{u}_2)e^{2t} = 2te^{2t}\mathbf{u}_1 + e^{2t}\mathbf{A}\mathbf{u}_2$$

$$\Rightarrow \qquad (\mathbf{u}_1 + 2\mathbf{u}_2)e^{2t} = e^{2t}\mathbf{A}\mathbf{u}_2 .$$

By dividing both sides of this equation by the nonzero term e^{2t}, we obtain

$$\mathbf{u}_1 + 2\mathbf{u}_2 = \mathbf{A}\mathbf{u}_2 \qquad \Rightarrow \qquad (\mathbf{A} - 2\mathbf{I})\mathbf{u}_2 = \mathbf{u}_1 .$$

Since all of these steps are reversible, if a vector $\mathbf{u}_2$ satisfies this last equation then $\mathbf{x}_2(t) = te^{2t}\mathbf{u}_1 + e^{2t}\mathbf{u}_2$ will be a solution to the system $\mathbf{x}' = \mathbf{A}\mathbf{x}$. Now we can use the formula $(\mathbf{A} - 2\mathbf{I})\mathbf{u}_2 = \mathbf{u}_1$ to find the vector $\mathbf{u}_2 = \mathrm{col}(v_1, v_2, v_3)$. Hence, we solve the equation

$$(\mathbf{A} - 2\mathbf{I})\mathbf{u}_2 = \begin{bmatrix} 0 & 1 & 6 \\ 0 & 0 & 5 \\ 0 & 0 & 0 \end{bmatrix} \begin{bmatrix} v_1 \\ v_2 \\ v_3 \end{bmatrix} = \begin{bmatrix} 1 \\ 0 \\ 0 \end{bmatrix}.$$

This equation is equivalent to the system $v_2 + 6v_3 = 1$, $5v_3 = 0$, which implies that $v_2 = 1, v_3 = 0$. Therefore, the vector $\mathbf{u}_2 = \mathrm{col}(0,1,0)$ will satisfy the equation $(\mathbf{A} - 2\mathbf{I})\mathbf{u}_2 = \mathbf{u}_1$ and, thus,

$$\mathbf{x}_2(t) = te^{2t}\begin{bmatrix} 1 \\ 0 \\ 0 \end{bmatrix} + e^{2t}\begin{bmatrix} 0 \\ 1 \\ 0 \end{bmatrix},$$

will be a second solution to the given system. We can see by inspection $\mathbf{x}_2(t)$ and $\mathbf{x}_1(t) = e^{2t}\mathbf{u}_1$ are linearly independent.

(d) To find a third linearly independent solution to this system we will try to find a solution of the form $\mathbf{x}_3(t) = \frac{t^2}{2}e^{2t}\mathbf{u}_1 + te^{2t}\mathbf{u}_2 + e^{2t}\mathbf{u}_3$, where $\mathbf{u}_3$ is a constant vector that we must find, and $\mathbf{u}_1$ and $\mathbf{u}_2$ are the vectors that we found in parts (b) and (c), respectively. To find the vector $\mathbf{u}_3$, we will proceed as we did in part (c) above. We will first show that $\mathbf{x}_3(t)$ will be a solution to the given system if and only if the vector $\mathbf{u}_3$ satisfies the equation $(\mathbf{A} - 2\mathbf{I})\mathbf{u}_3 = \mathbf{u}_2$. To do this we observe that

$$\mathbf{x}_3'(t) = te^{2t}\mathbf{u}_1 + t^2e^{2t}\mathbf{u}_1 + e^{2t}\mathbf{u}_2 + 2te^{2t}\mathbf{u}_2 + 2e^{2t}\mathbf{u}_3 .$$

Also, using the facts that

$$(\mathbf{A} - 2\mathbf{I})\mathbf{u}_1 = \mathbf{0} \qquad \Rightarrow \qquad \mathbf{A}\mathbf{u}_1 = 2\mathbf{u}_1, \tag{5}$$

and

$$(\mathbf{A} - 2\mathbf{I})\mathbf{u}_2 = \mathbf{u}_1 \qquad \Rightarrow \qquad \mathbf{u}_1 + 2\mathbf{u}_2 = \mathbf{A}\mathbf{u}_2 \tag{6}$$

we have

$$\mathbf{A}\mathbf{x}_3(t) = \mathbf{A}\left[\frac{t^2}{2}e^{2t}\mathbf{u}_1 + te^{2t}\mathbf{u}_2 + e^{2t}\mathbf{u}_3\right]$$

$$= \mathbf{A}\left(\frac{t^2}{2}e^{2t}\mathbf{u}_1\right) + \mathbf{A}\left(te^{2t}\mathbf{u}_2\right) + \mathbf{A}\left(e^{2t}\mathbf{u}_3\right), \qquad \text{distributive property}$$

$$= \frac{t^2}{2}e^{2t}(\mathbf{A}\mathbf{u}_1) + te^{2t}(\mathbf{A}\mathbf{u}_2) + e^{2t}(\mathbf{A}\mathbf{u}_3), \qquad \text{associative property}$$

$$= \frac{t^2}{2}e^{2t}(2\mathbf{u}_1) + te^{2t}(\mathbf{u}_1 + 2\mathbf{u}_2) + e^{2t}(\mathbf{A}\mathbf{u}_3), \qquad \text{(5) and (6) above}$$

$$= t^2e^{2t}\mathbf{u}_1 + te^{2t}\mathbf{u}_1 + 2te^{2t}\mathbf{u}_2 + e^{2t}\mathbf{A}\mathbf{u}_3 .$$

Therefore, for $\mathbf{x}_3(t)$ to satisfy the given system, we must have

$$\mathbf{x}_3'(t) = \mathbf{A}\mathbf{x}_3(t)$$

$$\Rightarrow \quad te^{2t}\mathbf{u}_1 + t^2e^{2t}\mathbf{u}_1 + e^{2t}\mathbf{u}_2 + 2te^{2t}\mathbf{u}_2 + 2e^{2t}\mathbf{u}_3 = t^2e^{2t}\mathbf{u}_1 + te^{2t}\mathbf{u}_1 + 2te^{2t}\mathbf{u}_2 + e^{2t}\mathbf{A}\mathbf{u}_3$$

$$\Rightarrow \quad e^{2t}\mathbf{u}_2 + 2e^{2t}\mathbf{u}_3 = e^{2t}\mathbf{A}\mathbf{u}_3$$

$$\Rightarrow \quad \mathbf{u}_2 + 2\mathbf{u}_3 = \mathbf{A}\mathbf{u}_3$$

$$\Rightarrow \quad (\mathbf{A} - 2\mathbf{I})\mathbf{u}_3 = \mathbf{u}_2 .$$

Again since these steps are reversible, we see that if a vector $\mathbf{u}_3$ satisfies the equation $(\mathbf{A} - 2\mathbf{I})\mathbf{u}_3 = \mathbf{u}_2$, then the vector $\mathbf{x}_3(t) = \frac{t^2}{2}e^{2t}\mathbf{u}_1 + te^{2t}\mathbf{u}_2 + e^{2t}\mathbf{u}_3$ will be a third linearly independent solution to the given system. Thus, we can use this equation to find the vector $\mathbf{u}_3 = \operatorname{col}(v_1, v_2, v_3)$. Hence, we solve

$$(\mathbf{A} - 2\mathbf{I})\mathbf{u}_3 = \begin{bmatrix} 0 & 1 & 6 \\ 0 & 0 & 5 \\ 0 & 0 & 0 \end{bmatrix}\begin{bmatrix} v_1 \\ v_2 \\ v_3 \end{bmatrix} = \begin{bmatrix} 0 \\ 1 \\ 0 \end{bmatrix}.$$

This equation is equivalent to the system $v_2 + 6v_3 = 0$, $5v_3 = 1$, which implies that $v_3 = \frac{1}{5}$, $v_2 = \frac{-6}{5}$. Therefore, if we let $\mathbf{u}_3 = \operatorname{col}\left(0, \frac{-6}{5}, \frac{1}{5}\right)$, then

$$\mathbf{x}_3(t) = \frac{t^2}{2}e^{2t}\begin{bmatrix}1\\0\\0\end{bmatrix} + te^{2t}\begin{bmatrix}0\\1\\0\end{bmatrix} + e^{2t}\begin{bmatrix}0\\-\frac{6}{5}\\\frac{1}{5}\end{bmatrix},$$

will be a third solution to the given system and we see by inspection that this solution is linearly independent from the solutions $\mathbf{x}_1(t)$ and $\mathbf{x}_2(t)$.

(e) Notice that

$$(\mathbf{A}-2\mathbf{I})^3\mathbf{u}_3 = (\mathbf{A}-2\mathbf{I})^2[(\mathbf{A}-2\mathbf{I})\mathbf{u}_3] = (\mathbf{A}-2\mathbf{I})^2\mathbf{u}_2 = (\mathbf{A}-2\mathbf{I})(\mathbf{A}-2\mathbf{I})\mathbf{u}_2 = (\mathbf{A}-2\mathbf{I})\mathbf{u}_1 = \mathbf{0}.$$

43. According to Problem 42, we will look for solutions of the form $\mathbf{x}(t) = t^r\mathbf{u}$, where r is an eigenvalue for the coefficient matrix and $\mathbf{u}$ is an associated eigenvector. To find the eigenvalues for this matrix, we solve the equation

$$|\mathbf{A}-r\mathbf{I}| = \begin{vmatrix}1-r & 3\\-1 & 5-r\end{vmatrix} = 0$$

$$\Rightarrow \quad (1-r)(5-r)+3=0 \quad \Rightarrow \quad r^2-6r+8=0 \quad \Rightarrow \quad (r-2)(r-4)=0.$$

Therefore, the coefficient matrix has the eigenvalues $r = 2, 4$. Since these are distinct eigenvalues, Theorem 6 on page 562 of the text assures us that their associated eigenvectors will be linearly independent. To find an eigenvector $\mathbf{u} = \operatorname{col}(u_1, u_2)$ associated with the eigenvalue $r = 2$, we solve the system

$$(\mathbf{A}-2\mathbf{I})\mathbf{u} = \begin{bmatrix}-1 & 3\\-1 & 3\end{bmatrix}\begin{bmatrix}u_1\\u_2\end{bmatrix} = \begin{bmatrix}0\\0\end{bmatrix},$$

which is equivalent to the equation $-u_1 + 3u_2 = 0$. Thus, if we let $u_2 = 1$ then, in order to satisfy this equation, we must have $u_1 = 3$. Hence, we see that the vector $\mathbf{u}_1 = \operatorname{col}(u_1, u_2) = \operatorname{col}(3,1)$ will be an eigenvector for the coefficient matrix of the given system associated with the eigenvalue $r = 2$. Therefore, according to Problem 42, one solution to this system will be given by

$$\mathbf{x}_1(t) = t^2\mathbf{u}_1 = t^2\begin{bmatrix}3\\1\end{bmatrix}.$$

To find an eigenvector associated with the eigenvalue $r = 4$, we solve the equation

$$(\mathbf{A}-4\mathbf{I})\mathbf{u} = \begin{bmatrix}-3 & 3\\-1 & 1\end{bmatrix}\begin{bmatrix}u_1\\u_2\end{bmatrix} = \begin{bmatrix}0\\0\end{bmatrix},$$

which is equivalent to the equation $u_1 = u_2$. Thus, if we let $u_2 = 1$, then we must have $u_1 = 1$ and so an eigenvector associated with the eigenvalue $r = 4$ will be given by the vector $\mathbf{u}_2 = \text{col}(u_1, u_2) = \text{col}(1,1)$. Therefore, another solution to the given system will be

$$\mathbf{x}_2(t) = t^4\mathbf{u}_2 = t^4\begin{bmatrix} 1 \\ 1 \end{bmatrix}.$$

Clearly the solutions $\mathbf{x}_1(t)$ and $\mathbf{x}_2(t)$ are linearly independent. So the general solution to the given system with $t > 0$ will be

$$\mathbf{x}(t) = c_1 t^2\begin{bmatrix} 3 \\ 1 \end{bmatrix} + c_2 t^4\begin{bmatrix} 1 \\ 1 \end{bmatrix} = c_1\begin{bmatrix} 3t^2 \\ t^2 \end{bmatrix} + c_2\begin{bmatrix} t^4 \\ t^4 \end{bmatrix}.$$

EXERCISES 9.6: Complex Eigenvalues, page 573

3. To find the eigenvalues for the matrix **A**, we solve the characteristic equation given by

$$|\mathbf{A} - r\mathbf{I}| = \begin{vmatrix} 1-r & 2 & -1 \\ 0 & 1-r & 1 \\ 0 & -1 & 1-r \end{vmatrix} = 0$$

$$\Rightarrow \quad (1-r)\begin{vmatrix} 1-r & 1 \\ -1 & 1-r \end{vmatrix} - 0 + 0 = 0$$

$$\Rightarrow \quad (1-r)\left[(1-r)^2 + 1\right] = 0 \qquad \Rightarrow \quad (1-r)\left(r^2 - 2r + 2\right) = 0.$$

By this equation and the quadratic formula, we see that the roots to the characteristic equation and, therefore, the eigenvalues for the matrix **A** are $r = 1$, and $r = 1 \pm i$. To find an eigenvector $\mathbf{u} = \text{col}(u_1, u_2, u_3)$ associated with the real eigenvalue $r = 1$, we solve the system

$$(\mathbf{A} - \mathbf{I})\mathbf{u} = \begin{bmatrix} 0 & 2 & -1 \\ 0 & 0 & 1 \\ 0 & -1 & 0 \end{bmatrix}\begin{bmatrix} u_1 \\ u_2 \\ u_3 \end{bmatrix} = \begin{bmatrix} 0 \\ 0 \\ 0 \end{bmatrix},$$

which implies that $u_2 = 0$, $u_3 = 0$. Therefore, we can set u_1 arbitrarily to any value, say $u_1 = s$. Then the vectors

$$\mathbf{u} = \text{col}(u_1, u_2, u_3) = \text{col}(s,0,0) = s\,\text{col}(1,0,0),$$

will satisfy the above equation and, therefore, be eigenvectors for the matrix **A**. Hence, if we set $s = 1$, we see that one eigenvector associated with the eigenvalue $r = 1$ will be the vector $\mathbf{u}_1 = \text{col}(1,0,0)$. Therefore, one solution to the given system will be

$$\mathbf{x}_1(t) = e^t \mathbf{u}_1 = e^t \begin{bmatrix} 1 \\ 0 \\ 0 \end{bmatrix}.$$

In order to find an eigenvector $\mathbf{z} = \text{col}(z_1, z_2, z_3)$ associated with the complex eigenvalue $r = 1 + i$, we solve the equation

$$[\mathbf{A} - (1+i)\mathbf{I}]\mathbf{z} = \begin{bmatrix} -i & 2 & -1 \\ 0 & -i & 1 \\ 0 & -1 & i \end{bmatrix} \begin{bmatrix} z_1 \\ z_2 \\ z_3 \end{bmatrix} = \begin{bmatrix} 0 \\ 0 \\ 0 \end{bmatrix}.$$

This equation is equivalent to the system

$$-iz_1 + 2z_2 - z_3 = 0 \qquad \text{and} \qquad -iz_2 + z_3 = 0.$$

Thus, if we let $z_2 = s$, then we see that $z_3 = is$ and

$$-iz_1 = -2z_2 + z_3 = -2s + is \qquad \Rightarrow \qquad (i)(-iz_1) = (i)(-2s + is)$$

$$\Rightarrow \qquad z_1 = -2is - s = -s - 2is,$$

where we have used the fact that $i^2 = -1$. Hence, eigenvectors associated with the eigenvalue $r = 1 + i$ will be $\mathbf{z} = s\,\text{col}(-1 - 2i, 1, i)$. By taking $s = 1$, we see that one eigenvector associated with this eigenvalue will be the vector

$$\mathbf{z}_1 = \begin{bmatrix} -1-2i \\ 1 \\ i \end{bmatrix} = \begin{bmatrix} -1 \\ 1 \\ 0 \end{bmatrix} + i \begin{bmatrix} -2 \\ 0 \\ 1 \end{bmatrix}.$$

Thus, by the notation on page 569 of the text, we have $\alpha = 1$, $\beta = 1$, $\mathbf{a} = \text{col}(-1,1,0)$, and $\mathbf{b} = \text{col}(-2,0,1)$. Therefore, according to formulas (6) and (7) on page 570 of the text, two more linearly independent solutions to the given system will be given by

$$\mathbf{x}_2(t) = (e^t \cos t)\mathbf{a} - (e^t \sin t)\mathbf{b} \qquad \text{and} \qquad \mathbf{x}_3(t) = (e^t \sin t)\mathbf{a} + (e^t \cos t)\mathbf{b}.$$

Hence, the general solution to the system given in this problem will be

$$\mathbf{x}(t) = c_1\mathbf{x}_2(t) + c_2\mathbf{x}_3(t) + c_3\mathbf{x}_1(t)$$

$$= c_1 e^t \cos t \begin{bmatrix} -1 \\ 1 \\ 0 \end{bmatrix} - c_1 e^t \sin t \begin{bmatrix} -2 \\ 0 \\ 1 \end{bmatrix} + c_2 e^t \sin t \begin{bmatrix} -1 \\ 1 \\ 0 \end{bmatrix} + c_2 e^t \cos t \begin{bmatrix} -2 \\ 0 \\ 1 \end{bmatrix} + c_3 e^t \begin{bmatrix} 1 \\ 0 \\ 0 \end{bmatrix}.$$

7. In order to find a fundamental matrix for this system, we must first find three linearly independent solutions. Thus, we seek the eigenvalues for the matrix **A** by solving the characteristic equation given by

$$|\mathbf{A}-r\mathbf{I}| = \begin{vmatrix} -r & 0 & 1 \\ 0 & -r & -1 \\ 0 & 1 & -r \end{vmatrix} = 0$$

$$\Rightarrow \quad -r\begin{vmatrix} -r & -1 \\ 1 & -r \end{vmatrix} - 0 + 0 = 0$$

$$\Rightarrow \quad -r\left(r^2+1\right) = 0 .$$

Hence, the eigenvalues for the matrix **A** will be $r = 0, \pm i$. To find an eigenvector $\mathbf{u} = \operatorname{col}(u_1, u_2, u_3)$ associated with the real eigenvalue $r = 0$, we solve the equation

$$(\mathbf{A} - 0\mathbf{I})\mathbf{u} = \begin{bmatrix} 0 & 0 & 1 \\ 0 & 0 & -1 \\ 0 & 1 & 0 \end{bmatrix}\begin{bmatrix} u_1 \\ u_2 \\ u_3 \end{bmatrix} = \begin{bmatrix} 0 \\ 0 \\ 0 \end{bmatrix},$$

which is equivalent to the system $u_3 = 0$, $u_2 = 0$. Thus, if we let u_1 have the arbitrary value $u_1 = s$, then the vectors

$$\mathbf{u} = \operatorname{col}(u_1, u_2, u_3) = \operatorname{col}(s, 0, 0) = s\operatorname{col}(1, 0, 0),$$

will satisfy this equation and will, therefore, be eigenvectors for the matrix **A** associated with the eigenvalue $r = 0$. Hence, by letting $s = 1$, we find that one of these eigenvectors will be the vector $\mathbf{u} = \operatorname{col}(1, 0, 0)$. Thus, one solution to the given system will be

$$\mathbf{x}_1(t) = e^0\mathbf{u}_1 = \begin{bmatrix} 1 \\ 0 \\ 0 \end{bmatrix}.$$

To find two more linearly independent solutions for this system, we will first look for an eigenvector associated with the complex eigenvalue $r = i$. That is, we seek a vector $\mathbf{z} = \operatorname{col}(z_1, z_2, z_3)$ which satisfies the equation

$$(\mathbf{A} - i\mathbf{I})\mathbf{z} = \begin{bmatrix} -i & 0 & 1 \\ 0 & -i & -1 \\ 0 & 1 & -i \end{bmatrix}\begin{bmatrix} z_1 \\ z_2 \\ z_3 \end{bmatrix} = \begin{bmatrix} 0 \\ 0 \\ 0 \end{bmatrix},$$

which is equivalent to the system

$$iz_1 = z_3, \qquad \text{and} \qquad iz_2 = -z_3 .$$

Thus, if we let z_3 be any arbitrary value, say $z_3 = is$, (which means that we must have $z_1 = s$ and $z_2 = -s$), then we see that the vectors given by

$$\mathbf{z} = \operatorname{col}(z_1, z_2, z_3) = \operatorname{col}(s, -s, is) = s\operatorname{col}(1, -1, i),$$

will be eigenvectors for the matrix **A** associated with the eigenvalue $r = i$. Therefore, by letting $s = 1$, we find that one of these eigenvectors will be the vector

$$\mathbf{z} = \begin{bmatrix} 1 \\ -1 \\ i \end{bmatrix} = \begin{bmatrix} 1 \\ -1 \\ 0 \end{bmatrix} + i\begin{bmatrix} 0 \\ 0 \\ 1 \end{bmatrix}.$$

From this, by the notation given on page 569 of the text, we see that $\alpha = 0$, $\beta = 1$, $\mathbf{a} = \text{col}(1,-1,0)$, and $\mathbf{b} = \text{col}(0,0,1)$. Therefore, by formulas (6) and (7) on page 570 of the text, two more linearly independent solutions for this system will be

$$\mathbf{x}_2(t) = (\cos t)\,\mathbf{a} - (\sin t)\,\mathbf{b} = \begin{bmatrix} \cos t \\ -\cos t \\ 0 \end{bmatrix} - \begin{bmatrix} 0 \\ 0 \\ \sin t \end{bmatrix} = \begin{bmatrix} \cos t \\ -\cos t \\ \sin t \end{bmatrix},$$

and

$$\mathbf{x}_3(t) = (\sin t)\,\mathbf{a} + (\cos t)\,\mathbf{b} = \begin{bmatrix} \sin t \\ -\sin t \\ 0 \end{bmatrix} + \begin{bmatrix} 0 \\ 0 \\ \cos t \end{bmatrix} = \begin{bmatrix} \sin t \\ -\sin t \\ \cos t \end{bmatrix}.$$

Finally, since a fundamental matrix for the system given in this problem must have three columns which are linearly independent solutions of the system, we see that such a fundamental matrix will be given by the matrix

$$\mathbf{X}(t) = \begin{bmatrix} 1 & \cos t & \sin t \\ 0 & -\cos t & -\sin t \\ 0 & -\sin t & \cos t \end{bmatrix}.$$

17. We will assume that $t > 0$. According to Problem 42 in Exercises 9.5, a solution to this Cauchy-Euler system will have the form $\mathbf{x}(t) = t^r\mathbf{u}$, where r is an eigenvalue for the coefficient matrix of the system and $\mathbf{u}$ is an eigenvector associated with this eigenvalue. Therefore, we first must find the eigenvalues for this matrix by solving the characteristic equation given by

$$|\mathbf{A} - r\mathbf{I}| = \begin{vmatrix} -1-r & -1 & 0 \\ 2 & -1-r & 1 \\ 0 & 1 & -1-r \end{vmatrix} = 0$$

$$\Rightarrow \quad (-1-r)\begin{vmatrix} -1-r & 1 \\ 1 & -1-r \end{vmatrix} + \begin{vmatrix} 2 & 1 \\ 0 & -1-r \end{vmatrix} = 0$$

$$\Rightarrow \quad (-1-r)\left[(-1-r)^2 - 1\right] + 2(-1-r) = 0 \qquad \Rightarrow \qquad (1+r)\left(r^2 + 2r + 2\right) = 0.$$

From this equation and by using the quadratic formula, we see that the eigenvalues for this coefficient matrix will be $r=-1,-1\pm i$. The eigenvectors associated with the real eigenvalue $r=-1$ will be the vectors $\mathbf{u}=\operatorname{col}(u_1,u_2,u_3)$ which satisfy the equation

$$(\mathbf{A}+\mathbf{I})\mathbf{u}=\begin{bmatrix}0 & -1 & 0\\ 2 & 0 & 1\\ 0 & 1 & 0\end{bmatrix}\begin{bmatrix}u_1\\ u_2\\ u_3\end{bmatrix}=\begin{bmatrix}0\\ 0\\ 0\end{bmatrix},$$

which is equivalent to the system $u_2=0$, $2u_1+u_3=0$. Thus, by letting $u_1=1$ (which means that $u_3=-2$), we see that the vector

$$\mathbf{u}=\operatorname{col}(u_1,u_2,u_3)=\operatorname{col}(1,0,-2),$$

satisfies this equation and is, therefore, an eigenvector of the coefficient matrix associated with the eigenvalue $r=-1$. Hence, according to Problem 42 of Exercises 9.5, we see that a solution to this Cauchy-Euler system will be given by

$$\mathbf{x}_1(t)=t^{-1}\mathbf{u}_1=t^{-1}\begin{bmatrix}1\\ 0\\ -2\end{bmatrix}=\begin{bmatrix}t^{-1}\\ 0\\ -2t^{-1}\end{bmatrix}.$$

To find the eigenvectors $\mathbf{z}=\operatorname{col}(z_1,z_2,z_3)$ associated with the complex eigenvalue $r=-1+i$, we solve the equation

$$[\mathbf{A}-(-1+i)\mathbf{I}]\mathbf{z}=\begin{bmatrix}-i & -1 & 0\\ 2 & -i & 1\\ 0 & 1 & -i\end{bmatrix}\begin{bmatrix}z_1\\ z_2\\ z_3\end{bmatrix}=\begin{bmatrix}0\\ 0\\ 0\end{bmatrix}$$

$$\Rightarrow\quad \begin{bmatrix}-i & -1 & 0\\ 0 & i & 1\\ 0 & 0 & 0\end{bmatrix}\begin{bmatrix}z_1\\ z_2\\ z_3\end{bmatrix}=\begin{bmatrix}0\\ 0\\ 0\end{bmatrix},$$

which is equivalent to the system $-iz_1-z_2=0$, $iz_2+z_3=0$. Thus, if we let $z_1=1$, we must let $z_2=-i$ and $z_3=-1$ in order to satisfy this system. Therefore, one eigenvector for the coefficient matrix associated with the eigenvalue $r=-1+i$ will be the vector $\mathbf{z}=\operatorname{col}(1,-i,-1)$ and another solution to this system will be $\mathbf{x}(t)=t^{-1+i}\mathbf{z}$. We would like to find real solutions to this problem. Therefore, we note that by Euler's formula we have

$$t^{-1+i}=t^{-1}\cdot t^{i}=t^{-1}e^{i\ln t}=t^{-1}[\cos(\ln t)+i\sin(\ln t)],$$

where we have made use of our assumption that $t>0$. Hence, the solution that we have just found becomes

$$\mathbf{x}(t)=t^{-1+i}\mathbf{z}=t^{-1}[\cos(\ln t)+i\sin(\ln t)]\mathbf{z}$$

$$= t^{-1}[\cos(\ln t) + i\sin(\ln t)]\begin{bmatrix} 1 \\ -i \\ -1 \end{bmatrix}$$

$$= \begin{bmatrix} t^{-1}\cos(\ln t) \\ t^{-1}\sin(\ln t) \\ -t^{-1}\cos(\ln t) \end{bmatrix} + i\begin{bmatrix} t^{-1}\sin(\ln t) \\ -t^{-1}\cos(\ln t) \\ -t^{-1}\sin(\ln t) \end{bmatrix}.$$

Thus, by Lemma 2 (adapted to systems) on page 193 of the text we see that two more linearly independent solutions to this Cauchy-Euler system will be

$$\mathbf{x}_2(t) = \begin{bmatrix} t^{-1}\cos(\ln t) \\ t^{-1}\sin(\ln t) \\ -t^{-1}\cos(\ln t) \end{bmatrix} \quad \text{and} \quad \mathbf{x}_3(t) = \begin{bmatrix} t^{-1}\sin(\ln t) \\ -t^{-1}\cos(\ln t) \\ -t^{-1}\sin(\ln t) \end{bmatrix},$$

and, hence, a general solution will be given by

$$\mathbf{x}(t) = c_1\begin{bmatrix} t^{-1} \\ 0 \\ -2t^{-1} \end{bmatrix} + c_2\begin{bmatrix} t^{-1}\cos(\ln t) \\ t^{-1}\sin(\ln t) \\ -t^{-1}\cos(\ln t) \end{bmatrix} + c_3\begin{bmatrix} t^{-1}\sin(\ln t) \\ -t^{-1}\cos(\ln t) \\ -t^{-1}\sin(\ln t) \end{bmatrix}.$$

EXERCISES 9.7: Nonhomogeneous Linear Systems, page 579

3. We must first find the general solution to the corresponding homogeneous system. Therefore, we first find the eigenvalues for the coefficient matrix **A** by solving the characteristic equation given by

$$|\mathbf{A} - r\mathbf{I}| = \begin{vmatrix} 1-r & -2 & 2 \\ -2 & 1-r & 2 \\ 2 & 2 & 1-r \end{vmatrix} = 0$$

$$\Rightarrow \quad (1-r)\begin{vmatrix} 1-r & 2 \\ 2 & 1-r \end{vmatrix} + 2\begin{vmatrix} -2 & 2 \\ 2 & 1-r \end{vmatrix} + 2\begin{vmatrix} -2 & 1-r \\ 2 & 2 \end{vmatrix} = 0$$

$$\Rightarrow \quad (1-r)\left[(1-r)^2 - 4\right] + 2\left[-2(1-r) - 4\right] + 2\left[-4 - 2(1-r)\right] = 0$$

$$\Rightarrow \quad (1-r)\left(r^2 - 2r - 3\right) + 4(2r - 6) = 0$$

$$\Rightarrow \quad (1-r)(r-3)(r+1) + 8(r-3) = 0$$

$$\Rightarrow \quad (r-3)\left(r^2 - 9\right) = 0$$

$$\Rightarrow \qquad (r-3)(r-3)(r+3)=0.$$

Thus, the eigenvalues for the matrix **A** are r = 3, –3, where r = 3 is an eigenvalue of multiplicity two. Notice that, even though the matrix **A** has only two distinct eigenvalues, we are still guaranteed three linearly independent eigenvectors because **A** is a 3 × 3 real symmetric matrix. To find an eigenvector associated with the eigenvalue r = –3, we must find a vector $\mathbf{u}=\mathrm{col}(u_1,u_2,u_3)$ which satisfies the system

$$(\mathbf{A}+3\mathbf{I})\mathbf{u}=\begin{bmatrix}4 & -2 & 2\\ -2 & 4 & 2\\ 2 & 2 & 4\end{bmatrix}\begin{bmatrix}u_1\\ u_2\\ u_3\end{bmatrix}=\begin{bmatrix}0\\ 0\\ 0\end{bmatrix}$$

$$\Rightarrow \quad \begin{bmatrix}1 & 0 & 1\\ 0 & 1 & 1\\ 0 & 0 & 0\end{bmatrix}\begin{bmatrix}u_1\\ u_2\\ u_3\end{bmatrix}=\begin{bmatrix}0\\ 0\\ 0\end{bmatrix},$$

which is equivalent to the system $u_1+u_3=0$, $u_2+u_3=0$. Hence, by letting $u_3=-1$, we must have $u_1=u_2=1$, and so the vector $\mathbf{u}_1=\mathrm{col}(1,1,-1)$ will then satisfy the above system. Therefore, this vector is an eigenvector for the matrix **A** associated with the eigenvalue r = –3. Thus, one solution to the corresponding homogeneous system is given by

$$\mathbf{x}_1(t)=e^{-3t}\mathbf{u}_1=e^{-3t}\begin{bmatrix}1\\ 1\\ -1\end{bmatrix}.$$

To find eigenvectors $\mathbf{u}=\mathrm{col}(u_1,u_2,u_3)$ associated with the eigenvalue r = 3, we solve the equation given by

$$(\mathbf{A}-3\mathbf{I})\mathbf{u}=\begin{bmatrix}-2 & -2 & 2\\ -2 & -2 & 2\\ 2 & 2 & -2\end{bmatrix}\begin{bmatrix}u_1\\ u_2\\ u_3\end{bmatrix}=\begin{bmatrix}0\\ 0\\ 0\end{bmatrix},$$

which is equivalent to the equation $u_1+u_2-u_3=0$. Thus, if we let $u_3=s$ and $u_2=v$, then we must have $u_1=s-v$. Hence, solutions to the above equation and, therefore, eigenvectors for **A** associated with the eigenvalue $r=-3$ will be the vectors

$$\mathbf{u}=\begin{bmatrix}s-v\\ v\\ s\end{bmatrix}=s\begin{bmatrix}1\\ 0\\ 1\end{bmatrix}+v\begin{bmatrix}-1\\ 1\\ 0\end{bmatrix},$$

where s and v are arbitrary scalars. Therefore, letting s = 1 and v = 0 yields the eigenvector $\mathbf{u}_2=\mathrm{col}(1,0,1)$. Similarly, by letting s = 0 and v = 1, we obtain the eigenvector $\mathbf{u}_3=\mathrm{col}(-1,1,0)$, which we can see by inspection is linearly independent from $\mathbf{u}_2$. Hence, two more solutions to the corresponding homogeneous system which are linearly independent from each other and from $\mathbf{x}_1(t)$ are given by

$$\mathbf{x}_2(t) = e^{3t}\mathbf{u}_2 = e^{3t}\begin{bmatrix} 1 \\ 0 \\ 1 \end{bmatrix} \quad \text{and} \quad \mathbf{x}_3(t) = e^{3t}\mathbf{u}_3 = e^{3t}\begin{bmatrix} -1 \\ 1 \\ 0 \end{bmatrix}.$$

Thus, the general solution to the corresponding homogeneous system will be

$$\mathbf{x}_h(t) = c_1 e^{-3t}\begin{bmatrix} 1 \\ 1 \\ -1 \end{bmatrix} + c_2 e^{3t}\begin{bmatrix} 1 \\ 0 \\ 1 \end{bmatrix} + c_3 e^{3t}\begin{bmatrix} -1 \\ 1 \\ 0 \end{bmatrix}.$$

To find a particular solution to the nonhomogeneous system, we note that

$$\boldsymbol{f}(t) = \begin{bmatrix} 2e^t \\ 4e^t \\ -2e^t \end{bmatrix} = e^t\begin{bmatrix} 2 \\ 4 \\ -2 \end{bmatrix} = e^t\mathbf{g},$$

where $\mathbf{g} = \text{col}(2,4,-2)$. Therefore, we will assume that a particular solution to the nonhomogeneous system will have the form $\mathbf{x}_p(t) = e^t\mathbf{a}$, where $\mathbf{a} = \text{col}(a_1, a_2, a_3)$ is a constant vector which must be determined. Hence, we see that $\mathbf{x}'_p(t) = e^t\mathbf{a}$. By substituting $\mathbf{x}_p(t)$ into the given system, we obtain

$$e^t\mathbf{a} = \mathbf{A}\mathbf{x}_p(t) + \boldsymbol{f}(t) = \mathbf{A}e^t\mathbf{a} + e^t\mathbf{g} = e^t\mathbf{A}\mathbf{a} + e^t\mathbf{g}.$$

Therefore, we have

$$e^t\mathbf{a} = e^t\mathbf{A}\mathbf{a} + e^t\mathbf{g} \quad \Rightarrow \quad \mathbf{a} = \mathbf{A}\mathbf{a} + \mathbf{g} \quad \Rightarrow \quad (\mathbf{I} - \mathbf{A})\mathbf{a} = \mathbf{g}$$

$$\Rightarrow \quad \begin{bmatrix} 0 & 2 & -2 \\ 2 & 0 & -2 \\ -2 & -2 & 0 \end{bmatrix}\begin{bmatrix} a_1 \\ a_2 \\ a_3 \end{bmatrix} = \begin{bmatrix} 2 \\ 4 \\ -2 \end{bmatrix}.$$

The last equation above can be solved by either performing elementary row operations on the augmented matrix or by solving the system

$$2a_2 - 2a_3 = 2,$$
$$2a_1 - 2a_3 = 4,$$
$$-2a_1 - 2a_2 = -2.$$

Either way, we obtain $a_1 = 1$, $a_2 = 0$, and $a_3 = -1$. Thus, a particular solution to the nonhomogeneous system will be given by

$$\mathbf{x}_p(t) = e^t\mathbf{a} = e^t\begin{bmatrix} 1 \\ 0 \\ -1 \end{bmatrix},$$

and so the general solution to the nonhomogeneous system will be

$$\mathbf{x}(t) = \mathbf{x}_h(t) + \mathbf{x}_p(t)$$

$$= c_1e^{-3t}\begin{bmatrix} 1 \\ 1 \\ -1 \end{bmatrix} + c_2e^{3t}\begin{bmatrix} 1 \\ 0 \\ 1 \end{bmatrix} + c_3e^{3t}\begin{bmatrix} -1 \\ 1 \\ 0 \end{bmatrix} + e^t\begin{bmatrix} 1 \\ 0 \\ -1 \end{bmatrix}.$$

13. We must first find a fundamental matrix for the corresponding homogeneous system $\mathbf{x}' = \mathbf{A}\mathbf{x}$. To this end, we first find the eigenvalues of the matrix $\mathbf{A}$ by solving the characteristic equation given by

$$|\mathbf{A} - r\mathbf{I}| = \begin{vmatrix} 2-r & 1 \\ -3 & -2-r \end{vmatrix} = 0$$

$$\Rightarrow \quad (2-r)(-2-r)+3=0 \quad \Rightarrow \quad r^2 - 1 = 0.$$

Thus, the eigenvalues of the coefficient matrix $\mathbf{A}$ are $r = \pm 1$. The eigenvectors associated with the eigenvalue $r = 1$ are the vectors $\mathbf{u} = \operatorname{col}(u_1, u_2)$ which satisfy the equation

$$(\mathbf{A} - \mathbf{I})\mathbf{u} = \begin{bmatrix} 1 & 1 \\ -3 & -3 \end{bmatrix}\begin{bmatrix} u_1 \\ u_2 \end{bmatrix} = \begin{bmatrix} 0 \\ 0 \end{bmatrix}.$$

This equation is equivalent to the equation $u_1 + u_2 = 0$. Therefore, if we let $u_1 = 1$, then we have $u_2 = -1$, so one eigenvector of the matrix $\mathbf{A}$ associated with the eigenvalue $r = 1$ is the vector $\mathbf{u}_1 = \operatorname{col}(1, -1)$. Hence, one solution of the corresponding homogeneous system is given by

$$\mathbf{x}_1(t) = e^t\mathbf{u}_1 = e^t\begin{bmatrix} 1 \\ -1 \end{bmatrix} = \begin{bmatrix} e^t \\ -e^t \end{bmatrix}.$$

To find an eigenvector associated with the eigenvalue $r = -1$, we solve the equation

$$(\mathbf{A} + \mathbf{I})\mathbf{u} = \begin{bmatrix} 3 & 1 \\ -3 & -1 \end{bmatrix}\begin{bmatrix} u_1 \\ u_2 \end{bmatrix} = \begin{bmatrix} 0 \\ 0 \end{bmatrix},$$

which is equivalent to the equation $3u_1 + u_2 = 0$. Since $u_1 = 1$ and $u_2 = -3$ satisfy this equation, one eigenvector for the matrix $\mathbf{A}$ associated with the eigenvalue $r = -1$ is the vector $\mathbf{u}_2 = \operatorname{col}(1, -3)$. Thus, another linearly independent solution of the corresponding homogeneous system is

$$\mathbf{x}_2(t) = e^{-t}\mathbf{u}_2 = e^{-t}\begin{bmatrix} 1 \\ -3 \end{bmatrix} = \begin{bmatrix} e^{-t} \\ -3e^{-t} \end{bmatrix}.$$

Hence, the general solution of the homogeneous system is given by

$$\mathbf{x}_h(t) = c_1 e^{t}\begin{bmatrix} 1 \\ -1 \end{bmatrix} + c_2 e^{-t}\begin{bmatrix} 1 \\ -3 \end{bmatrix},$$

and a fundamental matrix is

$$\mathbf{X}(t) = \begin{bmatrix} e^{t} & e^{-t} \\ -e^{t} & -3e^{-t} \end{bmatrix}.$$

To find the inverse matrix $\mathbf{X}^{-1}(t)$, we will perform row-reduction on the matrix $[\mathbf{X}(t)|\mathbf{I}]$. Thus, we have

$$[\mathbf{X}(t)|\mathbf{I}] = \left[\begin{array}{cc|cc} e^{t} & e^{-t} & 1 & 0 \\ -e^{t} & -3e^{-t} & 0 & 1 \end{array}\right]$$

$$\Rightarrow \quad \left[\begin{array}{cc|cc} e^{t} & e^{-t} & 1 & 0 \\ 0 & -2e^{-t} & 1 & 1 \end{array}\right]$$

$$\Rightarrow \quad \left[\begin{array}{cc|cc} e^{t} & 0 & \frac{3}{2} & \frac{1}{2} \\ 0 & e^{-t} & \frac{-1}{2} & \frac{-1}{2} \end{array}\right]$$

$$\Rightarrow \quad \left[\begin{array}{cc|cc} 1 & 0 & \frac{3}{2}e^{-t} & \frac{1}{2}e^{-t} \\ 0 & 1 & \frac{-1}{2}e^{t} & \frac{-1}{2}e^{t} \end{array}\right].$$

Therefore, we see that

$$\mathbf{X}^{-1}(t) = \begin{bmatrix} \frac{3}{2}e^{-t} & \frac{1}{2}e^{-t} \\ \frac{-1}{2}e^{t} & \frac{-1}{2}e^{t} \end{bmatrix}.$$

Hence, we have

$$\mathbf{X}^{-1}(t)\boldsymbol{f}(t) = \begin{bmatrix} \frac{3}{2}e^{-t} & \frac{1}{2}e^{-t} \\ \frac{-1}{2}e^{t} & \frac{-1}{2}e^{t} \end{bmatrix}\begin{bmatrix} 2e^{t} \\ 4e^{t} \end{bmatrix} = \begin{bmatrix} 5 \\ -3e^{2t} \end{bmatrix},$$

and so we have

$$\int \mathbf{X}^{-1}(t)f(t)\,dt = \begin{bmatrix} \int 5dt \\ -3\int e^{2t}\,dt \end{bmatrix} = \begin{bmatrix} 5t \\ \frac{-3}{2}e^{2t} \end{bmatrix},$$

where we have taken the constants of integration to be zero. Thus, by equation (10) on page 577 of the text, we see that

$$\mathbf{x}_p(t) = \begin{bmatrix} e^t & e^{-t} \\ -e^t & -3e^{-t} \end{bmatrix} \begin{bmatrix} 5t \\ \frac{-3}{2}e^{2t} \end{bmatrix} = \begin{bmatrix} 5te^t - \frac{3}{2}e^t \\ -5te^t + \frac{9}{2}e^t \end{bmatrix}.$$

Therefore, by adding $\mathbf{x}_h(t)$ and $\mathbf{x}_p(t)$ we obtain

$$\mathbf{x}(t) = c_1 e^t \begin{bmatrix} 1 \\ -1 \end{bmatrix} + c_2 e^{-t} \begin{bmatrix} 1 \\ -3 \end{bmatrix} + \begin{bmatrix} 5te^t - \frac{3}{2}e^t \\ -5te^t + \frac{9}{2}e^t \end{bmatrix}.$$

15. We must first find a fundamental matrix for the associated homogeneous system. We will do this by finding the solutions derived from the eigenvalues and the associated eigenvectors for the coefficient matrix **A**. Therefore, we find these eigenvalues by solving the characteristic equation given by

$$|\mathbf{A} - r\mathbf{I}| = \begin{vmatrix} -4-r & 2 \\ 2 & -1-r \end{vmatrix} = 0$$

$$\Rightarrow \quad (-4-r)(-1-r) - 4 = 0 \quad \Rightarrow \quad r^2 + 5r = 0.$$

Thus, the eigenvalues for the matrix **A** are $r = -5,\ 0$. An eigenvector for this matrix associated with the eigenvalue $r = 0$ is the vector $\mathbf{u} = \mathrm{col}(u_1, u_2)$ which satisfies the equation

$$\mathbf{Au} = \begin{bmatrix} -4 & 2 \\ 2 & -1 \end{bmatrix} \begin{bmatrix} u_1 \\ u_2 \end{bmatrix} = \begin{bmatrix} 0 \\ 0 \end{bmatrix}.$$

This equation is equivalent to the equation $2u_1 = u_2$. Therefore, if we let $u_1 = 1$ and $u_2 = 2$, then the vector $\mathbf{u}_1 = \mathrm{col}(1,2)$ satisfies this equation and is, therefore, an eigenvector for the matrix **A** associated with the eigenvalue $r = 0$. Hence, one solution to the homogeneous system is given by

$$\mathbf{x}_1(t) = e^0 \mathbf{u}_1 = \begin{bmatrix} 1 \\ 2 \end{bmatrix}.$$

To find an eigenvector associated with the eigenvalue $r = -5$, we solve the equation

$$(\mathbf{A}+5\mathbf{I})\mathbf{u}=\begin{bmatrix}1 & 2\\ 2 & 4\end{bmatrix}\begin{bmatrix}u_1\\ u_2\end{bmatrix}=\begin{bmatrix}0\\ 0\end{bmatrix},$$

which is equivalent to the equation $u_1+2u_2=0$. Thus, by letting $u_2=1$ and $u_1=-2$, the vector $\mathbf{u}_2=\operatorname{col}(u_1,u_2)=\operatorname{col}(-2,1)$ satisfies this equation and is, therefore, an eigenvector for $\mathbf{A}$ associated with the eigenvalue $r=-5$. Hence, since the two eigenvalues of $\mathbf{A}$ are distinct, we see that another linearly independent solution to the corresponding homogeneous system is given by

$$\mathbf{x}_2(t)=e^{-5t}\mathbf{u}_2=e^{-5t}\begin{bmatrix}-2\\ 1\end{bmatrix}=\begin{bmatrix}-2e^{-5t}\\ e^{-5t}\end{bmatrix}.$$

By combining these two solutions, we see that a general solution to the homogeneous system is

$$\mathbf{x}_h(t)=c_1\begin{bmatrix}1\\ 2\end{bmatrix}+c_2e^{-5t}\begin{bmatrix}-2\\ 1\end{bmatrix}$$

and a fundamental matrix for this system is the matrix

$$\mathbf{X}(t)=\begin{bmatrix}1 & -2e^{-5t}\\ 2 & e^{-5t}\end{bmatrix}.$$

We will use equation (10) on page 577 of the text to find a particular solution to the nonhomogeneous system. Thus, we need to find the inverse matrix $\mathbf{X}^{-1}(t)$. This can be done, for example, by performing row-reduction on the matrix $[\mathbf{X}(t)|\mathbf{I}]$ to obtain the matrix $[\mathbf{I}|\mathbf{X}^{-1}(t)]$. In this way, we find that the required inverse matrix is given by

$$\mathbf{X}^{-1}(t)=\begin{bmatrix}\frac{1}{5} & \frac{2}{5}\\ \frac{-2}{5}e^{5t} & \frac{1}{5}e^{5t}\end{bmatrix}.$$

Therefore, we have

$$\mathbf{X}^{-1}(t)\boldsymbol{f}(t)=\begin{bmatrix}\frac{1}{5} & \frac{2}{5}\\ \frac{-2}{5}e^{5t} & \frac{1}{5}e^{5t}\end{bmatrix}\begin{bmatrix}t^{-1}\\ 4+2t^{-1}\end{bmatrix}=\begin{bmatrix}t^{-1}+\frac{8}{5}\\ \frac{4}{5}e^{5t}\end{bmatrix}.$$

From this we see that

$$\int\mathbf{X}^{-1}(t)\boldsymbol{f}(t)\,dt=\begin{bmatrix}\int\left(t^{-1}+\frac{8}{5}\right)dt\\ \int\frac{4}{5}e^{5t}\,dt\end{bmatrix}=\begin{bmatrix}\ln|t|+\frac{8}{5}t\\ \frac{4}{25}e^{5t}\end{bmatrix},$$

where we have taken the constants of integration to be zero. Hence, by equation (10) on page 577 of the text, we obtain

$$\mathbf{x}_p(t) = \mathbf{X}(t)\int \mathbf{X}^{-1}(t)\boldsymbol{f}(t)\,dt =$$

$$= \begin{bmatrix} 1 & -2e^{-5t} \\ 2 & e^{-5t} \end{bmatrix} \begin{bmatrix} \ln|t| + \frac{8}{5}t \\ \frac{4}{25}e^{5t} \end{bmatrix} = \begin{bmatrix} \ln|t| + \frac{8}{5}t - \frac{8}{25} \\ 2\ln|t| + \frac{16}{5}t + \frac{4}{25} \end{bmatrix}.$$

Adding $\mathbf{x}_h(t)$ and $\mathbf{x}_p(t)$ yields the general solution to the nonhomogeneous system given by

$$\mathbf{x}(t) = c_1 \begin{bmatrix} 1 \\ 2 \end{bmatrix} + c_2 e^{-5t} \begin{bmatrix} -2 \\ 1 \end{bmatrix} + \begin{bmatrix} \ln|t| + \frac{8}{5}t - \frac{8}{25} \\ 2\ln|t| + \frac{16}{5}t + \frac{4}{25} \end{bmatrix}.$$

21. We will find the solution to this initial value problem by using equation (13) on page 578 of the text. Therefore, we must first find a fundamental matrix for the associated homogeneous system. This means that we must find the eigenvalues and corresponding eigenvectors for the coefficient matrix of this system by solving the characteristic equation

$$|\mathbf{A} - r\mathbf{I}| = \begin{vmatrix} -r & 2 \\ -1 & 3-r \end{vmatrix} = 0$$

$$\Rightarrow \quad -r(3-r)+2=0 \quad \Rightarrow \quad r^2 - 3r + 2 = 0 \quad \Rightarrow \quad (r-2)(r-1)=0.$$

Hence, $r = 1, 2$ are the eigenvalues for this matrix. To find an eigenvector $\mathbf{u} = \text{col}(u_1, u_2)$ for this coefficient matrix associated with the eigenvalue $r = 1$, we solve the system

$$(\mathbf{A} - \mathbf{I})\mathbf{u} = \begin{bmatrix} -1 & 2 \\ -1 & 2 \end{bmatrix} \begin{bmatrix} u_1 \\ u_2 \end{bmatrix} = \begin{bmatrix} 0 \\ 0 \end{bmatrix}.$$

This system is equivalent to the equation $u_1 = 2u_2$. Thus, $u_1 = 2$ and $u_2 = 1$ is a set of values which satisfies this equation and, therefore, the vector $\mathbf{u}_1 = \text{col}(2,1)$ is an eigenvector for the coefficient matrix corresponding to the eigenvalue $r = 1$. Hence, one solution to the homogeneous system is given by

$$\mathbf{x}_1(t) = e^t \mathbf{u}_1 = e^t \begin{bmatrix} 2 \\ 1 \end{bmatrix} = \begin{bmatrix} 2e^t \\ e^t \end{bmatrix}.$$

Similarly, by solving the equation

$$(\mathbf{A} - 2\mathbf{I})\mathbf{u} = \begin{bmatrix} -2 & 2 \\ -1 & 1 \end{bmatrix} \begin{bmatrix} u_1 \\ u_2 \end{bmatrix} = \begin{bmatrix} 0 \\ 0 \end{bmatrix},$$

we find that one eigenvector for the coefficient matrix associated with the eigenvalue $r = 2$ is $\mathbf{u}_2 = \text{col}(u_1, u_2) = \text{col}(1,1)$. Thus, another linearly independent solution to the associated homogeneous problem is given by

$$\mathbf{x}_2(t) = e^{2t}\mathbf{u}_2 = e^{2t}\begin{bmatrix} 1 \\ 1 \end{bmatrix} = \begin{bmatrix} e^{2t} \\ e^{2t} \end{bmatrix}.$$

By combining these two solutions, we obtain a general solution to the homogeneous system

$$\mathbf{x}_h(t) = c_1 e^t \begin{bmatrix} 2 \\ 1 \end{bmatrix} + c_2 e^{2t} \begin{bmatrix} 1 \\ 1 \end{bmatrix},$$

and the fundamental matrix

$$\mathbf{X}(t) = \begin{bmatrix} 2e^t & e^{2t} \\ e^t & e^{2t} \end{bmatrix}.$$

In order to use equation (13) on page 578 of the text, we must also find the inverse of the fundamental matrix. One way of doing this is to perform row-reduction on the matrix $[\mathbf{X}(t)|\mathbf{I}]$ to obtain the matrix $[\mathbf{I}|\mathbf{X}^{-1}(t)]$. Thus, we find that

$$\mathbf{X}^{-1}(t) = \begin{bmatrix} e^{-t} & -e^{-t} \\ -e^{-2t} & 2e^{-2t} \end{bmatrix}.$$

From this we see that

$$\mathbf{X}^{-1}(s)\mathbf{f}(s) = \begin{bmatrix} e^{-s} & -e^{-s} \\ -e^{-2s} & 2e^{-2s} \end{bmatrix}\begin{bmatrix} e^s \\ -e^s \end{bmatrix} = \begin{bmatrix} 2 \\ -3e^{-s} \end{bmatrix}.$$

(a) Using the initial condition $\mathbf{x}(0) = \begin{bmatrix} 5 \\ 4 \end{bmatrix}$, and $t_0 = 0$, we have

$$\mathbf{X}^{-1}(0) = \begin{bmatrix} 1 & -1 \\ -1 & 2 \end{bmatrix}.$$

Therefore

$$\int_{t_0}^{t} \mathbf{X}^{-1}(s)\mathbf{f}(s)\,ds = \int_0^t \mathbf{X}^{-1}(s)\mathbf{f}(s)\,ds = \begin{bmatrix} \int_0^t 2\,ds \\ -\int_0^t 3e^{-s}\,ds \end{bmatrix} = \begin{bmatrix} 2t \\ 3e^{-t} - 3 \end{bmatrix},$$

from which it follows that

$$\mathbf{X}(t)\int_{t_0}^{t}\mathbf{X}^{-1}(s)\boldsymbol{f}(s)\,ds=\begin{bmatrix}2e^{t} & e^{2t}\\ e^{t} & e^{2t}\end{bmatrix}\begin{bmatrix}2t\\ 3e^{-t}-3\end{bmatrix}=\begin{bmatrix}4te^{t}+3e^{t}-3e^{2t}\\ 2te^{t}+3e^{t}-3e^{2t}\end{bmatrix}.$$

We also find that

$$\mathbf{X}(t)\mathbf{X}^{-1}(t_0)\mathbf{x}_0=\begin{bmatrix}2e^{t} & e^{2t}\\ e^{t} & e^{2t}\end{bmatrix}\begin{bmatrix}1 & -1\\ -1 & 2\end{bmatrix}\begin{bmatrix}5\\ 4\end{bmatrix}$$

$$=\begin{bmatrix}2e^{t} & e^{2t}\\ e^{t} & e^{2t}\end{bmatrix}\begin{bmatrix}1\\ 3\end{bmatrix}=\begin{bmatrix}2e^{t}+3e^{2t}\\ e^{t}+3e^{2t}\end{bmatrix}.$$

Hence, by substituting these expressions into equation (13) on page 578 of the text, we obtain the solution to this initial value problem given by

$$\mathbf{x}(t)=\mathbf{X}(t)\mathbf{X}^{-1}(t_0)\mathbf{x}_0+\mathbf{X}(t)\int_{t_0}^{t}\mathbf{X}^{-1}(s)\boldsymbol{f}(s)\,ds$$

$$=\begin{bmatrix}2e^{t}+3e^{2t}\\ e^{t}+3e^{2t}\end{bmatrix}+\begin{bmatrix}4te^{t}+3e^{t}-3e^{2t}\\ 2te^{t}+3e^{t}-3e^{2t}\end{bmatrix}=\begin{bmatrix}4te^{t}+5e^{t}\\ 2te^{t}+4e^{t}\end{bmatrix}.$$

(b) Using the initial condition $\mathbf{x}(1)=\begin{bmatrix}0\\ 1\end{bmatrix}$, and $t_0=1$, we have

$$\mathbf{X}^{-1}(1)=\begin{bmatrix}e^{-1} & -e^{-1}\\ -e^{-2} & 2e^{-2}\end{bmatrix}.$$

Therefore

$$\int_{t_0}^{t}\mathbf{X}^{-1}(s)\boldsymbol{f}(s)\,ds=\int_{1}^{t}\mathbf{X}^{-1}(s)\boldsymbol{f}(s)\,ds=\begin{bmatrix}\int_1^t 2\,ds\\ -\int_1^t 3e^{-s}\,ds\end{bmatrix}=\begin{bmatrix}2t-2\\ 3e^{-t}-3e^{-1}\end{bmatrix},$$

from which it follows that

$$\mathbf{X}(t)\int_{t_0}^{t}\mathbf{X}^{-1}(s)\boldsymbol{f}(s)\,ds=\begin{bmatrix}2e^{t} & e^{2t}\\ e^{t} & e^{2t}\end{bmatrix}\begin{bmatrix}2t-2\\ 3e^{-t}-3e^{-1}\end{bmatrix}$$

$$=\begin{bmatrix}4te^{t}-4e^{t}+3e^{t}-3e^{2t-1}\\ 2te^{t}-2e^{t}+3e^{t}-3e^{2t-1}\end{bmatrix}=\begin{bmatrix}4te^{t}-e^{t}-3e^{2t-1}\\ 2te^{t}+e^{t}-3e^{2t-1}\end{bmatrix}.$$

We also find that

$$\mathbf{X}(t)\mathbf{X}^{-1}(t_0)\mathbf{x}_0 = \begin{bmatrix} 2e^t & e^{2t} \\ e^t & e^{2t} \end{bmatrix} \begin{bmatrix} e^{-1} & -e^{-1} \\ -e^{-2} & 2e^{-2} \end{bmatrix} \begin{bmatrix} 0 \\ 1 \end{bmatrix}$$

$$= \begin{bmatrix} 2e^t & e^{2t} \\ e^t & e^{2t} \end{bmatrix} \begin{bmatrix} -e^{-1} \\ 2e^{-2} \end{bmatrix} = \begin{bmatrix} -2e^{t-1} + 2e^{2t-2} \\ -e^{t-1} + 2e^{2t-2} \end{bmatrix}.$$

Hence, by substituting these expressions into equation (13) on page 578 of the text, we obtain the solution to this initial value problem given by

$$\mathbf{x}(t) = \mathbf{X}(t)\mathbf{X}^{-1}(t_0)\mathbf{x}_0 + \mathbf{X}(t)\int_{t_0}^{t} \mathbf{X}^{-1}(s)\mathbf{f}(s)\,ds$$

$$= \begin{bmatrix} -2e^{t-1} + 2e^{2t-2} \\ -e^{t-1} + 2e^{2t-2} \end{bmatrix} + \begin{bmatrix} 4te^t - e^t - 3e^{2t-1} \\ 2te^t + e^t - 3e^{2t-1} \end{bmatrix}$$

$$= \begin{bmatrix} -2e^{t-1} + 2e^{2t-2} + 4te^t - e^t - 3e^{2t-1} \\ -e^{t-1} + 2e^{2t-2} + 2te^t + e^t - 3e^{2t-1} \end{bmatrix}.$$

(c) Using the initial condition $\mathbf{x}(5) = \begin{bmatrix} 1 \\ 0 \end{bmatrix}$, and $t_0 = 5$, we have

$$\mathbf{X}^{-1}(5) = \begin{bmatrix} e^{-5} & -e^{-5} \\ -e^{-10} & 2e^{-10} \end{bmatrix}.$$

Therefore

$$\int_{t_0}^{t} \mathbf{X}^{-1}(s)\mathbf{f}(s)\,ds = \int_{5}^{t} \mathbf{X}^{-1}(s)\mathbf{f}(s)\,ds = \begin{bmatrix} \int_5^t 2\,ds \\ -\int_5^t 3e^{-s}\,ds \end{bmatrix} = \begin{bmatrix} 2t - 10 \\ 3e^{-t} - 3e^{-5} \end{bmatrix},$$

from which it follows that

$$\mathbf{X}(t)\int_{t_0}^{t} \mathbf{X}^{-1}(s)\mathbf{f}(s)\,ds = \begin{bmatrix} 2e^t & e^{2t} \\ e^t & e^{2t} \end{bmatrix} \begin{bmatrix} 2t - 10 \\ 3e^{-t} - 3e^{-5} \end{bmatrix}$$

$$= \begin{bmatrix} 4te^t - 20e^t + 3e^t - 3e^{2t-5} \\ 2te^t - 10e^t + 3e^t - 3e^{2t-5} \end{bmatrix} = \begin{bmatrix} 4te^t - 17e^t - 3e^{2t-5} \\ 2te^t - 7e^t - 3e^{2t-5} \end{bmatrix}.$$

We also find that

$$\mathbf{X}(t)\mathbf{X}^{-1}(t_0)\mathbf{x}_0 = \begin{bmatrix} 2e^t & e^{2t} \\ e^t & e^{2t} \end{bmatrix}\begin{bmatrix} e^{-5} & -e^{-5} \\ -e^{-10} & 2e^{-10} \end{bmatrix}\begin{bmatrix} 1 \\ 0 \end{bmatrix}$$

$$= \begin{bmatrix} 2e^t & e^{2t} \\ e^t & e^{2t} \end{bmatrix}\begin{bmatrix} e^{-5} \\ -e^{-10} \end{bmatrix} = \begin{bmatrix} 2e^{t-5} - e^{2t-10} \\ e^{t-5} - 2e^{2t-10} \end{bmatrix}.$$

Hence, by substituting these expressions into equation (13) on page 578 of the text, we obtain the solution to this initial value problem given by

$$\mathbf{x}(t) = \mathbf{X}(t)\mathbf{X}^{-1}(t_0)\mathbf{x}_0 + \mathbf{X}(t)\int_{t_0}^{t} \mathbf{X}^{-1}(s)\boldsymbol{f}(s)\,ds$$

$$= \begin{bmatrix} 2e^{t-5} - e^{2t-10} \\ e^{t-5} - 2e^{2t-10} \end{bmatrix} + \begin{bmatrix} 4te^t - 17e^t - 3e^{2t-5} \\ 2te^t - 7e^t - 3e^{2t-5} \end{bmatrix} = \begin{bmatrix} 2e^{t-5} - e^{2t-10} + 4te^t - 17e^t - 3e^{2t-5} \\ e^{t-5} - 2e^{2t-10} + 2te^t - 7e^t - 3e^{2t-5} \end{bmatrix}.$$

25. **(a)** We will find a fundamental solutions set for the corresponding homogeneous system by deriving solutions using the eigenvalues and associated eigenvectors for the coefficient matrix. Therefore, we first solve the characteristic equation

$$|\mathbf{A} - r\mathbf{I}| = \begin{vmatrix} -r & 1 \\ -2 & 3-r \end{vmatrix} = 0$$

$$\Rightarrow \quad -r(3-r)+2=0 \quad \Rightarrow \quad r^2 - 3r + 2 = 0 \quad \Rightarrow \quad (r-2)(r-1)=0.$$

Thus, we see that the eigenvalues for the coefficient matrix of this problem are $r = 1, 2$. Since these eigenvalues are distinct, the associated eigenvectors will be linearly independent, and so the solutions derived from these eigenvectors will also be linearly independent. Wc find an eigenvector for this matrix associated with the eigenvalue $r = 1$ by solving the equation

$$(\mathbf{A} - \mathbf{I})\mathbf{u} = \begin{bmatrix} -1 & 1 \\ -2 & 2 \end{bmatrix}\begin{bmatrix} u_1 \\ u_2 \end{bmatrix} = \begin{bmatrix} 0 \\ 0 \end{bmatrix}.$$

Since the vector $\mathbf{u}_1 = \text{col}(u_1, u_2) = \text{col}(1,1)$ satisfies this equation, we see that this vector is one such eigenvector and so one solution to the homogeneous problem is given by

$$\mathbf{x}_1(t) = e^t\mathbf{u}_1 = e^t\begin{bmatrix} 1 \\ 1 \end{bmatrix}.$$

To find an eigenvector associated with the eigenvalue $r = 2$, we solve the equation

$$(\mathbf{A} - 2\mathbf{I})\mathbf{u} = \begin{bmatrix} -2 & 1 \\ -2 & 1 \end{bmatrix}\begin{bmatrix} u_1 \\ u_2 \end{bmatrix} = \begin{bmatrix} 0 \\ 0 \end{bmatrix}.$$

The vector $\mathbf{u_2} = \text{col}(u_1, u_2) = \text{col}(1,2)$ is one vector which satisfies this equation and so it is one eigenvector of the coefficient matrix associated with the eigenvalue $r = 2$. Thus, another linearly independent solution to the corresponding homogeneous problem is given by

$$\mathbf{x_2}(t) = e^{2t}\mathbf{u_2} = e^{2t}\begin{bmatrix} 1 \\ 2 \end{bmatrix},$$

and a fundamental solution set for this homogeneous system is the set

$$\{e^t\mathbf{u_1}, e^{2t}\mathbf{u_2}\}, \text{ where } \mathbf{u_1} = \text{col}(1,1) \text{ and } \mathbf{u_2} = \text{col}(1,2).$$

(b) If we assume that $\mathbf{x}_p(t) = te^t\mathbf{a}$ for some constant vector $\mathbf{a} = \text{col}(a_1, a_2)$, then we have

$$\mathbf{x}'_p(t) = te^t\mathbf{a} + e^t\mathbf{a} = \begin{bmatrix} te^t a_1 \\ te^t a_2 \end{bmatrix} + \begin{bmatrix} e^t a_1 \\ e^t a_2 \end{bmatrix} = \begin{bmatrix} te^t a_1 + e^t a_1 \\ te^t a_2 + e^t a_2 \end{bmatrix}.$$

We also have

$$\begin{bmatrix} 0 & 1 \\ -2 & 3 \end{bmatrix}\mathbf{x}_p(t) + f(t) = \begin{bmatrix} 0 & 1 \\ -2 & 3 \end{bmatrix}\begin{bmatrix} te^t a_1 \\ te^t a_2 \end{bmatrix} + \begin{bmatrix} e^t \\ 0 \end{bmatrix}$$

$$= \begin{bmatrix} te^t a_2 + e^t \\ -2te^t a_1 + 3te^t a_2 \end{bmatrix}.$$

Thus, if $\mathbf{x}_p(t) = te^t\mathbf{a}$ is to satisfy this system, we must have

$$\begin{bmatrix} te^t a_1 + e^t a_1 \\ te^t a_2 + e^t a_2 \end{bmatrix} = \begin{bmatrix} te^t a_2 + e^t \\ -2te^t a_1 + 3te^t a_2 \end{bmatrix}$$

which means that

$$te^t a_1 + e^t a_1 = te^t a_2 + e^t,$$

$$te^t a_2 + e^t a_2 = -2te^t a_1 + 3te^t a_2.$$

By dividing out the term e^t and equating coefficients, this system becomes the system

$$a_1 = a_2, \qquad a_1 = 1,$$

$$a_2 = -2a_1 + 3a_2, \qquad a_2 = 0.$$

Since this set of equations implies that $1 = a_1 = a_2 = 0$, which is of course impossible, we see that this system has no solutions. Therefore, we cannot find a vector $\mathbf{a}$ for which $\mathbf{x}_p(t) = te^t\mathbf{a}$ is a particular solution to this problem.

(c) Assuming that

$$\mathbf{x}_p(t) = te^t\mathbf{a} + e^t\mathbf{b} = \begin{bmatrix} te^t a_1 \\ te^t a_2 \end{bmatrix} + \begin{bmatrix} e^t b_1 \\ e^t b_2 \end{bmatrix} = \begin{bmatrix} te^t a_1 + e^t b_1 \\ te^t a_2 + e^t b_2 \end{bmatrix},$$

where $\mathbf{a} = \operatorname{col}(a_1, a_2)$ and $\mathbf{b} = \operatorname{col}(b_1, b_2)$ are two constant vectors, implies that

$$\mathbf{x}'_p(t) = te^t\mathbf{a} + e^t\mathbf{a} + e^t\mathbf{b} = \begin{bmatrix} te^t a_1 + e^t a_1 + e^t b_1 \\ te^t a_2 + e^t a_2 + e^t b_2 \end{bmatrix}.$$

We also see that

$$\begin{bmatrix} 0 & 1 \\ -2 & 3 \end{bmatrix}\mathbf{x}_p(t) + \boldsymbol{f}(t) = \begin{bmatrix} 0 & 1 \\ -2 & 3 \end{bmatrix}\begin{bmatrix} te^t a_1 + e^t b_1 \\ te^t a_2 + e^t b_2 \end{bmatrix} + \begin{bmatrix} e^t \\ 0 \end{bmatrix}$$

$$= \begin{bmatrix} te^t a_2 + e^t b_2 + e^t \\ -2te^t a_1 - 2e^t b_1 + 3te^t a_2 + 3e^t b_2 \end{bmatrix}.$$

Thus, for $\mathbf{x}_p(t)$ to satisfy the differential equation we must have

$$\begin{bmatrix} te^t a_1 + e^t a_1 + e^t b_1 \\ te^t a_2 + e^t a_2 + e^t b_2 \end{bmatrix} = \begin{bmatrix} te^t a_2 + e^t b_2 + e^t \\ -2te^t a_1 - 2e^t b_1 + 3te^t a_2 + 3e^t b_2 \end{bmatrix}, \tag{7}$$

which implies the system of equations given by

$$te^t a_1 + e^t a_1 + e^t b_1 = te^t a_2 + e^t b_2 + e^t,$$

$$te^t a_2 + e^t a_2 + e^t b_2 = -2te^t a_1 - 2e^t b_1 + 3te^t a_2 + 3e^t b_2.$$

Dividing each equation by e^t and equating the coefficients in the resulting equations yields the system

$$\begin{aligned} a_1 &= a_2, \\ a_1 + b_1 &= b_2 + 1, \\ a_2 &= -2a_1 + 3a_2, \\ a_2 + b_2 &= -2b_1 + 3b_2. \end{aligned} \tag{8}$$

Taking the pair of equations on the right and simplifying yields the system

$$\begin{aligned} b_1 - b_2 &= 1 - a_1, \\ 2b_1 - 2b_2 &= -a_2. \end{aligned} \tag{9}$$

By multiplying the first of these equations by 2, we obtain the system

$$2b_1 - 2b_2 = 2 - 2a_1,$$

$$2b_1 - 2b_2 = -a_2,$$

which when subtracted yields $2 - 2a_1 + a_2 = 0$. Applying the first equation in (8) (the equation $a_1 = a_2$) to this equation yields $a_1 = a_2 = 2$. By substituting these values for a_1 and a_2 into equation (9) above we see that both equations reduce to the equation

$$b_2 = b_1 + 1.$$

(Note also that the remaining equation in (8) reduces to the first equation in that set.) Thus, b_1 is free to be any value, say $b_1 = s$, and the set of values $a_1 = a_2 = 2$, $b_1 = s$, $b_2 = s + 1$, satisfies all of the equations given in (8) and, hence, the system given in (7). Therefore, particular solutions to the nonhomogeneous equation given in this problem are

$$\mathbf{x}_p(t) = te^t \begin{bmatrix} 2 \\ 2 \end{bmatrix} + e^t \begin{bmatrix} s \\ s+1 \end{bmatrix}$$

$$\mathbf{x}_p(t) = te^t \begin{bmatrix} 2 \\ 2 \end{bmatrix} + e^t \begin{bmatrix} 0 \\ 1 \end{bmatrix} + se^t \begin{bmatrix} 1 \\ 1 \end{bmatrix}.$$

But, since the vector $\mathbf{u} = e^t \mathrm{col}(1,1)$ is a solution to the corresponding homogeneous system, this term can be incorporated into the solution $\mathbf{x}_h(t)$ and we obtain one particular solution to this problem given by

$$\mathbf{x}_p(t) = te^t \begin{bmatrix} 2 \\ 2 \end{bmatrix} + e^t \begin{bmatrix} 0 \\ 1 \end{bmatrix}.$$

(d) To find the general solution to the nonhomogeneous system given in this problem, we first form the solution to the corresponding homogeneous system using the fundamental solution set found in part (a). Thus, we have

$$\mathbf{x}_h(t) = c_1 e^t \begin{bmatrix} 1 \\ 1 \end{bmatrix} + c_2 e^{2t} \begin{bmatrix} 1 \\ 2 \end{bmatrix}.$$

By adding the solution found in part (c) to this solution, we obtain the general solution given by

$$\mathbf{x}(t) = c_1 e^t \begin{bmatrix} 1 \\ 1 \end{bmatrix} + c_2 e^{2t} \begin{bmatrix} 1 \\ 2 \end{bmatrix} + te^t \begin{bmatrix} 2 \\ 2 \end{bmatrix} + e^t \begin{bmatrix} 0 \\ 1 \end{bmatrix}.$$

EXERCISES 9.8: The Matrix Exponential Function, page 590

3. **(a)** From the characteristic equation, $|\mathbf{A}-r\mathbf{I}|=0$, we obtain

$$|\mathbf{A}-r\mathbf{I}|=\begin{vmatrix} 2-r & 1 & -1 \\ -3 & -1-r & 1 \\ 9 & 3 & -4-r \end{vmatrix}=0$$

$$\Rightarrow \quad (2-r)\begin{vmatrix} -1-r & 1 \\ 3 & -4-r \end{vmatrix} - \begin{vmatrix} -3 & 1 \\ 9 & -4-r \end{vmatrix} - \begin{vmatrix} -3 & -1-r \\ 9 & 3 \end{vmatrix}=0$$

$$\Rightarrow \quad (2-r)[(-1-r)(-4-r)-3]-[-3(-4-r)-9]-[-9-9(-1-r)]=0$$

$$\Rightarrow \quad r^3+3r^2+3r+1=0$$

$$\Rightarrow \quad (r+1)^3=0.$$

Therefore, for the matrix **A**, $r=-1$ is an eigenvalue of multiplicity three. Thus, by the Cayley-Hamilton theorem as stated on page 585 of the text, we have

$$(\mathbf{A}+\mathbf{I})^3=\mathbf{0},$$

(so that $r=-1$ and $k=3$).

(b) In order to find $e^{\mathbf{A}t}$, we first notice (as was done in the text on page 585) that

$$e^{\mathbf{A}t}=e^{[-\mathbf{I}+(\mathbf{A}+\mathbf{I})]t}, \quad \text{commutative and associative properties of matrix addition}$$

$$=e^{-\mathbf{I}t}\cdot e^{(\mathbf{A}+\mathbf{I})t}, \quad \text{property (d) on page 583 of the text [since } (\mathbf{A}+\mathbf{I})\mathbf{I} = \mathbf{I}(\mathbf{A}+\mathbf{I})]$$

$$=e^{-t}\mathbf{I}e^{(\mathbf{A}+\mathbf{I})t}, \quad \text{property (e) on page 583 of the text}$$

$$=e^{-t}e^{(\mathbf{A}+\mathbf{I})t}.$$

Therefore, to find $e^{\mathbf{A}t}$ we need only to find $e^{(\mathbf{A}+\mathbf{I})t}$ then multiply the resulting expression by e^{-t}. By formula (2) on page 582 of the text and using the fact that $(\mathbf{A}+\mathbf{I})^3=0$ (which implies that $(\mathbf{A}+\mathbf{I})^n=0$ for $n\geq 3$), we have

$$e^{(\mathbf{A}+\mathbf{I})t}=\mathbf{I}+(\mathbf{A}+\mathbf{I})t+(\mathbf{A}+\mathbf{I})^2\left(\frac{t^2}{2}\right)+\cdots+(\mathbf{A}+\mathbf{I})^n\left(\frac{t^n}{n!}\right)+\cdots \quad (10)$$

$$= \mathbf{I} + (\mathbf{A} + \mathbf{I})t + (\mathbf{A} + \mathbf{I})^2 \left(\frac{t^2}{2} \right).$$

Since

$$(\mathbf{A} + \mathbf{I})^2 = \begin{bmatrix} 3 & 1 & -1 \\ -3 & 0 & 1 \\ 9 & 3 & -3 \end{bmatrix} \begin{bmatrix} 3 & 1 & -1 \\ -3 & 0 & 1 \\ 9 & 3 & -3 \end{bmatrix} = \begin{bmatrix} -3 & 0 & 1 \\ 0 & 0 & 0 \\ -9 & 0 & 3 \end{bmatrix},$$

equation (10) becomes

$$e^{(\mathbf{A}+\mathbf{I})t} = \begin{bmatrix} 1 & 0 & 0 \\ 0 & 1 & 0 \\ 0 & 0 & 1 \end{bmatrix} + \begin{bmatrix} 3t & t & -t \\ -3t & 0 & t \\ 9t & 3t & -3t \end{bmatrix} + \begin{bmatrix} -3\frac{t^2}{2} & 0 & \frac{t^2}{2} \\ 0 & 0 & 0 \\ -9\frac{t^2}{2} & 0 & 3\frac{t^2}{2} \end{bmatrix}$$

$$= \begin{bmatrix} 1 + 3t - \frac{3}{2}t^2 & t & -t + \frac{1}{2}t^2 \\ -3t & 1 & t \\ 9t - \frac{9}{2}t^2 & 3t & 1 - 3t + \frac{3}{2}t^2 \end{bmatrix}.$$

Hence, we have

$$e^{\mathbf{A}t} = e^{-t} \begin{bmatrix} 1 + 3t - \frac{3}{2}t^2 & t & -t + \frac{1}{2}t^2 \\ -3t & 1 & t \\ 9t - \frac{9}{2}t^2 & 3t & 1 - 3t + \frac{3}{2}t^2 \end{bmatrix}.$$

9. By equation (6) on page 586 of the text, we see that $e^{\mathbf{A}t} = \mathbf{X}(t)\mathbf{X}^{-1}(0)$, where $\mathbf{X}(t)$ is a fundamental matrix for the system $\mathbf{x}' = \mathbf{A}\mathbf{x}$. We will construct this fundamental matrix from three linearly independent solutions derived from the eigenvalues and associated eigenvectors for the matrix $\mathbf{A}$. Thus, we solve the characteristic equation

$$|\mathbf{A} - r\mathbf{I}| = \begin{vmatrix} -r & 1 & 0 \\ 0 & -r & 1 \\ 1 & -1 & 1-r \end{vmatrix} = 0$$

$$\Rightarrow \quad -r\begin{vmatrix} -r & 1 \\ -1 & 1-r \end{vmatrix} - \begin{vmatrix} 0 & 1 \\ 1 & 1-r \end{vmatrix} = 0$$

$$\Rightarrow \quad -r[-r(1-r)+1]+1=0$$

$$\Rightarrow \quad r^3 - r^2 + r - 1 = 0$$

$$\Rightarrow \quad (r-1)(r^2+1)=0.$$

Therefore, the eigenvalues of the matrix **A** are $r = 1, \pm i$. To find an eigenvector $\mathbf{u} = \operatorname{col}(u_1, u_2, u_3)$ associated with the eigenvalue $r = 1$, we solve the system,

$$(\mathbf{A} - \mathbf{I})\mathbf{u} = \mathbf{0}$$

$$\Rightarrow \quad \begin{bmatrix} -1 & 1 & 0 \\ 0 & -1 & 1 \\ 1 & -1 & 0 \end{bmatrix} \begin{bmatrix} u_1 \\ u_2 \\ u_3 \end{bmatrix} = \begin{bmatrix} 0 \\ 0 \\ 0 \end{bmatrix}$$

$$\Rightarrow \quad \begin{bmatrix} -1 & 0 & 1 \\ 0 & -1 & 1 \\ 0 & 0 & 0 \end{bmatrix} \begin{bmatrix} u_1 \\ u_2 \\ u_3 \end{bmatrix} = \begin{bmatrix} 0 \\ 0 \\ 0 \end{bmatrix}.$$

This system is equivalent to the system $u_1 = u_3$, $u_2 = u_3$. Hence, u_3 is free to be any arbitrary value, say $u_3 = 1$. Then $u_1 = u_2 = 1$, and so the vector $\mathbf{u} = \operatorname{col}(1,1,1)$ is an eigenvector associated with $r = 1$. Hence, one solution to the system $\mathbf{x}' = \mathbf{A}\mathbf{x}$ is given by

$$\mathbf{x}_1(t) = e^t \mathbf{u} = e^t \begin{bmatrix} 1 \\ 1 \\ 1 \end{bmatrix} = \begin{bmatrix} e^t \\ e^t \\ e^t \end{bmatrix}.$$

Since the eigenvalue $r = i$ is complex, we want to find two more linearly independent solutions for the system $\mathbf{x}' = \mathbf{A}\mathbf{x}$ derived from the eigenvectors associated with this eigenvalue. These eigenvectors ($\mathbf{z} = \operatorname{col}(z_1, z_2, z_3)$) must satisfy the equation

$$(\mathbf{A} - i\mathbf{I})\mathbf{z} = \begin{bmatrix} -i & 1 & 0 \\ 0 & -i & 1 \\ 1 & -1 & 1-i \end{bmatrix} \begin{bmatrix} z_1 \\ z_2 \\ z_3 \end{bmatrix} = \begin{bmatrix} 0 \\ 0 \\ 0 \end{bmatrix},$$

which is equivalent to the system $z_1 = -z_3$, $z_2 = -iz_3$. Thus, one solution to this system is $z_3 = 1$, $z_1 = -1$, and $z_2 = -i$ and so one eigenvector for **A** associated with the eigenvalue $r = i$ is given by

$$\mathbf{z} = \begin{bmatrix} z_1 \\ z_2 \\ z_3 \end{bmatrix} = \begin{bmatrix} -1 \\ -i \\ 1 \end{bmatrix} = \begin{bmatrix} -1 \\ 0 \\ 1 \end{bmatrix} + i \begin{bmatrix} 0 \\ -1 \\ 0 \end{bmatrix}.$$

By the notation on page 569 of the text, this means that $\alpha = 0$, $\beta = 1$, $\mathbf{a} = \operatorname{col}(-1,0,1)$ and $\mathbf{b} = \operatorname{col}(0,-1,0)$. Therefore, by equations (6) and (7) on page 570 of the text we see that two more linearly independent solutions to the system $\mathbf{x}' = \mathbf{A}\mathbf{x}$ are given by

$$\mathbf{x}_2(t) = e^{0 \cdot t}(\cos t)\mathbf{a} - e^{0 \cdot t}(\sin t)\mathbf{b} = \begin{bmatrix} -\cos t \\ 0 \\ \cos t \end{bmatrix} - \begin{bmatrix} 0 \\ -\sin t \\ 0 \end{bmatrix} = \begin{bmatrix} -\cos t \\ \sin t \\ \cos t \end{bmatrix},$$

$$\mathbf{x}_3(t) = e^{0 \cdot t}(\sin t)\,\mathbf{a} + e^{0 \cdot t}(\cos t)\,\mathbf{b} = \begin{bmatrix} -\sin t \\ 0 \\ \sin t \end{bmatrix} + \begin{bmatrix} 0 \\ -\cos t \\ 0 \end{bmatrix} = \begin{bmatrix} -\sin t \\ -\cos t \\ \sin t \end{bmatrix}.$$

Thus, a fundamental matrix for this system is

$$\mathbf{X}(t) = \begin{bmatrix} e^t & -\cos t & -\sin t \\ e^t & \sin t & -\cos t \\ e^t & \cos t & \sin t \end{bmatrix}$$

$$\Rightarrow \quad \mathbf{X}(0) = \begin{bmatrix} 1 & -1 & 0 \\ 1 & 0 & -1 \\ 1 & 1 & 0 \end{bmatrix}.$$

To find the inverse of the matrix $\mathbf{X}(0)$ we can, for example, perform row-reduction on the matrix $[\mathbf{X}(t)|\mathbf{I}]$ to obtain the matrix $[\mathbf{I}|\mathbf{X}^{-1}(t)]$. Thus, we see that

$$\mathbf{X}^{-1}(0) = \begin{bmatrix} \frac{1}{2} & 0 & \frac{1}{2} \\ -\frac{1}{2} & 0 & \frac{1}{2} \\ \frac{1}{2} & -1 & \frac{1}{2} \end{bmatrix}.$$

Hence, we obtain

$$e^{\mathbf{A}t} = \mathbf{X}(t)\mathbf{X}^{-1}(0) = \begin{bmatrix} e^t & -\cos t & -\sin t \\ e^t & \sin t & -\cos t \\ e^t & \cos t & \sin t \end{bmatrix} \begin{bmatrix} \frac{1}{2} & 0 & \frac{1}{2} \\ -\frac{1}{2} & 0 & \frac{1}{2} \\ \frac{1}{2} & -1 & \frac{1}{2} \end{bmatrix}$$

$$= \frac{1}{2}\begin{bmatrix} e^t + \cos t - \sin t & 2\sin t & e^t - \cos t - \sin t \\ e^t - \sin t - \cos t & 2\cos t & e^t + \sin t - \cos t \\ e^t - \cos t + \sin t & -2\sin t & e^t + \cos t + \sin t \end{bmatrix}.$$

11. The first step in finding $e^{\mathbf{A}t}$ using a fundamental matrix for the system $\mathbf{x}' = \mathbf{A}\mathbf{x}$ is to find the eigenvalues for the matrix **A**. Thus, we solve the characteristic equation

$$|\mathbf{A} - r\mathbf{I}| = \begin{vmatrix} 5-r & -4 & 0 \\ 1 & -r & 2 \\ 0 & 2 & 5-r \end{vmatrix} = 0$$

$$\Rightarrow \quad (5-r)\begin{vmatrix} -r & 2 \\ 2 & 5-r \end{vmatrix} + 4\begin{vmatrix} 1 & 2 \\ 0 & 5-r \end{vmatrix} = 0$$

$$\Rightarrow \quad (5-r)\left[-r(5-r)-4\right] + 4(5-r) = 0$$

$$\Rightarrow \quad r(r-5)^2 = 0 .$$

Therefore, the eigenvalues of **A** are $r = 0,5$, with $r = 5$ an eigenvalue of multiplicity two. Next we must find the eigenvectors and generalized eigenvectors for the matrix **A** and from these vectors derive three linearly independent solutions of the system $\mathbf{x}' = \mathbf{A}\mathbf{x}$. To find the eigenvector associated with the eigenvalue $r = 0$, we solve the equation

$$\mathbf{A}\mathbf{u} = \begin{bmatrix} 5 & -4 & 0 \\ 1 & 0 & 2 \\ 0 & 2 & 5 \end{bmatrix}\begin{bmatrix} u_1 \\ u_2 \\ u_3 \end{bmatrix} = \begin{bmatrix} 0 \\ 0 \\ 0 \end{bmatrix}$$

$$\Rightarrow \quad \begin{bmatrix} 1 & 0 & 2 \\ 0 & 2 & 5 \\ 0 & 0 & 0 \end{bmatrix}\begin{bmatrix} u_1 \\ u_2 \\ u_3 \end{bmatrix} = \begin{bmatrix} 0 \\ 0 \\ 0 \end{bmatrix}.$$

This equation is equivalent to the system $u_1 = -2u_3$, $2u_2 = -5u_3$ and one solution to this system is $u_3 = 2$, $u_1 = -4$, $u_2 = -5$. Therefore, one eigenvector of the matrix **A** associated with the eigenvalue $r = 0$ is given by the vector

$$\mathbf{u}_1 = \operatorname{col}(u_1, u_2, u_3) = \operatorname{col}(-4,-5,2),$$

and so one solution to the system $\mathbf{x}' = \mathbf{A}\mathbf{x}$ is

$$\mathbf{x}_1(t) = e^0\mathbf{u}_1 = \begin{bmatrix} -4 \\ -5 \\ 2 \end{bmatrix}.$$

To find an eigenvector associated with the eigenvalue $r = 5$, we solve the equation

$$(\mathbf{A}-5\mathbf{I})\mathbf{u}=\begin{bmatrix}0 & -4 & 0\\ 1 & -5 & 2\\ 0 & 2 & 0\end{bmatrix}\begin{bmatrix}u_1\\ u_2\\ u_3\end{bmatrix}=\begin{bmatrix}0\\ 0\\ 0\end{bmatrix},$$

which is equivalent to the system $u_2=0,\ u_1=-2u_3$. One solution to this system is $u_3=1$, $u_1=-2$, $u_2=0$. Thus, one eigenvector of the matrix **A** associated with the eigenvalue $r=5$ is the vector

$$\mathbf{u}_2=\operatorname{col}(u_1,u_2,u_3)=\operatorname{col}(-2,0,1),$$

and so another linearly independent solution to the system $\mathbf{x}'=\mathbf{A}\mathbf{x}$ is given by

$$\mathbf{x}_2(t)=e^{5t}\mathbf{u}_2=e^{5t}\begin{bmatrix}-2\\ 0\\ 1\end{bmatrix}=\begin{bmatrix}-2e^{5t}\\ 0\\ e^{5t}\end{bmatrix}.$$

Since $r=5$ is an eigenvalue of multiplicity two, we can find a generalized eigenvector (with $k=2$) associated with the eigenvalue $r=5$ which will be linearly independent from the vector $\mathbf{u}_2$ found above. Thus, we solve the equation

$$(\mathbf{A}-5\mathbf{I})^2\mathbf{u}=\mathbf{0}. \tag{11}$$

Because

$$(\mathbf{A}-5\mathbf{I})^2=\begin{bmatrix}0 & -4 & 0\\ 1 & -5 & 2\\ 0 & 2 & 0\end{bmatrix}\begin{bmatrix}0 & -4 & 0\\ 1 & -5 & 2\\ 0 & 2 & 0\end{bmatrix}=\begin{bmatrix}-4 & 20 & -8\\ -5 & 25 & -10\\ 2 & -10 & 4\end{bmatrix},$$

we see that equation (11) becomes

$$\begin{bmatrix}-4 & 20 & -8\\ -5 & 25 & -10\\ 2 & -10 & 4\end{bmatrix}\begin{bmatrix}u_1\\ u_2\\ u_3\end{bmatrix}=\begin{bmatrix}0\\ 0\\ 0\end{bmatrix}$$

$$\Rightarrow\quad\begin{bmatrix}-1 & 5 & -2\\ 0 & 0 & 0\\ 0 & 0 & 0\end{bmatrix}\begin{bmatrix}u_1\\ u_2\\ u_3\end{bmatrix}=\begin{bmatrix}0\\ 0\\ 0\end{bmatrix}.$$

This equation is equivalent to the equation $-u_1+5u_2-2u_3=0$ and is, therefore, satisfied if we let $u_2=s$ and $u_3=v$ and $u_1=5s-2v$ for any values of s and v. Hence, solutions to equation (11) are given by

$$\mathbf{u} = \begin{bmatrix} u_1 \\ u_2 \\ u_3 \end{bmatrix} = \begin{bmatrix} 5s - 2v \\ s \\ v \end{bmatrix} = s\begin{bmatrix} 5 \\ 1 \\ 0 \end{bmatrix} + v\begin{bmatrix} -2 \\ 0 \\ 1 \end{bmatrix}.$$

Notice that the vectors $v\,\text{col}(-2,0,1)$ are the eigenvectors that we found above associated with the eigenvalue $r = 5$. Since we are looking for a vector which satisfies equation (11) and is linearly independent from this eigenvector we will choose $s = 1$ and $v = 0$. Thus, a generalized eigenvector for the matrix $\mathbf{A}$ associated with the eigenvalue $r = 5$ and linearly independent of the eigenvector $\mathbf{u}_2$ is given by

$$\mathbf{u}_3 = \text{col}(5,1,0).$$

Hence, by formula (8) on page 587 of the text, we see that another linearly independent solution to the system $\mathbf{x}' = \mathbf{A}\mathbf{x}$ is given by

$$\mathbf{x}_3(t) = e^{\mathbf{A}t}\mathbf{u}_3 = e^{5t}\{\mathbf{u}_3 + t(\mathbf{A} - 5\mathbf{I})\mathbf{u}_3\}$$

$$= e^{5t}\begin{bmatrix} 5 \\ 1 \\ 0 \end{bmatrix} + te^{5t}\begin{bmatrix} 0 & -4 & 0 \\ 1 & -5 & 2 \\ 0 & 2 & 0 \end{bmatrix}\begin{bmatrix} 5 \\ 1 \\ 0 \end{bmatrix}$$

$$= e^{5t}\begin{bmatrix} 5 \\ 1 \\ 0 \end{bmatrix} + te^{5t}\begin{bmatrix} -4 \\ 0 \\ 2 \end{bmatrix} = \begin{bmatrix} 5e^{5t} - 4te^{5t} \\ e^{5t} \\ 2te^{5t} \end{bmatrix},$$

where we have used the fact that, by our choice of $\mathbf{u}_3$, $(\mathbf{A} - 5\mathbf{I})^2\mathbf{u}_3 = \mathbf{0}$ and so $(\mathbf{A} - 5\mathbf{I})^n\mathbf{u}_3 = \mathbf{0}$ for $n \geq 2$. (This is the reason why we used the generalized eigenvector to calculate $\mathbf{x}_3(t)$. The Cayley-Hamilton theorem, as given on page 585 of the text, states that $\mathbf{A}$ satisfies its characteristic equation, which in this case means that $\mathbf{A}(\mathbf{A} - 5\mathbf{I})^2 = \mathbf{0}$. However, we cannot assume from this fact that $(\mathbf{A} - 5\mathbf{I})^2 = \mathbf{0}$ because in matrix multiplication it is possible for two nonzero matrices to have a zero product.)

Our last step is to find a fundamental matrix for the system $\mathbf{x}' = \mathbf{A}\mathbf{x}$ using the linearly independent solutions found above and then to use this fundamental matrix to calculate $e^{\mathbf{A}t}$. Thus, from these three solutions we obtain the fundamental matrix given by

$$\mathbf{X}(t) = \begin{bmatrix} -4 & -2e^{5t} & 5e^{5t} - 4te^{5t} \\ -5 & 0 & e^{5t} \\ 2 & e^{5t} & 2te^{5t} \end{bmatrix}$$

$$\Rightarrow \quad \mathbf{X}(0)=\begin{bmatrix}-4 & -2 & 5\\ -5 & 0 & 1\\ 2 & 1 & 0\end{bmatrix}.$$

We can find the inverse matrix $\mathbf{X}^{-1}(0)$ by (for example) performing row-reduction on the matrix $[\mathbf{X}(0)|\mathbf{I}]$ to obtain the matrix $[\mathbf{I}|\mathbf{X}^{-1}(0)]$. Thus, we find

$$\mathbf{X}^{-1}(0)=\frac{1}{25}\begin{bmatrix}1 & -5 & 2\\ -2 & 10 & 21\\ 5 & 0 & 10\end{bmatrix}.$$

Therefore, by formula (6) on page 586 of the text, we see that

$$e^{\mathbf{A}t}=\mathbf{X}(t)\mathbf{X}^{-1}(0)$$

$$=\frac{1}{25}\begin{bmatrix}-4 & -2e^{5t} & 5e^{5t}-4te^{5t}\\ -5 & 0 & e^{5t}\\ 2 & e^{5t} & 2te^{5t}\end{bmatrix}\begin{bmatrix}1 & -5 & 2\\ -2 & 10 & 21\\ 5 & 0 & 10\end{bmatrix}$$

$$=\frac{1}{25}\begin{bmatrix}-4+29e^{5t}-20te^{5t} & 20-20e^{5t} & -8+8e^{5t}-40te^{5t}\\ -5+5e^{5t} & 25 & -10+10e^{5t}\\ 2-2e^{5t}+10te^{5t} & -10+10e^{5t} & 4+21e^{5t}+20te^{5t}\end{bmatrix}.$$

17. We first calculate the eigenvalues for the matrix **A** by solving the characteristic equation

$$|\mathbf{A}-r\mathbf{I}|=\begin{vmatrix}-r & 1 & 0\\ 0 & -r & 1\\ -2 & -5 & -4-r\end{vmatrix}=0$$

$$\Rightarrow \quad -r\begin{vmatrix}-r & 1\\ -5 & -4-r\end{vmatrix}-\begin{vmatrix}0 & 1\\ -2 & -4-r\end{vmatrix}=0$$

$$\Rightarrow \quad -r[-r(-4-r)+5]-2=0$$

$$\Rightarrow \quad r^3+4r^2+5r+2=0.$$

$$\Rightarrow \quad (r+1)^2(r+2)=0.$$

Thus, the eigenvalues for **A** are $r=-1,-2$, with $r=-1$ an eigenvalue of multiplicity two. To find an eigenvector $\mathbf{u}=\text{col}(u_1,u_2,u_3)$ associated with the eigenvalue $r=-1$, we solve the equation

$$(\mathbf{A}+\mathbf{I})\mathbf{u} = \begin{bmatrix} 1 & 1 & 0 \\ 0 & 1 & 1 \\ -2 & -5 & -3 \end{bmatrix}\begin{bmatrix} u_1 \\ u_2 \\ u_3 \end{bmatrix} = \begin{bmatrix} 0 \\ 0 \\ 0 \end{bmatrix},$$

which is equivalent to the system $u_1 = u_3$, $u_2 = -u_3$. Therefore, by letting $u_3 = 1$ (so that $u_1 = 1$ and $u_2 = -1$), we see that one eigenvector for the matrix **A** associated with the eigenvalue $r = -1$ is the vector

$$\mathbf{u}_1 = \operatorname{col}(u_1, u_2, u_3) = \operatorname{col}(1,-1,1).$$

Hence, one solution to the system $\mathbf{x}' = \mathbf{A}\mathbf{x}$ is given by

$$\mathbf{x}_1(t) = e^{-t}\mathbf{u}_1 = e^{-t}\begin{bmatrix} 1 \\ -1 \\ 1 \end{bmatrix}.$$

Since $r = -1$ s an eigenvalue of multiplicity two, we can find a generalized eigenvector associated with this eigenvalue (with k = 2) which will be linearly independent from the vector $\mathbf{u}_1 = \operatorname{col}(1,-1,1)$. To do this, we solve the equation

$$(\mathbf{A}+\mathbf{I})^2\mathbf{u} = \mathbf{0}$$

$$\Rightarrow \quad \begin{bmatrix} 1 & 1 & 0 \\ 0 & 1 & 1 \\ -2 & -5 & -3 \end{bmatrix}\begin{bmatrix} 1 & 1 & 0 \\ 0 & 1 & 1 \\ -2 & -5 & -3 \end{bmatrix}\begin{bmatrix} u_1 \\ u_2 \\ u_3 \end{bmatrix} = \begin{bmatrix} 0 \\ 0 \\ 0 \end{bmatrix}$$

$$\Rightarrow \quad \begin{bmatrix} 1 & 2 & 1 \\ -2 & -4 & -2 \\ 4 & 8 & 4 \end{bmatrix}\begin{bmatrix} u_1 \\ u_2 \\ u_3 \end{bmatrix} = \begin{bmatrix} 0 \\ 0 \\ 0 \end{bmatrix}$$

$$\Rightarrow \quad \begin{bmatrix} 1 & 2 & 1 \\ 0 & 0 & 0 \\ 0 & 0 & 0 \end{bmatrix}\begin{bmatrix} u_1 \\ u_2 \\ u_3 \end{bmatrix} = \begin{bmatrix} 0 \\ 0 \\ 0 \end{bmatrix},$$

which is equivalent to the equation $u_1 + 2u_2 + u_3 = 0$. This equation will be satisfied if we let $u_3 = s$, $u_2 = v$, and $u_1 = -2v - s$ for any values of s and v. Thus, generalized eigenvectors associated with the eigenvalue $r = -1$ are given by

$$\mathbf{u} = \begin{bmatrix} u_1 \\ u_2 \\ u_3 \end{bmatrix} = \begin{bmatrix} -2v - s \\ v \\ s \end{bmatrix} = s\begin{bmatrix} -1 \\ 0 \\ 1 \end{bmatrix} + v\begin{bmatrix} -2 \\ 0 \\ 1 \end{bmatrix}.$$

Hence, by letting $s = 2$ and $v = -1$, we find one such generalized eigenvector to be the vector

$$\mathbf{u}_2 = \text{col}(0,-1,2),$$

which we see by inspection is linearly independent from $\mathbf{u}_1$. Therefore, by equation (8) on page 587 of the text, we obtain a second linearly independent solution of the system $\mathbf{x}' = \mathbf{A}\mathbf{x}$ given by

$$\mathbf{x}_2(t) = e^{\mathbf{A}t}\mathbf{u}_2 = e^{-t}\{\mathbf{u}_2 + t(\mathbf{A}+\mathbf{I})\mathbf{u}_2\}$$

$$= e^{-t}\begin{bmatrix} 0 \\ -1 \\ 2 \end{bmatrix} + te^{-t}\begin{bmatrix} 1 & 1 & 0 \\ 0 & 1 & 1 \\ -2 & -5 & -3 \end{bmatrix}\begin{bmatrix} 0 \\ -1 \\ 2 \end{bmatrix}$$

$$= e^{-t}\begin{bmatrix} 0 \\ -1 \\ 2 \end{bmatrix} + te^{-t}\begin{bmatrix} -1 \\ 1 \\ -1 \end{bmatrix} = e^{-t}\begin{bmatrix} -t \\ -1+t \\ 2-t \end{bmatrix}.$$

In order to obtain a third linearly independent solution to this system, we will find an eigenvector associated with the eigenvalue $r = -2$ by solving the equation

$$(\mathbf{A}+2\mathbf{I})\mathbf{u} = \begin{bmatrix} 2 & 1 & 0 \\ 0 & 2 & 1 \\ -2 & -5 & -2 \end{bmatrix}\begin{bmatrix} u_1 \\ u_2 \\ u_3 \end{bmatrix} = \begin{bmatrix} 0 \\ 0 \\ 0 \end{bmatrix},$$

This equation is equivalent to the system $2u_1 + u_2 = 0$, $2u_2 + u_3 = 0$. One solution to this system is given by $u_1 = 1$, $u_2 = -2$, and $u_3 = 4$. Thus, one eigenvector associated with the eigenvalue $r = -2$ is the vector

$$\mathbf{u}_3 = \text{col}(u_1, u_2, u_3) = \text{col}(1,-2,4),$$

and another linearly independent solution to this system is given by

$$\mathbf{x}_3(t) = e^{-2t}\mathbf{u}_3 = e^{-2t}\begin{bmatrix} 1 \\ -2 \\ 4 \end{bmatrix}.$$

Hence, by combining the three linearly independent solutions that we have just found, we see that a general solution to this system is

$$\mathbf{x}(t) = c_1e^{-t}\begin{bmatrix} 1 \\ -1 \\ 1 \end{bmatrix} + c_2e^{-t}\begin{bmatrix} -t \\ -1+t \\ 2-t \end{bmatrix} + c_3e^{-2t}\begin{bmatrix} 1 \\ -2 \\ 4 \end{bmatrix}.$$

23. In Problem 3, we found that

$$e^{\mathbf{A}t} = e^{-t}\begin{bmatrix} 1+3t-\frac{3}{2}t^2 & t & -t+\frac{1}{2}t^2 \\ -3t & 1 & t \\ 9t-\frac{9}{2}t^2 & 3t & 1-3t+\frac{3}{2}t^2 \end{bmatrix}.$$

In order to use the variation of parameters formula (equation (13) on page 589 of the text), we need to find expressions for $e^{\mathbf{A}(t-t_0)}\mathbf{x}_0$ and $\int_0^t e^{\mathbf{A}(t-s)} f(s)ds$, where we have used the fact that $t_0 = 0$. Thus, we first notice that

$$\int_0^t e^{\mathbf{A}(t-s)} f(s)ds = \int_0^t e^{\mathbf{A}t-\mathbf{A}s} f(s)ds = e^{\mathbf{A}t}\int_0^t e^{-\mathbf{A}s} f(s)ds .$$

Since $\boldsymbol{f}(s) = \mathrm{col}(0, s, 0)$, we observe that

$$e^{-\mathbf{A}s} f(s) = e^{s}\begin{bmatrix} 1-3s-\frac{3}{2}s^2 & -s & s+\frac{1}{2}s^2 \\ 3s & 1 & -s \\ -9s-\frac{9}{2}s^2 & -3s & 1+3s+\frac{3}{2}s^2 \end{bmatrix}\begin{bmatrix} 0 \\ s \\ 0 \end{bmatrix}$$

$$= e^{s}\begin{bmatrix} -s^2 \\ s \\ -3s^2 \end{bmatrix} = \begin{bmatrix} -e^s s^2 \\ e^s s \\ -3e^s s^2 \end{bmatrix}.$$

Therefore, we have

$$\int_0^t e^{\mathbf{A}(t-s)} f(s)ds = e^{\mathbf{A}t}\int_0^t e^{-\mathbf{A}s} f(s)ds$$

$$= e^{\mathbf{A}t}\begin{bmatrix} \int_0^t -e^s s^2\, ds \\ \int_0^t e^s s\, ds \\ \int_0^t -3e^s s^2\, ds \end{bmatrix} = e^{\mathbf{A}t}\begin{bmatrix} 2-e^t(t^2-2t+2) \\ 1+e^t(t-1) \\ 6-3e^t(t^2-2t+2) \end{bmatrix},$$

where we have used integration by parts to evaluate the three integrals above. Next, since $\mathbf{x}_0 = \mathrm{col}(0,3,0)$, we see that

$$e^{\mathbf{A}(t-0)}\mathbf{x}_0 = e^{\mathbf{A}t}\mathbf{x}_0 = e^{-t}\begin{bmatrix} 1+3t-\frac{3}{2}t^2 & t & -t+\frac{1}{2}t^2 \\ -3t & 1 & t \\ 9t-\frac{9}{2}t^2 & 3t & 1-3t+\frac{3}{2}t^2 \end{bmatrix}\begin{bmatrix} 0 \\ 3 \\ 0 \end{bmatrix} = e^{-t}\begin{bmatrix} 3t \\ 3 \\ 9t \end{bmatrix}.$$

Finally, substituting these expressions into the variation of parameters formula (13) on page 589 of the text, yields

$$\mathbf{x}(t) = \mathrm{e}^{\mathbf{A}(t-0)}\,\mathbf{x}_0 + \int_0^t \mathrm{e}^{\mathbf{A}(t-s)}\, f(s)ds = \mathrm{e}^{-t}\begin{bmatrix} 3t \\ 3 \\ 9t \end{bmatrix} + \mathrm{e}^{\mathbf{A}t}\begin{bmatrix} 2-\mathrm{e}^{t}(t^2-2t+2) \\ 1+\mathrm{e}^{t}(t-1) \\ 6-3\,\mathrm{e}^{t}(t^2-2t+2) \end{bmatrix},$$

where $\mathrm{e}^{\mathbf{A}t}$ is given above.

CHAPTER 10: Partial Differential Equations

EXERCISES 10.2: Method of Separation of Variables, page 611

5. To find the general solution to this equation, we first observe that the auxiliary equation associated with the corresponding homogeneous equation is given by $r^2 - 1 = 0$. This equation has roots $r = \pm 1$. Thus, the solution to the corresponding homogeneous equation is given by

$$y_h(x) = C_1 e^x + C_2 e^{-x}.$$

By the method of undetermined coefficients, we see that the form of a particular solution to the nonhomogeneous equation is

$$y_p(x) = A + Bx,$$

where we have used the fact that neither $y = 1$ nor $y = x$ is a solution to the corresponding homogeneous equation. To find *A* and *B*, we note that

$$y_p'(x) = B \qquad \text{and} \qquad y_p''(x) = 0.$$

By substituting these expressions into the original differential equation, we obtain

$$y_p''(x) - y_p(x) = -A - Bx = 1 - 2x.$$

By equating coefficients, we see that $A = -1$ and $B = 2$. Substituting these values into the equation for $y_p(x)$ yields

$$y_p(x) = -1 + 2x.$$

Thus, we see that

$$y(x) = y_h(x) + y_p(x) = C_1 e^x + C_2 e^{-x} - 1 + 2x.$$

Next we try to find C_1 and C_2 so that the solution *y*(*x*) will satisfy the boundary conditions. That is, we want to find C_1 and C_2 satisfying

$$y(0) = C_1 + C_2 - 1 = 0,$$

and

$$y(1) = C_1 e + C_2 e^{-1} + 1 = 1 + e.$$

From the first equation we see that

$$C_2 = 1 - C_1.$$

Substituting this expression for C_2 into the second equation and simplifying yields

$$e - e^{-1} = C_1(e - e^{-1}).$$

Thus, $C_1 = 1$ and $C_2 = 0$. Therefore,

$$y(x) = e^x + 2x - 1.$$

is the only solution to the boundary value problem.

13. First note that the auxiliary equation for this problem is $r^2 + \lambda = 0$. To find eigenvalues which yield nontrivial solutions we will consider the three cases $\lambda < 0$, $\lambda = 0$, and $\lambda > 0$.

Case 1, $\lambda < 0$: In this case the roots to the auxiliary equation are $\pm\sqrt{-\lambda}$ (where we note that $-\lambda$ is a positive number). Therefore, a general solution to the differential equation $y'' + \lambda y = 0$ is given by

$$y(x) = C_1 e^{\sqrt{-\lambda}x} + C_2 e^{-\sqrt{-\lambda}x}.$$

In order to apply the boundary conditions we need to find $y'(x)$. Thus, we have

$$y'(x) = \sqrt{-\lambda}C_1 e^{\sqrt{-\lambda}x} - \sqrt{-\lambda}C_2 e^{-\sqrt{-\lambda}x}.$$

By applying the boundary conditions we obtain

$$y(0) - y'(0) = C_1 + C_2 - \sqrt{-\lambda}C_1 + \sqrt{-\lambda}C_2 = 0$$

$$\Rightarrow \quad \left(1 - \sqrt{-\lambda}\right)C_1 + \left(1 + \sqrt{-\lambda}\right)C_2 = 0,$$

and

$$y(\pi) = C_1 e^{\sqrt{-\lambda}\pi} + C_2 e^{-\sqrt{-\lambda}\pi} = 0$$

$$\Rightarrow \quad C_2 = -C_1 e^{2\sqrt{-\lambda}\pi}$$

By combining these expressions, we observe that

$$\left(1 - \sqrt{-\lambda}\right)C_1 - \left(1 + \sqrt{-\lambda}\right)C_1 e^{2\sqrt{-\lambda}\pi} = 0$$

$$C_1\left[\left(1 - \sqrt{-\lambda}\right) - \left(1 + \sqrt{-\lambda}\right)e^{2\sqrt{-\lambda}\pi}\right] = 0. \qquad (1)$$

This last expression will be true if $C_1 = 0$ or if

$$e^{2\sqrt{-\lambda}\pi} = \frac{1 - \sqrt{-\lambda}}{1 + \sqrt{-\lambda}}.$$

But since $\sqrt{-\lambda} > 0$, we see that $e^{2\sqrt{-\lambda}\pi} > 1$ while $\dfrac{1 - \sqrt{-\lambda}}{1 + \sqrt{-\lambda}} < 1$. Therefore, the only way that equation (1) can be true is for $C_1 = 0$. This means that C_2 must also equal zero and so in this case we have only the trivial solution.

Case 2, $\lambda = 0$: In this case we are solving the differential equation $y'' = 0$. This equation has a general solution given by

$$y(x) = C_1 + C_2 x \quad \Rightarrow \quad y'(x) = C_2.$$

By applying the boundary conditions we obtain

$$y(0) - y'(0) = C_1 - C_2 = 0 \qquad \text{and} \qquad y(\pi) = C_1 + C_2\pi = 0\,.$$

Solving these equations simultaneously yields $C_1 = C_2 = 0$. Thus, we again find only the trivial solution.

Case 3, $\lambda > 0$: In this case the roots to the associated auxiliary equation are $r = \pm\sqrt{\lambda}i$. Therefore, the general solution is given by

$$y(x) = C_1 \cos\left(\sqrt{\lambda}x\right) + C_2 \sin\left(\sqrt{\lambda}x\right)$$

$$\Rightarrow \qquad y'(x) = -\sqrt{\lambda}C_1 \sin\left(\sqrt{\lambda}x\right) + \sqrt{\lambda}C_2 \cos\left(\sqrt{\lambda}x\right).$$

By applying the boundary conditions, we obtain

$$y(0) - y'(0) = C_1 - \sqrt{\lambda}C_2 = 0 \qquad \Rightarrow \qquad C_1 = \sqrt{\lambda}C_2\,,$$

and

$$y(\pi) = C_1 \cos\left(\sqrt{\lambda}\pi\right) + C_2 \sin\left(\sqrt{\lambda}\pi\right) = 0\,.$$

By combining these results, we obtain

$$C_2\left[\sqrt{\lambda}\cos\left(\sqrt{\lambda}\pi\right) + \sin\left(\sqrt{\lambda}\pi\right)\right] = 0\,.$$

Therefore, in order to obtain a solution other than the trivial solution, we must solve the equation

$$\sqrt{\lambda}\cos\left(\sqrt{\lambda}\pi\right) + \sin\left(\sqrt{\lambda}\pi\right) = 0\,.$$

By simplifying this equation becomes

$$\tan\left(\sqrt{\lambda}\pi\right) = -\sqrt{\lambda}\,.$$

To see that there exist values for $\lambda > 0$ which satisfy this equation, we examine the graphs of the equations $y = -x$ and $y = \tan \pi x$. For any values of $x > 0$ where these two graphs intersect, we set $\lambda = x^2$. These values for λ will be the eigenvalues that we seek. From the graph in Figure 10-A, we see that there are (countably) infinitely many such eigenvalues. These values satisfy the equations

$$\tan(\sqrt{\lambda_n}\pi) + \sqrt{\lambda_n} = 0\,.$$

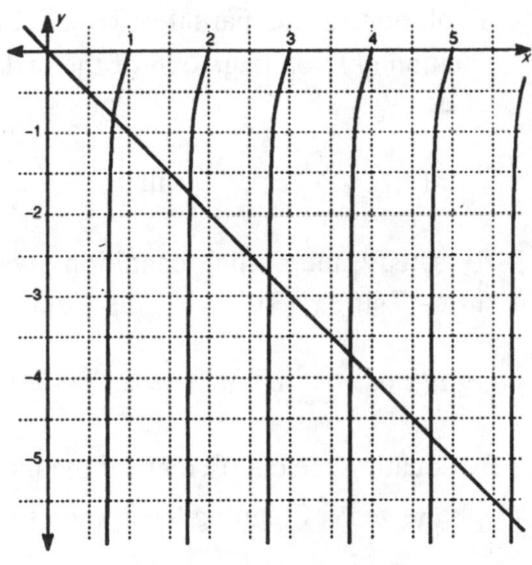

Figure 10-A. The intersections of the graphs of $y = -x$ and $y = \tan \pi x$, $x > 0$.

As n becomes large, we can also see from the graph that these eigenvalues approach the square of odd multiples of $\frac{1}{2}$. That is

$$\lambda_n \approx \frac{(2n-1)^2}{4},$$

when n is large. Corresponding to the eigenvalue λ_n we obtain the solutions

$$y_n(x) \approx C_{1,n} \cos\left(\sqrt{\lambda_n}\, x\right) + C_{2,n} \sin\left(\sqrt{\lambda_n}\, x\right)$$

$$= \sqrt{\lambda_n}\, C_{2,n} \cos\left(\sqrt{\lambda_n}\, x\right) + C_{2,n} \sin\left(\sqrt{\lambda_n}\, x\right) \qquad (\text{since } C_{1,n} = \sqrt{\lambda_n}\, C_{2,n})$$

$$\Rightarrow \qquad y_n(x) = C_n\left[\sqrt{\lambda_n} \cos\left(\sqrt{\lambda_n}\, x\right) + \sin\left(\sqrt{\lambda_n}\, x\right)\right], \qquad C_n \text{ arbitrary.}$$

17. We are solving the problem

$$\frac{\partial u(x,t)}{\partial t} = 3\frac{\partial^2 u(x,t)}{\partial t^2}, \qquad 0 < x < \pi, \qquad t > 0,$$

$$u(0, t) = u(\pi, t) = 0, \quad t > 0,$$

$$u(x,0) = \sin x - 7\sin 3x + \sin 5x \;.$$

A solution to this partial differential equation satisfying the first boundary condition is given in equation (11) on page 606 of the text. By letting $\beta = 3$ and $L = \pi$ in this equation we obtain the series

$$u(x,t) = \sum_{n=1}^{\infty} c_n\, e^{-3n^2 t} \sin nx \,. \tag{2}$$

To satisfy the initial condition, we let $t = 0$ in this equation and set the result equal to $\sin x - 7\sin 3x + \sin 5x$. This yields

$$u(x,0) = \sum_{n=1}^{\infty} c_n \sin nx = \sin x - 7\sin 3x + \sin 5x \;.$$

By equating the coefficients of the like terms, we see that $c_1 = 1$, $c_3 = -7$, and $c_5 = 1$ and all other c_n's are zero. Plugging these values into equation (2) gives the solution

$$u(x,t) = e^{-3(1)^2 t} \sin x - 7e^{-3(3)^2 t} \sin 3x + e^{-3(5)^2 t} \sin 5x$$

$$= e^{-3t} \sin x - 7e^{-27t} \sin 3x + e^{-75t} \sin 5x \,.$$

21. By letting $\alpha = 3$ and $L = \pi$ in formula (24) on page 609 of the text, we see that the solution we want will have the form

$$u(x,t) = \sum_{n=1}^{\infty} [a_n \cos 3nt + b_n \sin 3nt] \sin nx \;. \tag{3}$$

Therefore, we see that

$$\frac{\partial u}{\partial t} = \sum_{n=1}^{\infty} [-3na_n \sin 3nt + 3nb_n \cos 3nt] \sin nx \;.$$

In order for the solution to satisfy the initial conditions, we must find a_n and b_n such that

$$u(x,0) = \sum_{n=1}^{\infty} a_n \sin nx = 6 \sin 2x + 2 \sin 6x \;,$$

and

$$\frac{\partial u(x,0)}{\partial t} = \sum_{n=1}^{\infty} 3nb_n \sin nx = 11 \sin 9x - 14 \sin 15x \;.$$

From the first condition, we observe that we must have a term for $n = 2, 6$ and for these terms we want $a_2 = 6$ and $a_6 = 2$. All of the other a_n's must be zero. By comparing coefficients in the second condition, we see that we require

$$(3)(9)b_9 = 11 \quad \text{or} \quad b_9 = \frac{11}{27}, \qquad \text{and} \qquad (3)(15)b_{15} = -14 \quad \text{or} \quad b_{15} = \frac{-14}{45} \;.$$

We also see that all other values for b_n must be zero. Therefore, by substituting these values into equation (3) above, we obtain the solution of the vibrating string problem with $\alpha = 3$, $L = \pi$ and $f(x)$ and $g(x)$ as given. This solution is given by

$$u(x,t) = 6[\cos(3)(2)t] \sin 2x + 2[\cos(3)(6)t] \sin 6x$$
$$+ \frac{11}{27}[\sin(3)(9)t] \sin 9x - \frac{14}{45}[\sin(3)(15)t] \sin 15x \;.$$

Or by simplifying, we obtain

$$u(x,t) = 6 \cos 6t \sin 2x + 2 \cos 18t \sin 6x + \frac{11}{27} \sin 27t \sin 9x - \frac{14}{45} \sin 45t \sin 15x \;.$$

23. We know from equation (11) on page 606 of the text that a formal solution to the heat flow problem is given by

$$u(x,t) = \sum_{n=1}^{\infty} c_n e^{-2(n\pi)^2 t} \sin n\pi x \;, \tag{4}$$

where we have made the substitutions $\beta = 2$ and $L = 1$. For this function to be a solution to the problem it must satisfy the initial condition $u(x, 0) = f(x)$, $0 < x < 1$. Therefore, we let $t = 0$ in equation (4) above and set the result equal to $f(x)$ to obtain

$$u(x,0)=\sum_{n=1}^{\infty} c_n \sin n\pi x=\sum_{n=1}^{\infty} \frac{1}{n^2}\sin n\pi x .$$

By equating coefficients, we see that $c_n = n^{-2}$. Substituting these values of c_n into equation (4) yields the solution

$$u(x,t)=\sum_{n=1}^{\infty} n^{-2}e^{-2(n\pi)^2 t}\sin n\pi x .$$

EXERCISES 10.3: Fourier Series, page 627

5. Note that $f(-x)=e^{x}\cos(-3x)=e^{x}\cos 3x$. Since

$$f(-x)=e^{x}\cos 3x\neq e^{-x}\cos 3x=f(x)$$

unless $x=0$, we see that this function is not even. Similarly since

$$f(-x)=e^{x}\cos 3x\neq -e^{-x}\cos 3x=-f(x),$$

this function is also not odd.

13. For this problem $T=1$. Thus, by Definition 1 on page 618 of the text, the Fourier series for this function will be given by

$$\frac{a_0}{2}+\sum_{n=1}^{\infty}\left(a_n\cos n\pi x+b_n\sin n\pi x\right) \tag{5}$$

To compute a_0, we use equation (9) given in Definition 1 in the text noting that $\cos(0\cdot\pi x)=1$. Thus, we have

$$a_0=\int_{-1}^{1}x^2\,dx=\left.\frac{x^3}{3}\right|_{-1}^{1}=\frac{1}{3}+\frac{1}{3}=\frac{2}{3}.$$

To find a_n for $n = 1, 2, 3, \ldots$, we again use equation (9) on page 618 of the text. This yields

$$a_n=\int_{-1}^{1}x^2\cos n\pi x\,dx=2\int_{0}^{1}x^2\cos n\pi x\,dx,$$

where we have used the fact that $x^2\cos n\pi x$ is an even function. Thus, using integration by parts twice, we obtain

$$a_n=2\int_{0}^{1}x^2\cos n\pi x\,dx=2\left[\left.\frac{x^2}{n\pi}\sin n\pi x\right|_{0}^{1}-\frac{2}{n\pi}\int_{0}^{1}x\sin n\pi x\,dx\right]$$

$$= 2\left[\left(\frac{\sin n\pi}{n\pi} - 0\right) - \frac{2}{n\pi}\left(-\frac{x}{n\pi}\cos n\pi x\Big|_0^1 + \frac{1}{n\pi}\int_0^1 \cos n\pi x\, dx\right)\right]$$

$$= 2\left[0 + \frac{2}{n^2\pi^2}(\cos n\pi - 0) - \frac{2}{n^2\pi^2}\left(\frac{1}{n\pi}\sin n\pi x\Big|_0^1\right)\right]$$

$$= \frac{4}{n^2\pi^2}(-1)^n - \frac{4}{n^3\pi^3}(\sin n\pi - 0) = \frac{4}{n^2\pi^2}(-1)^n .$$

To calculate the b_n's, note that since x^2 is even and $\sin n\pi x$ is odd, their product is odd (see Problem 7 in this section of the text). Since $x^2 \cos n\pi x$ is also continuous, by Theorem 1 on page 614 of the text, we have

$$b_n = \int_{-1}^{1} x^2 \sin n\pi x\, dx = 0 .$$

By plugging these coefficients into equation (5) above, we see that the Fourier series associated with x^2 is given by

$$\frac{1}{3} + \sum_{n=1}^{\infty} \frac{4}{n^2\pi^2}(-1)^n \cos n\pi x .$$

21. We use Theorem 2 on page 624 of the text. Notice that $f(x) = x^2$ and $f(x) = 2x$ are continuous on $[-1, 1]$. Thus, the Fourier series for f converges to $f(x)$ for $-1 < x < 1$. Furthermore,

$$f(-1^+) = \lim_{x \to -1^+} x^2 = 1 \qquad \text{and} \qquad f(1^-) = \lim_{x \to 1^-} x^2 = 1 .$$

Hence,

$$\frac{1}{2}\left[f(-1^+) + f(1^-)\right] = \frac{1}{2}[1 + 1] = 1 ,$$

and so, by Theorem 2, the sum of the Fourier series equals 1 when $x = \pm 1$. Therefore, the Fourier series converges to

$$f(x) = x^2 \quad \text{for} \quad -1 \le x \le 1 .$$

Since the sum function must be periodic with period 2, the sum function is the 2-periodic extension of $f(x)$ which we can write as

$$g(x) = (x - 2n)^2 , \quad 2n - 1 \le x < 2n + 1, \quad n = 0, \pm 1, \pm 2, \ldots .$$

29. To calculate the coefficients of this expansion we use formula (20) on page 623 of the text. Thus we have

$$c_0 = \frac{\int_{-1}^{1} f(x)\,dx}{\|P_0\|^2} = \frac{0}{\|P_0\|^2} = 0\,,$$

where we have used the fact that $f(x)$ is an odd function. To find c_1 we first calculate the denominator to be

$$\|P_1\|^2 = \int_{-1}^{1} P_1^2(x)\,dx = \int_{-1}^{1} x^2\,dx = \left.\frac{x^3}{3}\right|_{-1}^{1} = \frac{2}{3}\,.$$

Therefore, we obtain

$$c_1 = \frac{3}{2}\int_{-1}^{1} f(x)P_1(x)\,dx = \frac{3}{2}2\int_{0}^{1} x\,dx = 3\left.\frac{x^2}{2}\right|_{0}^{1} = \frac{3}{2}\,.$$

Notice that in order to calculate the above integral, we used the fact that the product of the two odd functions $f(x)$ and $P_1(x)$ is even. To find c_2, we first observe that since $f(x)$ is odd and $P_2(x)$ is even their product is odd and so we have

$$\int_{-1}^{1} f(x)P_2(x)\,dx = 0\,.$$

Hence

$$c_2 = \frac{\int_{-1}^{1} f(x)P_2(x)\,dx}{\|P_2\|^2} = 0\,.$$

31. We need to show that

$$\int_{-\infty}^{\infty} H_n(x)H_m(x)\mathrm{e}^{-x^2}\,dx = 0\,,$$

for $m \neq n$ where m, n = 0, 1, 2. Therefore, we need to calculate several integrals. Let's begin with $m = 0$, $n = 2$. Here we see that

$$\int_{-\infty}^{\infty} H_0(x)H_2(x)\mathrm{e}^{-x^2}\,dx = \int_{-\infty}^{\infty}\left(4x^2 - 2\right)\mathrm{e}^{-x^2}\,dx$$

$$= \lim_{N\to\infty}\int_{0}^{N}\left(4x^2 - 2\right)\mathrm{e}^{-x^2}\,dx + \lim_{M\to\infty}\int_{-M}^{0}\left(4x^2 - 2\right)\mathrm{e}^{-x^2}\,dx\,.$$

We will first calculate the indefinite integral using integration by parts with the substitution

$$u = x \qquad dv = 2x\,\mathrm{e}^{-x^2}\,dx$$

$$du = dx \qquad v = -\mathrm{e}^{-x^2}\,.$$

That is we find

$$\int (4x^2-2)e^{-x^2}\,dx = 2\int 2x^2 e^{-x^2}\,dx - 2\int e^{-x^2}\,dx$$

$$= 2\left[-xe^{-x^2} + \int e^{-x^2}\,dx\right] - 2\int e^{-x^2}\,dx = -2xe^{-x^2} + C.$$

Substituting this result in for the integrals we are calculating and using l'Hopital's rule to find the limits, yields

$$\int_{-\infty}^{\infty} H_0(x)H_2(x)e^{-x^2}\,dx = \lim_{N\to\infty} -2xe^{-x^2}\Big|_0^N + \lim_{M\to\infty} -2xe^{-x^2}\Big|_{-M}^0$$

$$= \lim_{N\to\infty}\left[\frac{-2N}{e^{N^2}} + 0\right] + \lim_{M\to\infty}\left[0 - \frac{2M}{e^{M^2}}\right]$$

$$= \lim_{N\to\infty}\frac{-2N}{e^{N^2}} - \lim_{M\to\infty}\frac{2M}{e^{M^2}} = 0 + 0 = 0.$$

When $m = 0, n = 1$ and $m = 1, n = 2$, the integrals are, respectively,

$$\int_{-\infty}^{\infty} H_0(x)H_1(x)e^{-x^2}\,dx = \int_{-\infty}^{\infty} 2xe^{-x^2}\,dx\,,$$

and

$$\int_{-\infty}^{\infty} H_1(x)H_2(x)e^{-x^2}\,dx = \int_{-\infty}^{\infty} 2x(4x^2-2)e^{-x^2}\,dx\,.$$

In each case the integrands are odd functions and hence their integrals over symmetric intervals of the form $(-N, N)$ are zero. Since it is easy to show that the above improper integrals are convergent, we get

$$\int_{-\infty}^{\infty} = \lim_{N\to\infty}\int_{-N}^{N} = \lim_{N\to\infty} 0 = 0.$$

Since we have shown that the 3 integrals above are all equal to zero, the first three Hermite polynomials are orthogonal.

EXERCISES 10.4: Fourier Cosine and Sine Series, page 635

3. **(a)** The π-periodic extension $\tilde{f}(x)$ on the interval $(-\pi, \pi)$ is

$$\tilde{f}(x)=\begin{cases}0, & -\pi<x<-\dfrac{\pi}{2},\\ 1, & -\dfrac{\pi}{2}<x<0,\\ 0, & 0<x<\dfrac{\pi}{2},\\ 1, & \dfrac{\pi}{2}<x<\pi,\end{cases}$$

with $\tilde{f}(x+2\pi) = \tilde{f}(x)$. The graph of this function is given in Figure 10-B.

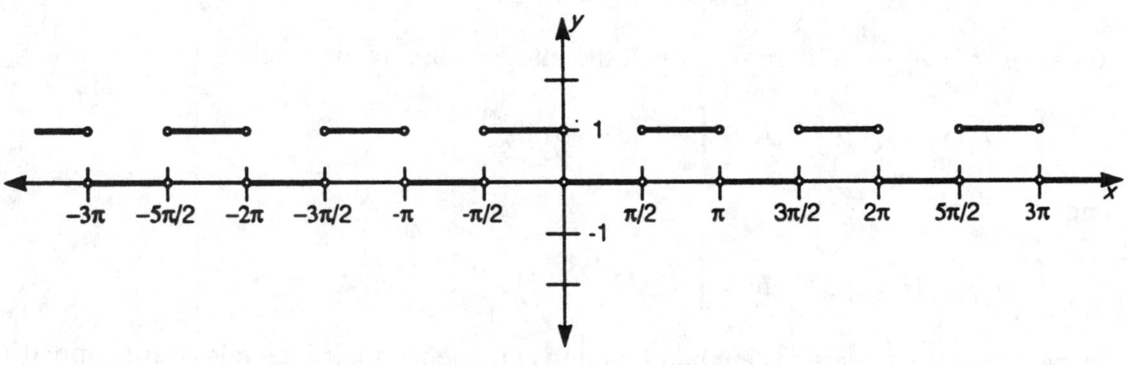

Figure 10-B. The graph of the π-periodic extension of f.

(b) Using the formula on page 631 of the text, the odd 2π-periodic extension f_o on the interval $(-\pi,\pi)$ is

$$f_o(x) = \begin{cases}-f(-x), & -\pi<x<0,\\ \\ f(x), & 0<x<\pi,\end{cases} = \begin{cases}-1, & -\pi<x<\dfrac{-\pi}{2},\\ 0, & \dfrac{-\pi}{2}<x<0,\\ 0, & 0<x<\dfrac{\pi}{2},\\ 1, & \dfrac{\pi}{2}<x<\pi,\end{cases}$$

with $f_o(x+2\pi) = f_o(x)$. The graph of f_o is given in Figure 10-C.

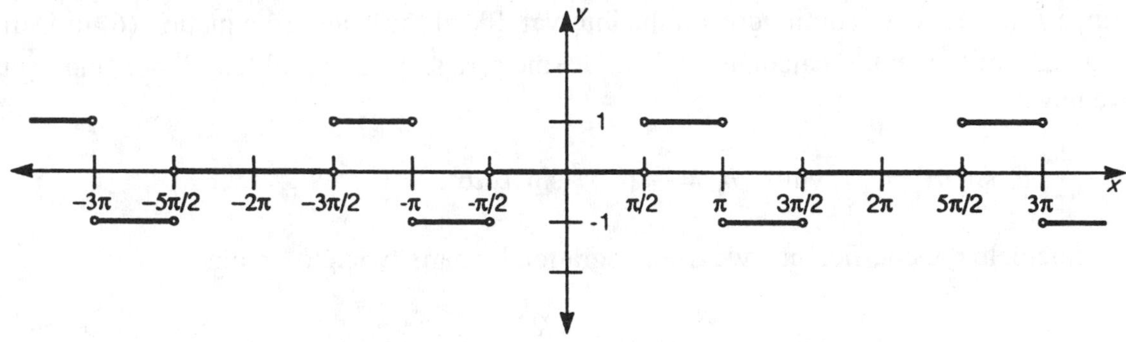

Figure 10-C. The graph of the odd 2π-periodic extension of f.

(c) Using the formula on page 632 of the text, the even 2π-periodic extension f_e on the interval $(-\pi, \pi)$ is

$$f_e(x) = \begin{cases} f(-x), & -\pi < x < 0, \\ f(x), & 0 < x < \pi, \end{cases} = \begin{cases} 1, & -\pi < x < \dfrac{-\pi}{2}, \\ 0, & \dfrac{-\pi}{2} < x < 0, \\ 0, & 0 < x < \dfrac{\pi}{2}, \\ 1, & \dfrac{\pi}{2} < x < \pi, \end{cases}$$

with $f_e(x + 2\pi) = f_e(x)$. The graph of this function is given in Figure 10-D.

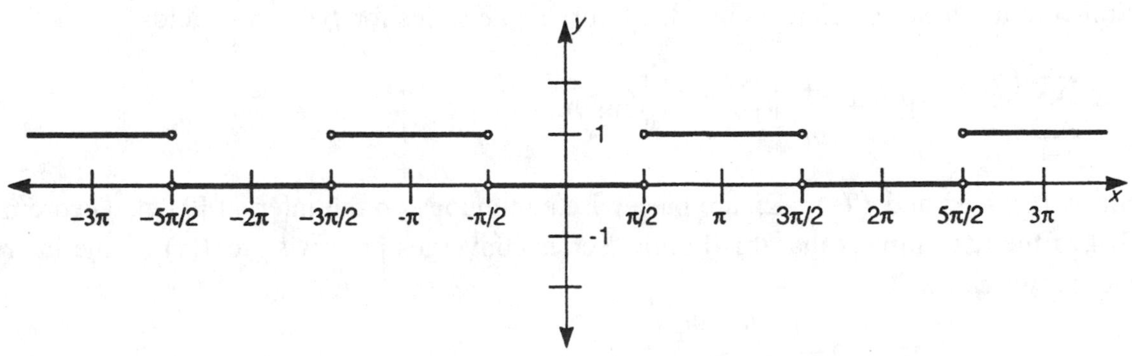

Figure 10-D. The graph of the even 2π-periodic extension of f.

7. Since f is piecewise continuous on the interval $[0, \pi]$, we can use equation (6) in Definition 2 on page 633 of the text to calculate its Fourier sine series. In this problem $T = \pi$ and $f(x) = x^2$. Thus we have

$$\sum_{n=1}^{\infty} b_n \sin nx\,, \qquad \text{with } b_n = \frac{2}{\pi}\int_0^{\pi} x^2 \sin nx\, dx\,.$$

To calculate the coefficients, we use integration by parts twice to obtain

$$\frac{\pi}{2} b_n = \int_0^{\pi} x^2 \sin nx\, dx = -\frac{x^2}{n}\cos nx\bigg|_0^{\pi} + \frac{2}{n}\int_0^{\pi} x\cos nx\, dx$$

$$= -\frac{\pi^2}{n}\cos n\pi + 0 + \frac{2}{n}\left[\frac{x}{n}\sin nx\bigg|_0^{\pi} - \frac{1}{n}\int_0^{\pi}\sin nx\, dx\right]$$

$$= -\frac{\pi^2}{n}\cos n\pi + \frac{2}{n}\left[0 - \frac{1}{n}\left(-\frac{1}{n}\cos nx\bigg|_0^{\pi}\right)\right]$$

$$= -\frac{\pi^2}{n}\cos n\pi + \frac{2}{n^3}(\cos n\pi - \cos 0),$$

where $n = 1,2,3, \ldots$. Since $\cos n\pi = 1$ if n is even and $\cos n\pi = -1$ if n is odd for $n = 1, 2, 3, \ldots$, we see that

$$\frac{\pi}{2} b_n = -\frac{\pi^2}{n}(-1)^n + \frac{2}{n^3}\left[(-1)^n - 1\right].$$

Therefore, for $n = 1, 2, 3, \ldots$, we have

$$b_n = \frac{2\pi}{n}(-1)^{n+1} + \frac{4}{\pi n^3}\left[(-1)^n - 1\right].$$

Substituting these coefficients into the Fourier sine series for $f(x) = x^2$, yields

$$\sum_{n=1}^{\infty}\left\{\frac{2\pi}{n}(-1)^{n+1} + \frac{4}{\pi n^3}\left[(-1)^n - 1\right]\right\}\sin nx\,.$$

Since $f(x) = x^2$ and $f'(x) = 2x$ are piecewise continuous on the interval $[0, \pi]$, Theorem 2 on page 624 of the text implies that this Fourier series converges pointwise to $f(x)$ on the interval $(0, \pi)$. Hence, we can write

$$f(x) = x^2 = \sum_{n=1}^{\infty}\left\{\frac{2\pi}{n}(-1)^{n+1} + \frac{4}{\pi n^3}\left[(-1)^n - 1\right]\right\}\sin nx\,,$$

for x in the interval $(0, \pi)$. But since the odd 2π-periodic extension of $f(x)$ is discontinuous at odd multiples of π, the Gibbs' phenomenon (see Problem 39 on page 630 of the text) occurs around these points, and so the convergence of this Fourier sine series is not uniform on $(0, \pi)$.

13. Since $f(x) = e^x$ is piecewise continuous on the interval [0, 1], we can use Definition 2 on page 633 of the text to find its Fourier cosine series. Therefore, we have

$$\frac{a_0}{2} + \sum_{n=1}^{\infty} a_n \cos n\pi x\,, \qquad \text{where } a_n = 2\int_0^1 e^x \cos n\pi x \, dx\,.$$

Using the fact that $\cos 0 = 1$, we find the coefficient a_0 to be

$$a_0 = 2\int_0^1 e^x \, dx = 2[e - 1]\,.$$

We will use integration by parts twice (or the table of integrals on the inside cover of the text) to calculate the integrals in the remaining coefficients. This yields

$$\int e^x \cos n\pi x \, dx = \frac{e^x(\cos n\pi x + n\pi \sin n\pi x)}{1 + n^2\pi^2}\,,$$

where $n = 1, 2, 3, \ldots$. Thus, the remaining coefficients are given by

$$a_n = 2\int_0^1 e^x \cos n\pi x \, dx = \left. \frac{2e^x(\cos n\pi x + n\pi \sin n\pi x)}{1 + n^2\pi^2} \right|_0^1$$

$$= \frac{2e(\cos n\pi)}{1 + n^2\pi^2} - \frac{2(1)}{1 + n^2\pi^2} = \frac{2\left[(-1)^n e - 1\right]}{1 + n^2\pi^2}\,, \qquad n = 1, 2, 3, \ldots,$$

where we have used the fact that $\cos n\pi = 1$ if n is even and $\cos n\pi = -1$ if n is odd. By substituting the above coefficients into the Fourier cosine series for f given above, we obtain

$$e^x = e - 1 + 2\sum_{n=1}^{\infty} \frac{(-1)^n e - 1}{1 + n^2\pi^2} \cos n\pi x\,,$$

for $0 < x < 1$. Note that we can say that e^x for $0 < x < 1$ equals its Fourier cosine series because this series converges uniformly. To see this, first notice that the even 2π-periodic extension of $f(x) = e^x$, $0 < x < 1$ is given by

$$f_e(x) = \begin{cases} e^{-x}, & -1 < x < 0, \\ e^x, & 0 < x < 1, \end{cases}$$

with $f_e(x + 2\pi) = f_e(x)$. Since this extension is continuous on $(-\infty, \infty)$ and $f_e'(x)$ is piecewise continuous on [–1, 1], Theorem 3 on page 625 of the text states that its Fourier series (which is the one we found above) converges uniformly to $f_e(x)$ on [–1, 1] and so it converges uniformly to $f(x) = e^x$ on (0, 1).

17. This problem is the same as the heat flow problem on page 604 of the text with $\beta = 5$, $L = \pi$ and $f(x) = 1 - \cos 2x$. Therefore, the formal solution to this problem is given in equations (11) and (12) on pages 606 and 607 of the text. Thus, the formal solution is

$$u(x,t) = \sum_{n=1}^{\infty} c_n e^{-5n^2 t} \sin nx, \qquad 0 < x < \pi,\ t > 0, \tag{6}$$

where

$$f(x) = 1 - \cos 2x = \sum_{n=1}^{\infty} c_n \sin nx .$$

Therefore, we must find the Fourier sine series for $1 - \cos 2x$. To do this, we can use equations (6) and (7) of Definition 2 on page 633 of the text. Hence, the coefficients are given by

$$c_n = \frac{2}{\pi} \int_0^{\pi} (1 - \cos 2x) \sin nx \, dx$$

$$= \frac{2}{\pi} \int_0^{\pi} \sin nx \, dx - \frac{2}{\pi} \int_0^{\pi} \cos 2x \sin nx \, dx, \qquad n = 1, 2, 3, \ldots .$$

Calculating the first integral above yields

$$\frac{2}{\pi} \int_0^{\pi} \sin nx \, dx = -\frac{2}{n\pi}(\cos n\pi - 1) = \frac{2}{n\pi}\left[1 - (-1)^n\right],$$

where we have used the fact that $\cos n\pi = 1$ if n is even and $\cos n\pi = -1$ if n is odd. To calculate the second integral, we use the fact that $2\cos\alpha \sin\beta = \sin(\beta - \alpha) + \sin(\alpha + \beta)$, to obtain

$$-\frac{2}{\pi} \int_0^{\pi} \cos 2x \sin nx \, dx = -\frac{1}{\pi}\left\{ \int_0^{\pi} \sin[(n-2)x] dx + \int_0^{\pi} \sin[(n+2)x] dx \right\}$$

$$= \frac{1}{\pi(n-2)}\{\cos[(n-2)\pi] - 1\} + \frac{1}{\pi(n+2)}\{\cos[(n+2)\pi] - 1\}$$

$$= \frac{1}{\pi(n-2)}\left[(-1)^n - 1\right] + \frac{1}{\pi(n+2)}\left[(-1)^n - 1\right].$$

Combining these two integrals yields

$$c_n = \frac{2}{n\pi}\left[1 - (-1)^n\right] + \frac{1}{\pi(n-2)}\left[(-1)^n - 1\right] + \frac{1}{\pi(n+2)}\left[(-1)^n - 1\right]$$

$$\Rightarrow \qquad c_n = \begin{cases} 0, & \text{if } n \text{ is even,} \\ \dfrac{4}{\pi n} - \dfrac{2}{\pi(n-2)} - \dfrac{2}{\pi(n+2)}, & \text{if } n \text{ is odd,} \end{cases}$$

for $n = 1, 2, 3, \ldots$. Hence, we obtain the formal solution to this problem by substituting these coefficients into equation (6) above and setting $n = 2k - 1$. Therefore, we have

$$u(x,t)=\frac{2}{\pi}\sum_{n=1}^{\infty}\left[\frac{2}{2k-1}-\frac{1}{2k+1}-\frac{1}{2k-3}\right]e^{-5(2k-1)^2t}\sin(2k-1)x.$$

EXERCISES 10.5: The Heat Equation, page 648

3. If we let $\beta = 3$, $L = \pi$ and $f(x) = x$, we see that this problem has the same form as the problem in Example 1 on page 637 of the text. Therefore, we can find the formal solution to this problem by substituting these values into equation (14) on page 639 of the text. Hence, we have

$$u(x,t)=\sum_{n=0}^{\infty}c_n e^{-n^2t}\cos nx, \qquad \text{where} \quad f(x)=\sum_{n=0}^{\infty}c_n\cos nx. \tag{7}$$

Thus, we must find the Fourier cosine series coefficients for $f(x) = x$, $0 < x < \pi$. (Note that the even 2π-extension for $f(x) = x$, $0 < x < \pi$, which is given by

$$f_e(x)=\begin{cases}-x, & \text{for } -\pi<x<0,\\ x, & \text{for } 0<x<\pi,\end{cases}$$

with $f_e(x+2\pi)=f_e(x)$, is continuous. Also note that its derivative is piecewise continuous on $[-\pi,\pi]$. Therefore, the Fourier series for this extension converges uniformly to f_e. This means that the equality sign in the second equation given in formula (7) above is justified for $0<x<\pi$.) To find the required Fourier series coefficients, we use equations (4) and (5) given in Definition 2 on page 633 of the text. Hence, we have

$$x=\frac{a_0}{2}+\sum_{n=1}^{\infty}a_n\cos nx,$$

(so that $c_0=\frac{a_0}{2}$ and $c_n=a_n$ for $n = 1, 2, 3, \ldots$) where

$$a_0=\frac{2}{\pi}\int_0^{\pi}x\,dx=\frac{2}{\pi}\cdot\left.\frac{x^2}{2}\right|_0^{\pi}=\pi, \quad \text{and} \qquad a_n=\frac{2}{\pi}\int_0^{\pi}x\cos nx\,dx,$$

for $n = 1, 2, 3, \ldots$. To calculate the second integral above we use integration by parts to obtain

$$a_n=\frac{2}{\pi}\int_0^{\pi}x\cos nx\,dx=\frac{2}{\pi}\left[\left.\frac{x}{n}\sin nx\right|_0^{\pi}-\frac{1}{n}\int_0^{\pi}\sin nx\,dx\right]$$

$$=\frac{2}{\pi}\left\{0-\frac{1}{n}\left[-\frac{1}{n}(\cos nx-1)\right]\right\}=\frac{2}{\pi n^2}(\cos n\pi-1)=\frac{2}{\pi n^2}\left[(-1)^n-1\right].$$

Combining these results yields

$$a_n = \begin{cases} \pi, & \text{if } n = 0, \\ \dfrac{-4}{\pi n^2}, & \text{if } n \text{ is odd}, \\ 0, & \text{if } n \text{ is even and } n \neq 0, \end{cases}$$

where n = 1, 2, 3, … . The formal solution for this problem is, therefore, found by substituting these coefficients into the first equation given in formula (7) above.(Recall that $c_0 = \dfrac{a_0}{2}$ and $c_n = a_n$ for n = 1, 2, 3, … .) Thus, we have

$$u(x,t) = \frac{\pi}{2} e^{-0} \cos 0 - \sum_{n=0}^{\infty} \frac{4}{\pi(2n+1)^2} e^{-3(2n+1)^2 t} \cos(2n+1)x$$

$$= \frac{\pi}{2} - \sum_{n=0}^{\infty} \frac{4}{\pi(2n+1)^2} e^{-3(2n+1)^2 t} \cos(2n+1)x \ .$$

7. This problem has nonhomogeneous boundary conditions and so has the same form as the problem in Example 2 on page 640 of the text. By comparing these two problems, we see that for this problem $\beta = 2$, $L = \pi$, $U_1 = 5$, $U_2 = 10$, and $f(x) = \sin 3x - \sin 5x$. To solve this problem, we assume that the solution consists of a steady state solution $v(x)$ and a transient solution $w(x, t)$. The steady state solution is given in equation (24) on page 641 of the text and is

$$v(x) = 5 + \frac{(10-5)x}{\pi} = 5 + \frac{5}{\pi} x \,.$$

The formal transient solution is given by equations (39) and (40) on page 643 of the text. By using these equations and making appropriate substitutions, we obtain

$$w(x,t) = \sum_{n=1}^{\infty} c_n e^{-2n^2 t} \sin nx \,, \tag{8}$$

where the coefficients (the c_n's) are given by

$$f(x) - v(x) = \sin 3x - \sin 5x - 5 - \frac{5}{\pi} x = \sum_{n=1}^{\infty} c_n \sin nx \,, \qquad 0 < x < \pi.$$

Therefore, we must find the Fourier sine series coefficients for the function $\sin 3x - \sin 5x - 5 - \dfrac{5x}{\pi}$ for $0 < x < \pi$. Since the function $\sin 3x - \sin 5x$ is already in the form of a sine series, we only need to find the Fourier sine series for $-5 - \dfrac{5x}{\pi}$ and then add $\sin 3x - \sin 5x$ to this series. The resulting coefficients are the ones that we need. (Note that the Fourier sine series for $-5 - 5x/\pi$ will converge pointwise but not uniformly to $-5 - 5x/\pi$ for $0 < x < \pi$.)

To find the desired Fourier series we use equations (6) and (7) in Definition 2 on page 633 of the text. Thus, with the appropriate substitutions, we have

$$-5-\frac{5x}{\pi}x=\sum_{n=1}^{\infty} b_n \sin nx, \quad \text{where} \quad b_n=\frac{2}{\pi}\int_0^{\pi}\left(-5-\frac{5x}{\pi}\right)\sin nx\,dx .$$

To find the b_n's, we will use integration by parts to obtain

$$\begin{aligned}
b_n &= -\frac{10}{\pi}\int_0^{\pi}\sin nx\,dx-\frac{10}{\pi^2}\int_0^{\pi}x\sin nx\,dx \\
&= \frac{10}{n\pi}(\cos n\pi-1)-\frac{10}{\pi^2}\left[-\frac{x}{n}\cos nx\Big|_0^{\pi}+\frac{1}{n}\int_0^{\pi}\cos nx\,dx\right] \\
&= \frac{10}{n\pi}(\cos n\pi-1)-\frac{10}{\pi^2}\left[-\frac{\pi}{n}\cos n\pi+0\right] \\
&= \frac{10}{n\pi}\cos n\pi-\frac{10}{n\pi}+\frac{10}{n\pi}\cos n\pi=\frac{10}{n\pi}(2\cos n\pi-1) \\
&= \frac{10}{n\pi}\left[2(-1)^n-1\right], \qquad n=1,2,3,\dots .
\end{aligned}$$

Thus, the Fourier sine series for $\sin 3x-\sin 5x-5-\frac{5x}{\pi}$ is given by

$$\begin{aligned}
\sin 3x-\sin 5x-5-\frac{5x}{\pi} &= \sin 3x-\sin 5x+\sum_{n=1}^{\infty}\frac{10}{n\pi}\left[2(-1)^n-1\right]\sin nx \\
&= \sin 3x-\sin 5x-\frac{30}{\pi}\sin x+\frac{10}{2\pi}\sin 2x-\frac{30}{3\pi}\sin 3x+\frac{10}{4\pi}\sin 4x \\
&\qquad -\frac{30}{5\pi}\sin 5x+\sum_{n=6}^{\infty}\frac{10}{n\pi}\left[2(-1)^n-1\right]\sin nx \\
&= -\frac{30}{\pi}\sin x+\frac{5}{\pi}\sin 2x+\left(1-\frac{10}{\pi}\right)\sin 3x+\frac{5}{2\pi}\sin 4x \\
&\qquad -\left(1+\frac{6}{\pi}\right)\sin 5x+\sum_{n=6}^{\infty}\frac{10}{n\pi}\left[2(-1)^n-1\right]\sin nx .
\end{aligned}$$

We therefore obtain the formal transient solution by taking the coefficients from this Fourier series and substituting them in for the c_n coefficients in equation (8) above. Thus, we find

$$w(x,t)=-\frac{30}{\pi}e^{-2t}\sin x+\frac{5}{\pi}e^{-2(2)^2t}\sin 2x+\left(1-\frac{10}{\pi}\right)e^{-2(3)^2t}\sin 3x+\frac{5}{2\pi}e^{-2(4)^2t}\sin 4x$$

$$-\left(1+\frac{6}{\pi}\right)e^{-2(5)^2t}\sin 5x+\sum_{n=6}^{\infty}\frac{10}{n\pi}\left[2(-1)^n-1\right]e^{-2n^2t}\sin nx\,,$$

and so the formal solution to the original problem is given by

$$u(x,t)=v(x)+w(x,t)$$

$$=5+\frac{5}{\pi}x-\frac{30}{\pi}e^{-2t}\sin x+\frac{5}{\pi}e^{-8t}\sin 2x+\left(1-\frac{10}{\pi}\right)e^{-18t}\sin 3x+\frac{5}{2\pi}e^{-32t}\sin 4x$$

$$-\left(1+\frac{6}{\pi}\right)e^{-50t}\sin 5x+\sum_{n=6}^{\infty}\frac{10}{n\pi}\left[2(-1)^n-1\right]e^{-2n^2t}\sin nx\,.$$

9. Notice that this problem is a nonhomogeneous partial differential equation and has the same form as the problem given in Example 3 on page 642 of the text. By comparing these problems, we see that here $\beta=1, P(x)=e^{-x}$, $L=\pi$, $U_1=U_2=0$, and $f(x)=\sin 2x$. As in Example 3, we will assume that the solution is the sum of a steady state solution $v(x)$ and a transient solution $w(x, t)$. The steady state solution is the solution to the boundary value problem

$$v''(x)=e^{-x}, \quad 0<x<\pi, \qquad v(0)=v(\pi)=0\,.$$

Thus the steady state solution can be found either by solving this ODE or by substituting the appropriate values into equation (35) given on page 642 of the text. By either method we find

$$v(x)=\frac{e^{-\pi}-1}{\pi}x-e^{-x}+1\,.$$

The formal transient solution is then given by equations (39) and (40) on page 643 of the text. By making the appropriate substitutions into this equation, we obtain

$$w(x,t)=\sum_{n=1}^{\infty}c_n e^{-n^2t}\sin nx\,, \tag{9}$$

where the c_n's are given by

$$f(x)-v(x)=\sin 2x-\frac{e^{-\pi}-1}{\pi}x-e^{-x}+1=\sum_{n=1}^{\infty}c_n\sin nx\,.$$

Hence, the problem is to find the Fourier sine coefficients for $f(x)-v(x)$. The first term, $f(x)=\sin 2x$, is already in the desired form. Therefore, the Fourier sine series for $f(x)-v(x)$ is

$$\sin 2x + \sum_{n=1}^{\infty} b_n \sin nx = b_1 \sin x + (b_2 + 1)\sin 2x + \sum_{n=3}^{\infty} b_n \sin nx ,$$

where the b_n's are the Fourier sine coefficients for $-v(x)$. This implies that if $n \neq 2$, then $c_n = b_n$ and if $n = 2$, then $c_2 = b_2 + 1$. The b_n coefficients are given by equation (7) on page 633 of the text. Thus, we have

$$b_n = \frac{2}{\pi} \int_0^{\pi} \left[-\frac{e^{-\pi} - 1}{\pi} x + e^{-x} - 1 \right] \sin nx \, dx$$

$$= \frac{2}{\pi}\left(-\frac{e^{-\pi} - 1}{\pi} \right) \int_0^{\pi} x \sin nx \, dx + \frac{2}{\pi} \int_0^{\pi} e^{-x} \sin nx \, dx - \frac{2}{\pi} \int_0^{\pi} \sin nx \, dx .$$

We will calculate each integral separately. The first integral is found by using integration by parts. This yields

$$\frac{2}{\pi}\left(-\frac{e^{-\pi} - 1}{\pi} \right) \int_0^{\pi} x \sin nx \, dx = \frac{-2(e^{-\pi} - 1)}{\pi^2} \left[-\frac{x}{n} \cos nx \Big|_0^{\pi} + \frac{1}{n} \int_0^{\pi} \cos nx \, dx \right]$$

$$= \frac{-2(e^{-\pi} - 1)}{\pi^2} \left[-\frac{\pi}{n} \cos n\pi + 0 + 0 \right] = \frac{2(e^{-\pi} - 1)}{\pi n} (-1)^n .$$

To find the second integral we use the table of integrals on the inside front cover of the text (or use integration by parts twice) to obtain

$$\frac{2}{\pi} \int_0^{\pi} e^{-x} \sin nx \, dx = \frac{2}{\pi} \left[\frac{-e^{-\pi} n \cos n\pi + n}{1 + n^2} \right] = \frac{2n}{(1+n^2)\pi} \left[e^{-\pi} (-1)^{n+1} + 1 \right].$$

The last integral is found to be

$$-\frac{2}{\pi} \int_0^{\pi} \sin nx \, dx = \frac{2}{n\pi} [\cos n\pi - 1] = \frac{2}{n\pi} [(-1)^n - 1].$$

By combining all of these results, we find that the Fourier coefficients for $-v(x)$ are given by

$$b_n = \frac{2(e^{-\pi} - 1)}{\pi n} (-1)^n + \frac{2n}{(1+n^2)\pi} \left[e^{-\pi} (-1)^{n+1} + 1 \right] + \frac{2}{n\pi} [(-1)^n - 1].$$

Therefore, the coefficients for the formal transient solution are

$$c_n = \begin{cases} \dfrac{2(e^{-\pi} - 1)}{\pi n} (-1)^n + \dfrac{2n}{(1+n^2)\pi} [e^{-\pi} (-1)^{n+1} + 1] + \dfrac{2}{n\pi} [(-1)^n - 1], & \text{if } n \neq 2, \\ \dfrac{e^{-\pi} - 1}{\pi} + \dfrac{4}{5\pi} (1 - e^{-\pi}) + 1, & \text{if } n = 2. \end{cases}$$

Since the formal solution to the PDE given in this problem is the sum of its steady state solution and its transient solution, we find this final solution to be

$$u(x,t) = v(x) + w(x,t) = \frac{e^{-\pi} - 1}{\pi} x - e^{-x} + 1 + \sum_{n=1}^{\infty} c_n e^{-n^2 t} \sin nx .$$

where the c_n's are given above.

11. Let $u(x,t) = X(x)T(t)$. Substituting $u(x,t) = X(x)T(t)$ into the PDE yields

$$T'(t)X(x) = 4X''(x)T(t) \qquad \Rightarrow \qquad \frac{T'(t)}{4T(t)} = \frac{X''(x)}{X(x)} = K,$$

where K is a constant. Substituting the solution $u(x,t) = X(x)T(t)$ into the boundary conditions, we obtain

$$X'(0)T(t) = 0, \qquad X(\pi)T(t) = 0, \quad t > 0.$$

Thus, we assume that $X'(0) = 0$ and $X(\pi) = 0$ since this allows the expressions above to be true for all $t > 0$ without implying that $u(x,t) \equiv 0$. Therefore, we have the two ODE's

$$X''(x) = KX(x), \qquad 0 < x < \pi,$$
$$X'(0) = X(\pi) = 0, \tag{10}$$

and

$$T'(t) = 4KT(t), \qquad t > 0. \tag{11}$$

To solve boundary value problem (10), we will examine three cases.

Case 1: Assume $K = 0$. Now equation (10) becomes $X''(0) = 0$. The solution is $X(x) = ax + b$, where a and b are arbitrary constants. To find these constants we use the boundary conditions in (10). Thus, we have

$$X'(0) = a = 0 \qquad \Rightarrow \qquad a = 0 \qquad \Rightarrow \qquad X(x) = b,$$

and so

$$X(\pi) = b = 0, \qquad \Rightarrow \qquad b = 0.$$

Therefore, in this case we have only the trivial solution.

Case 2: Assume $K > 0$. In this case the auxiliary equation for equation (10) is $r^2 - K = 0$. The roots to this equation are $r = \pm\sqrt{K}$. Thus, the solution is

$$X(x) = C_1 e^{\sqrt{K}x} + C_2 e^{-\sqrt{K}x},$$

where C_1 and C_2 are arbitrary constants. To find these constants we again use the boundary conditions in (10). We first note that

$$X'(x) = C_1\sqrt{K}e^{\sqrt{K}x} - C_2\sqrt{K}e^{-\sqrt{K}x}.$$

Therefore,

$$X'(0)=C_1\sqrt{K}-C_2\sqrt{K}=0 \qquad \Rightarrow \qquad C_1=C_2$$

$$\Rightarrow \qquad X(x)=C_1 e^{\sqrt{K}x}+C_1 e^{-\sqrt{K}x}.$$

The other boundary condition implies that

$$X(\pi)=C_1 e^{\sqrt{K}\pi}+C_1 e^{-\sqrt{K}\pi}=0 \qquad \Rightarrow \qquad C_1\left[e^{2\sqrt{K}\pi}+1\right]=0.$$

The only way that the final equation above can be zero is for C_1 to be zero. Therefore, we again obtain only the trivial solution.

<u>Case 3: Assume $K < 0$, so $-K > 0$.</u> Then the auxiliary equation for equation (10) is $r^2+K=0$ and its roots are $r=\pm i\sqrt{-K}$. Therefore, the solution is

$$X(x)=C_1\sin\left(\sqrt{-K}x\right)+C_2\cos\left(\sqrt{-K}x\right),$$

$$\Rightarrow \qquad X'(x)=C_1\sqrt{-K}\cos\left(\sqrt{-K}x\right)+C_2\sqrt{-K}\sin\left(\sqrt{-K}x\right),$$

Using the boundary condition $X'(0)=0$, we obtain

$$X'(0)=0=C_1\sqrt{-K}\cos 0+C_2\sqrt{-K}\sin 0=C_1\sqrt{-K} \qquad \Rightarrow \qquad C_1=0.$$

Hence, $X(x)=C_2\cos\left(\sqrt{-K}x\right)$. Applying the other boundary condition yields

$$X(\pi)=0=\cos\left(\sqrt{-K}\pi\right)$$

$$\Rightarrow \qquad \sqrt{-K}\pi=(2n+1)\frac{\pi}{2} \qquad \Rightarrow \qquad K=-\frac{(2n+1)^2}{4}, \qquad n=1,2,3,\ldots.$$

Therefore, nontrivial solutions to problem (10) above are given by

$$X_n(x)=c_n\cos\left(\frac{2n+1}{2}x\right), \quad n=1,2,3,\ldots.$$

By substituting the values of K into equation (11), we obtain

$$T'(t)=-(2n+1)^2T(t), \qquad t>0.$$

This is a separable differential equation, and, therefore, we find

$$\frac{1}{T}dt=-(2n+1)^2\,dt$$

$$\Rightarrow \qquad \ln|T|=-(2n+1)^2t+A$$

$$\Rightarrow \qquad T_n(t)=b_n e^{-(2n+1)^2t}, \qquad n=1,2,3,\ldots \quad (\text{where } b_n=\pm e^A).$$

Hence, by the superposition principle (and since $u_n(x,t)=X_n(x)T_n(t)$), we see that the formal solution to the original PDE is

$$u(x,t) = \sum_{n=0}^{\infty} b_n e^{-(2n+1)^2 t} c_n \cos\left(\frac{2n+1}{2}x\right) \tag{12}$$

$$= \sum_{n=0}^{\infty} a_n e^{-(2n+1)^2 t} \cos\left[\left(n+\frac{1}{2}\right)x\right], \qquad \text{where } a_n = b_n c_n .$$

To find the a_n's, we use the initial condition to obtain

$$u(x,0) = f(x) = \sum_{n=0}^{\infty} a_n \cos\left[\left(n+\frac{1}{2}\right)x\right]. \tag{13}$$

Therefore, the formal solution to this PDE is given by equation (12) where the a_n's are given by equation (13).

17. This problem is similar to the problem given in Example 4 on page 643 of the text with $\beta = 1$, $L = W = \pi$, and $f(x, y) = y$. The formal solution to this problem is given in equation (52) on page 645 of the text with its coefficients given on pages 645 and 646 in equations (54) and (55). By making appropriate substitutions in the first of these equations, we see that the formal solution to this problem is

$$u(x, y, t) = \sum_{m=0}^{\infty} \sum_{n=1}^{\infty} a_{mn} e^{-(m^2+n^2)t^2} \cos mx \sin ny . \tag{14}$$

We can find the coefficients, a_{0n}, n = 1, 2, 3, …, by using equation (54) on page 645 of the text with the appropriate substitutions. This yields

$$a_{0n} = \frac{2}{\pi^2} \int_0^{\pi} \int_0^{\pi} y \sin ny \, dx \, dy = \frac{2}{\pi^2} \int_0^{\pi} y \sin ny \left[\int_0^{\pi} dx \right] dy$$

$$= \frac{2}{\pi} \int_0^{\pi} y \sin ny \, dy \qquad \text{(use integration by parts)}$$

$$= \frac{2}{\pi} \left[-\frac{y}{n} \cos ny \Big|_0^{\pi} + \frac{1}{n} \int_0^{\pi} \cos ny \, dy \right]$$

$$= \frac{2}{\pi} \left[-\frac{\pi}{n} \cos n\pi + \left(\frac{1}{n^2} \sin ny \Big|_0^{\pi} \right) \right] = -\frac{2}{\pi} \left(\frac{\pi}{n} \cos n\pi \right) = \frac{2}{n} (-1)^{n+1} .$$

We will use equation (55) on page 646 of the text to find the other coefficients. Thus for $m \geq 1$ and $n \geq 1$, we have

$$a_{mn} = \frac{4}{\pi^2} \int_0^{\pi} \int_0^{\pi} y \cos mx \sin ny \, dx \, dy$$

$$= \frac{4}{\pi^2} \int_0^{\pi} y \sin ny \left(\int_0^{\pi} \cos mx \, dx \right) dy = \frac{4}{\pi^2} \int_0^{\pi} y \sin ny \,(0)\, dy = 0 .$$

The formal solution to this problem is found by substituting these coefficients into equation (14). To do this we first note that the coefficients for any terms containing $m \neq 0$ are zero. Hence, only terms containing $m = 0$ will appears in the summation. Therefore, the formal solution is given by

$$u(x, y, t) = \sum_{n=1}^{\infty} \frac{2}{n} (-1)^{n+1} e^{-n^2 t} \sin ny = 2 \sum_{n=1}^{\infty} \frac{(-1)^{n+1}}{n} e^{-n^2 t} \sin ny \; .$$

EXERCISES 10.6: The Wave Equation, page 660

1. This problem has the form of the problem given in equations (1)-(4) on page 649 of the text. Here, however, $\alpha = 1$, $L = 1$, and $f(x) = x(1 - x)$, and $g(x) = \sin 7\pi x$. This problem is consistent because

$$f(0) = 0 = f(1) \qquad \text{and} \qquad g(0) = \sin 0 = 0 = \sin 7\pi = g(1) .$$

The solution to this problem was derived in Section 10.2 of the text and given again in equation (5) on page 649 of the text. Making appropriate substitutions in equation (5) yields a formal solution given by

$$u(x,t) = \sum_{n=1}^{\infty} \left[a_n \cos n\pi t + b_n \sin n\pi t \right] \sin n\pi x. \tag{15}$$

To find the a_n's we note that they are the Fourier sine coefficients for $x(1 - x)$ and so are given by equation (7) on page 633 of the text. Thus, for $n = 1, 2, 3, \ldots$, we have

$$a_n = 2 \int_0^1 x(1-x) \sin n\pi x \, dx = 2 \left[\int_0^1 x \sin n\pi x \, dx - \int_0^1 x^2 \sin n\pi x \, dx \right].$$

We will use integration by parts to calculate these two integrals. This yields

$$\int_0^1 x \sin n\pi x \, dx = -\frac{1}{n\pi} \cos n\pi = -\frac{1}{n\pi} (-1)^n ,$$

and

$$\int_0^1 x^2 \sin n\pi x \, dx = -\frac{1}{n\pi} \cos n\pi - \frac{2}{n^2 \pi^2} \left(\frac{-1}{n\pi} \cos n\pi + 1 \right)$$

$$= -\frac{1}{n\pi} (-1)^n + \frac{2}{n^3 \pi^3} \left[(-1)^n - 1 \right].$$

Therefore, for $n = 1, 2, 3, \ldots$, we see that

$$a_n = 2\left\{-\frac{1}{n\pi}(-1)^n + \frac{1}{n\pi}(-1)^n - \frac{2}{n^3\pi^3}\left[(-1)^n - 1\right]\right\} = -\frac{4}{n^3\pi^3}\left[(-1)^n - 1\right].$$

This can also be expressed as

$$a_n = \begin{cases} 0, & \text{if } n \text{ is even,} \\ \dfrac{8}{n^3\pi^3}, & \text{if } n \text{ is odd.} \end{cases}$$

The b_n's were found in equation (7) on page 650. By making appropriate substitutions in this equations we have

$$\sin 7\pi x = \sum_{n=1}^{\infty} n\pi b_n \sin nx .$$

From this we see that for $n = 7$

$$7\pi b_7 = 1 \quad \Rightarrow \quad b_7 = \frac{1}{7\pi},$$

and for all other n's, $b_n = 0$. By substituting these coefficients into the formal solution given in equation (15) above, we obtain

$$u(x,t) = \frac{1}{7\pi}\sin 7\pi t \sin 7\pi x + \sum_{n=0}^{\infty} \frac{8}{\left[(2n+1)\pi\right]^3}\cos\left[(2n+1)\pi t\right]\sin\left[(2n+1)\pi x\right].$$

5. First we note that this problem is consistent because $g(0) = 0 = g(L)$ and $f(0) = 0 = f(L)$. The formal solution to this problem is given in equation (5) on page 649 of the text with the coefficients given in equations (6) and (7) on page 650. By equation (7), we see that

$$g(x) = 0 = \sum_{n=1}^{\infty} b_n \frac{n\pi\alpha}{L}\sin\left(\frac{n\pi x}{L}\right).$$

Thus, each term in this infinite series must be zero and so $b_n = 0$ for all n's. Therefore, the formal solution given in equation (5) on page 649 of the text becomes

$$u(x,t) = \sum_{n=1}^{\infty} a_n \cos\left(\frac{n\pi\alpha}{L}t\right)\sin\left(\frac{n\pi x}{L}\right). \tag{16}$$

To find the a_n's we note that by equation (6) on page 650 of the text these coefficients are the Fourier sine coefficients for $f(x)$. Therefore, by using equation (7) on page 633 of the text, for $n = 1, 2, 3, \ldots$, we have

$$a_n = \frac{2}{L}\int_0^L f(x)\sin\left(\frac{n\pi x}{L}\right)dx = \frac{2}{L}\left[\frac{h_0}{a}\int_0^a x\sin\left(\frac{n\pi x}{L}\right)dx + h_0\int_a^L \frac{L-x}{L-a}\sin\left(\frac{n\pi x}{L}\right)dx\right]$$

$$= \frac{2h_0}{L}\left[\frac{1}{a}\int_0^a x\sin\left(\frac{n\pi x}{L}\right)dx + \frac{L}{L-a}\int_a^L \sin\left(\frac{n\pi x}{L}\right)dx - \frac{1}{L-a}\int_a^L x\sin\left(\frac{n\pi x}{L}\right)dx\right].$$

By using integration by parts, we find

$$\int x\sin\left(\frac{n\pi x}{L}\right)dx = -\frac{xL}{n\pi}\cos\left(\frac{n\pi x}{L}\right) + \frac{L^2}{n^2\pi^2}\sin\left(\frac{n\pi x}{L}\right).$$

Therefore, for $n = 1, 2, 3, \ldots$, the coefficients become

$$a_n = \frac{2h_0}{L}\left[\frac{1}{a}\left(-\frac{aL}{n\pi}\cos\left(\frac{n\pi a}{L}\right) + \frac{L^2}{n^2\pi^2}\sin\left(\frac{n\pi a}{L}\right)\right) - \frac{L^2}{n\pi(L-a)}\left(\cos n\pi - \cos\left(\frac{n\pi a}{L}\right)\right)\right.$$

$$\left. - \frac{1}{L-a}\left(-\frac{L^2}{n\pi}\cos n\pi + \frac{aL}{n\pi}\cos\left(\frac{n\pi a}{L}\right) + \frac{L^2}{n^2\pi^2}\sin n\pi - \frac{L^2}{n^2\pi^2}\sin\left(\frac{n\pi a}{L}\right)\right)\right].$$

After simplifying, this becomes

$$a_n = \frac{2h_0 L^2}{n^2\pi^2 a(L-a)}\sin\left(\frac{n\pi a}{L}\right), \qquad n = 1, 2, 3, \ldots.$$

By substituting this result into equation (16) above, we obtain the formal solution to this problem given by

$$u(x,t) = \frac{2h_0 L^2}{\pi^2 a(L-a)}\sum_{n=1}^{\infty}\frac{1}{n^2}\sin\left(\frac{n\pi a}{L}\right)\cos\left(\frac{n\pi a t}{L}\right)\sin\left(\frac{n\pi x}{L}\right).$$

7. If we let $\alpha = 1$, $h(x,t) = tx$, $L = \pi$, $f(x) = \sin x$, and $g(x) = 5\sin 2x - 3\sin 5x$, then we see that this problem has the same form as the problem given in Example 1 on page 651 of the text. The formal solution to the problem in Example 1 is given in equation (16) on page 652 of the text. Therefore, with the appropriate substitutions, the formal solution to this problem is

$$u(x,t) = \sum_{n=1}^{\infty}\left\{a_n\cos nt + b_n\sin nt + \frac{1}{n}\int_0^t h_n(s)\sin[n(t-s)]\,ds\right\}\sin nx. \qquad (17)$$

The a_n's are shown in equation (14) on page 652 of the text to satisfy

$$\sin x = \sum_{n=1}^{\infty} a_n \sin nx.$$

Thus, the only nonzero term in this infinite series is the term for $n = 1$. Therefore, we see that $a_1 = 1$ and $a_n = 0$ for $n \neq 1$. The b_n's are given in equation (15) on page 652 of the text and so must satisfy

$$5\sin 2x - 3\sin 5x = \sum_{n=1}^{\infty} nb_n \sin nx,$$

which implies that

$$2b_2 = 5 \quad \Rightarrow \quad b_2 = \frac{5}{2}, \qquad \text{and} \qquad 5b_5 = -3 \quad \Rightarrow \quad b_5 = -\frac{3}{5},$$

and $b_n = 0$ for all other values of n. To calculate the integral given in the formal solution we must first find the functions $h_n(t)$. To do this, we note that in Example 1, the functions $h_n(t)$, $n = 1, 2, \ldots$, are the Fourier sine coefficients for $h(x, t) = tx$ with t fixed. These functions are given below equation (13) on page 652 of the text. (We will assume proper convergence of this series.) Thus, we have

$$h_n(t) = \frac{2}{\pi}\int_0^{\pi} tx \sin nx\, dx = \frac{2t}{\pi}\int_0^{\pi} x \sin nx\, dx$$

$$= \frac{2t}{\pi}\left[-\frac{\pi}{n}\cos n\pi + 0 + \frac{1}{n^2}\sin n\pi - \sin 0\right]$$

$$= -\frac{2t}{\pi}\cos n\pi = \frac{2t}{\pi}(-1)^{n+1}, \qquad n = 1, 2, 3, \ldots,$$

where we have used integration by parts to calculate this integral. Substituting this result into the integral in equation (17) above yields

$$\int_0^t h_n(s)\sin[n(t-s)]\,ds = \int_0^t \frac{2s}{n}(-1)^{n+1}\sin[n(t-s)]\,ds$$

$$= \frac{2}{n}(-1)^{n+1}\left[\frac{t}{n} - \frac{1}{n^2}\sin nt\right] = \frac{2}{n^3}(-1)^{n+1}(nt - \sin nt), \qquad \text{where } n = 1, 2, 3, \ldots .$$

By plugging the a_n's, the b_n's, and the result we just found into equation (17), we obtain the formal solution to this problem given by

$$u(x,t) = \cos t \sin x + \frac{5}{2}\sin 2t \sin 2x - \frac{3}{5}\sin 5t \sin 5x + \sum_{n=1}^{\infty}\frac{1}{n}\left[\frac{2}{n^3}(-1)^{n+1}(nt - \sin nt)\right]\sin nx$$

$$= \cos t \sin x + \frac{5}{2}\sin 2t \sin 2x - \frac{3}{5}\sin 5t \sin 5x + 2\sum_{n=1}^{\infty}\left[\frac{(-1)^{n+1}}{n^3}\left(t - \frac{\sin nt}{n}\right)\right]\sin nx .$$

11. We will assume that a solution to this problem has the form $u(x,t) = X(x)T(t)$. Substituting this expression into the partial differential equations yields

$$X(x)T''(t) + X(x)T'(t) + X(x)T(t) = \alpha^2 X''(x)T(t) .$$

Dividing this equation by $\alpha^2 X(x)T(t)$ yields

$$\frac{T''(t) + T'(t) + T(t)}{\alpha^2 T(t)} = \frac{X''(x)}{X(x)} .$$

Since these two expressions must be equal for all x in $(0, L)$ **and** all $t > 0$, they can not vary. Therefore, they must both equal a constant, say K. This gives us the two ordinary differential equations

$$\frac{T''(t)+T'(t)+T(t)}{\alpha^2 T(t)} = K \qquad \Rightarrow \qquad T''(t)+T'(t)+\left(1-\alpha^2 K\right)T(t)=0, \tag{18}$$

and

$$\frac{X''(x)}{X(x)} = K \qquad \Rightarrow \qquad X''(x)-KX(x)=0. \tag{19}$$

Substituting $u(x,t)=X(x)T(t)$ into the boundary conditions, $u(0,t)=u(L,t)=0$, $t>0$, we obtain

$$X(0)T(t)=0=X(L)T(t), \quad t>0.$$

Since we are seeking a nontrivial solution to the partial differential equation, we do not want $T(t)\equiv 0$. Therefore, for the above equation to be zero, we must have $X(0)=X(L)=0$. Combining this fact with equation (19) above yields the boundary value problem given by

$$X''(x)-KX(x)=0, \quad \text{with } X(0)=X(L)=0.$$

This problem was solved in Section 10.2 of the text. There we found that for $K=-\left(\frac{n\pi}{L}\right)^2$, $n=1, 2, 3,\ldots,$ we obtain nonzero solutions of the form

$$X_n(x) = A_n \sin\left(\frac{n\pi x}{L}\right), \quad n = 1, 2, 3, \ldots . \tag{20}$$

Plugging these values of K into equation (18) above yields the family of linear ordinary differential equations with constant coefficients given by

$$T''(t)+T'(t)+\left(1+\frac{\alpha^2 n^2\pi^2}{L^2}\right)T(t)=0, \quad n=1, 2, 3, \ldots . \tag{21}$$

The auxiliary equations associated with these ODE's are

$$r^2+r+\left(1+\frac{\alpha^2 n^2\pi^2}{L^2}\right)=0.$$

By using the quadratic formula, we obtain the roots to these auxiliary equations. Thus, we have

$$r=\frac{-1\pm\sqrt{1-4\left(1+\dfrac{\alpha^2 n^2\pi^2}{L^2}\right)}}{2}=-\frac{1}{2}\pm\frac{\sqrt{L^2-4L^2-4\alpha^2 n^2\pi^2}}{2L}$$

$$=-\frac{1}{2}\pm\frac{\sqrt{3L^2+4\alpha^2 n^2\pi^2}}{2L}i, \qquad n=1, 2, 3, \ldots .$$

Hence, the solutions to the linear equations given in equation (21) above are

$$T_n(t) = e^{-t/2}\left[B_n \cos\left(\frac{\sqrt{3L^2+4\alpha^2 n^2\pi^2}}{2L}t\right) + C_n \sin\left(\frac{\sqrt{3L^2+4\alpha^2 n^2\pi^2}}{2L}t\right)\right],$$

for $n = 1, 2, 3, \ldots$. By letting

$$\beta_n = \frac{\sqrt{3L^2+4\alpha^2 n^2\pi^2}}{2L}, \qquad (22)$$

for $n = 1, 2, 3, \ldots$, this family of solutions can be more easily written as

$$T_n(t) = e^{-t/2}[B_n \cos\beta_n t + C_n \sin\beta_n t].$$

Substituting the solutions we have just found and the solutions given in equation (20) above into $u(x,t) = X(x)T(t)$, yields solutions to the original partial differential equation given by

$$u_n(x,t) = X_n(x)T_n(t) = e^{-t/2}[A_n B_n \cos\beta_n t + A_n C_n \sin\beta_n t]\sin\left(\frac{n\pi x}{L}\right), \qquad n = 1, 2, 3, \ldots .$$

By the superposition principle, we see that solutions to the PDE will have the form

$$u(x,t) = \sum_{n=1}^{\infty} e^{-t/2}[a_n \cos\beta_n t + b_n \sin\beta_n t]\sin\left(\frac{n\pi x}{L}\right),$$

where β_n is given in equation (22) above, $a_n = A_n B_n$, and $b_n = A_n C_n$. To find the coefficients a_n and b_n, we use the initial conditions $u(x,0) = f(x)$ and $\dfrac{\partial u(x,0)}{\partial t} = 0$. Therefore, since

$$\frac{\partial u(x,t)}{\partial t} = \sum_{n=1}^{\infty}\left\{\left(-\frac{1}{2}\right)e^{-t/2}[a_n \cos\beta_n t + b_n \sin\beta_n t]\sin\left(\frac{n\pi x}{L}\right)\right.$$

$$\left. + e^{-t/2}[-a_n\beta_n \sin\beta_n t + b_n\beta_n \cos\beta_n t]\sin\left(\frac{n\pi x}{L}\right)\right\},$$

we have

$$\frac{\partial u(x,0)}{\partial t} = 0 = \sum_{n=1}^{\infty}\left\{-\frac{a_n}{2} + b_n\beta_n\right\}\sin\left(\frac{n\pi x}{L}\right).$$

Hence, each term in this infinite series must be zero which implies that

$$-\frac{a_n}{2} + b_n\beta_n = 0 \qquad \Rightarrow \qquad b_n = \frac{a_n}{2\beta_n}, \qquad n = 1, 2, 3, \ldots .$$

Thus, we can write

$$u(x,t) = \sum_{n=1}^{\infty} a_n\, e^{-t/2}\left[\cos\beta_n t + \frac{1}{2\beta_n}\sin\beta_n t\right]\sin\left(\frac{n\pi x}{L}\right), \qquad (23)$$

where β_n is given above in equation (22). To find the a_n's, we use the remaining initial condition to obtain

$$u(x,0) = f(x) = \sum_{n=1}^{\infty} a_n \sin\left(\frac{n\pi x}{L}\right).$$

Therefore, the a_n's are the Fourier sine coefficients of $f(x)$ and so satisfy

$$a_n = \frac{2}{L}\int_0^L f(x)\sin\left(\frac{n\pi x}{L}\right)dx. \qquad (24)$$

Combining all of these results, we see that a formal solution to the telegraph problem is given by equation (23) where β_n and a_n are given in equation (22) and (24) respectively.

15. This problem has the form of the problem solved in Example 2 on page 655 of the text with $f(x) = g(x) = x$. There it was found that d'Alembert's formula given in equation (32) on page 655 of the text is a solution to this problem. By making the appropriate substitutions in this equation (and noting that $f(x+\alpha t) = x+\alpha t$ and $f(x-\alpha t) = x-\alpha t$), we obtain the solution

$$u(x,t) = \frac{1}{2}[x+\alpha t + x - \alpha t] + \frac{1}{2\alpha}\int_{x-\alpha t}^{x+\alpha t} s\,ds = x + \frac{1}{2\alpha}\left[\left.\frac{s^2}{2}\right|_{x-\alpha t}^{x+\alpha t}\right]$$

$$= x + \frac{1}{4\alpha}\left[(x+\alpha t)^2 - (x-\alpha t)^2\right] = x + \frac{1}{4\alpha}[4\alpha t x] = x + tx.$$

EXERCISES 10.7: Laplace's Equation, page 673

3. To solve this problem using separation of variables, we will assume that a solution has the form $u(x,y) = X(x)Y(y)$. Making this substitution into the partial differential equation yields

$$X''(x)Y(y) + X(x)Y''(y) = 0.$$

By dividing the above equation by $X(x)Y(y)$, we obtain

$$\frac{X''(x)}{X(x)} + \frac{Y''(y)}{Y(y)} = 0.$$

Since this equation must be true for $0 < x < \pi$ and $0 < y < \pi$, there must be a constant K such that

$$\frac{X''(x)}{X(x)} = -\frac{Y''(y)}{Y(y)} = K, \qquad 0 < x < \pi \text{ and } 0 < y < \pi.$$

This leads to the two ordinary differential equations given by

$$X''(x) - KX(x) = 0, \qquad (25)$$

and

$$Y''(y) + KY(y) = 0. \tag{26}$$

By making the substitution $u(x, y) = X(x)Y(y)$ into the first boundary conditions $u(0, y) = u(\pi, y) = 0$, we obtain

$$X(0)Y(y) = X(\pi)Y(y) = 0.$$

Since we do not want the trivial solution which would be obtained if we let $Y(y) \equiv 0$, these boundary conditions imply that

$$X(0) = X(\pi) = 0.$$

Combining these boundary conditions with equation (25) above yields the boundary value problem

$$X''(x) - KX(x) = 0, \qquad \text{with} \quad X(0) = X(\pi) = 0.$$

To solve this problem, we will consider three cases.

Case 1: $K = 0$. For this case, the differential equation becomes $X''(x) = 0$, which has solutions $X(x) = A + Bx$. By applying the first of the boundary conditions, we obtain

$$X(0) = 0 = A \qquad \Rightarrow \qquad X(x) = Bx.$$

The second boundary condition yields

$$X(\pi) = B\pi = 0 \qquad \Rightarrow \qquad B = 0.$$

Thus, in this case we obtain only the trivial solution.

Case 2: $K > 0$. In this case, the auxiliary equation associated with this differential equation is $r^2 - K = 0$, which has the real roots $r = \pm\sqrt{K}$. Thus, solutions to this problem are given by

$$X(x) = Ae^{\sqrt{K}x} + Be^{-\sqrt{K}x}.$$

Applying the boundary conditions, yields

$$X(0) = 0 = A + B \;\Rightarrow\; \qquad A = -B \qquad \Rightarrow \qquad X(x) = -Be^{\sqrt{K}x} + Be^{-\sqrt{K}x},$$

and

$$X(\pi) = 0 = -Be^{\sqrt{K}\pi} + Be^{-\sqrt{K}\pi} \qquad \Rightarrow \qquad -B\left(e^{2\sqrt{K}\pi} - 1\right) = 0.$$

This last expression is true only if $K = 0$ or if $B = 0$. Since we are assuming that $K > 0$, we must have $B = 0$ which means that $A = -B = 0$. Therefore, in this case we again find only the trivial solution.

Case 3: $K < 0$. The auxiliary equation associated with the differential equation in this case has the complex valued roots $r = \pm\sqrt{-K}i$, (where $-K > 0$). Therefore, solutions to the ODE for this case are given by

$$X(x) = A\cos\left(\sqrt{-K}x\right) + B\sin\left(\sqrt{-K}x\right).$$

By applying the boundary conditions, we obtain

$$X(0) = 0 = A \qquad \Rightarrow \qquad X(x) = B\sin\left(\sqrt{-K}x\right),$$

and

$$X(\pi) = 0 = B\sin\left(\sqrt{-K}\pi\right) \qquad \Rightarrow \qquad \sqrt{-K} = n \qquad \Rightarrow \qquad K = -n^2, \qquad n = 1, 2, 3, \ldots,$$

where we have assumed that $B \neq 0$ since this would lead to the trivial solution. Therefore, nontrivial solutions $X_n(x) = B_n \sin nx$ are obtained when $K = -n^2$, $n = 1, 2, 3 \ldots$.
To solve the differential equation given in equation (26) above, we use these values for *K*. This yields the family of linear ordinary differential equations given by

$$Y''(y) - n^2 Y(y) = 0, \qquad n = 1, 2, 3, \ldots.$$

The auxiliary equations associated with these ODE's are $r^2 - n^2 = 0$, which have the real roots $r = \pm n$, $n = 1, 2, 3, \ldots$. Hence, the solutions to this family of differential equations are given by

$$Y_n(y) = C_n e^{ny} + D_n e^{-ny}, \qquad n = 1, 2, 3, \ldots.$$

With the substitutions $K_{1,n} = C_n + D_n$ and $K_{2,n} = C_n - D_n$, so that

$$\frac{K_{1,n} + K_{2,n}}{2} = C_n \quad \text{and} \quad \frac{K_{1,n} - K_{2,n}}{2} = D_n,$$

we see that these solutions can be written as

$$Y_n(y) = \frac{K_{1,n}e^{ny} + K_{1,n}e^{-ny}}{2} - \frac{K_{2,n}e^{ny} + K_{2,n}e^{-ny}}{2} = K_{1,n}\cosh ny + K_{2,n}\sinh ny.$$

This last expression can in turn be written as

$$Y_n(y) = A_n \sinh(ny + D_n),$$

where $A_n = K_{2,n}^2 - K_{1,n}^2$, and $D_n = \tanh^{-1}\left(\dfrac{K_{1,n}}{K_{2,n}}\right)$. (See Problem 18.)

The last boundary condition $u(x,\pi) = X(x)Y(\pi) = 0$ implies that $Y(\pi) = 0$ (since we do not want the trivial solution). Therefore, by substituting π into the solutions just found, we obtain

$$Y_n(\pi) = A_n \sinh(n\pi + D_n) = 0.$$

Since we do not want $A_n = 0$, this implies that $\sinh(n\pi + D_n) = 0$. This will be true only if $n\pi + D_n = 0$ or in other words if $D_n = -n\pi$. Substituting these expressions for D_n into the family of solutions we found for *Y*, yields

$$Y_n(y) = A_n \sinh(ny - n\pi).$$

Therefore, substituting the solutions just found for *X*(*x*) and *Y*(*y*) into $u_n(x,y) = X_n(x)Y_n(y)$ we see that $u_n(x,y) = a_n \sin nx \sinh(ny - n\pi)$, where $a_n = A_n B_n$. By the superposition principle, a formal solution to the original partial differential equation is given by

$$u(x,y)=\sum_{n=1}^{\infty} a_n \sin nx \sinh(ny-n\pi)\,. \tag{27}$$

In order to find an expression for the coefficients a_n, we will apply the remaining boundary condition, $u(x,0)=f(x)$. From this condition, we obtain

$$u(x,0)=f(x)=\sum_{n=1}^{\infty} a_n \sin nx \sinh(-n\pi)\,,$$

which implies that $a_n \sinh(-n\pi)$ are the coefficients of the Fourier sine series of $f(x)$. Therefore, by equation (7) on page 633 of the text, we see that (with $T=\pi$)

$$a_n \sinh(-n\pi)=\frac{2}{\pi}\int_0^{\pi} f(x)\sin nx\,dx$$

$$\Rightarrow \qquad a_n=\frac{2}{\pi\sinh(-n\pi)}\int_0^{\pi} f(x)\sin nx\,dx\,.$$

Thus, a formal solution to this ODE is given in equation (27) with the a_n's given by the equation above.

5. This problem has two nonhomogeneous boundary conditions, and, therefore, we will solve two PDE problems, one for each of these boundary conditions. These problems are

$$\frac{\partial^2 u}{\partial x^2}+\frac{\partial^2 u}{\partial y^2}=0\,; \qquad 0<x<\pi, \qquad 0<y<1,$$

$$\frac{\partial u(0,y)}{\partial x}=\frac{\partial u(\pi,y)}{\partial x}=0\,, \quad 0\le y\le 1,$$

$$u(x,0)=\cos x-\cos 3x\,, \quad 0\le x\le \pi,$$

$$u(x,1)=0\,, \quad 0\le x\le \pi,$$

and

$$\frac{\partial^2 u}{\partial x^2}+\frac{\partial^2 u}{\partial y^2}=0\,; \qquad 0<x<\pi, \qquad 0<y<1,$$

$$\frac{\partial u(0,y)}{\partial x}=\frac{\partial u(\pi,y)}{\partial x}=0\,, \quad 0\le y\le 1,$$

$$u(x,0)=0\,, \quad 0\le x\le \pi,$$

$$u(x,1)=\cos 2x\,, \quad 0\le x\le \pi.$$

If u_1 and u_2 are solutions to the first and second problems respectively, then $u = u_1 + u_2$ will be a solution to the original problem. To see this notice that

$$\frac{\partial^2 u}{\partial x^2} + \frac{\partial^2 u}{\partial y^2} = \frac{\partial^2 u_1}{\partial x^2} + \frac{\partial^2 u_2}{\partial x^2} + \frac{\partial^2 u_1}{\partial y^2} + \frac{\partial^2 u_2}{\partial y^2} = 0 + 0 = 0,$$

$$\frac{\partial u(0, y)}{\partial x} = \frac{\partial u_1(0, y)}{\partial x} + \frac{\partial u_2(0, y)}{\partial x} = 0 + 0 = 0,$$

$$\frac{\partial u(\pi, y)}{\partial x} = \frac{\partial u_1(\pi, y)}{\partial x} + \frac{\partial u_2(\pi, y)}{\partial x} = 0 + 0 = 0,$$

$$u(x,0) = u_1(x,0) + u_2(x,0) = \cos x - \cos 3x + 0 = \cos x - \cos 3x,$$

$$u(x,1) = u_1(x,1) + u_2(x,1) = 0 + \cos 2x = \cos 2x.$$

This is an application of the superposition principle.

The first of these two problems has the form of the problem given in Example 1 on page 663 of the text with $a = \pi$, $b = 1$, and $f(x) = \cos x - \cos 3x$. A formal solution to this problem is given in equation (10) on page 665 of the text. Thus, by making the appropriate substitutions, we find that a formal solution to the first problem is

$$u_1(x, y) = E_0(y-1) + \sum_{n=1}^{\infty} E_n \cos nx \sinh(ny - n).$$

To find the coefficients E_n, we use the nonhomogeneous boundary condition

$$u(x,0) = \cos x - \cos 3x.$$

Thus, we have

$$u_1(x,0) = \cos x - \cos 3x = -E_0 + \sum_{n=1}^{\infty} E_n \cos nx \sinh(-n).$$

From this we see that for $n = 1$,

$$E_1 \sinh(-1) = 1 \quad \Rightarrow \quad E_1 = \frac{1}{\sinh(-1)},$$

and for $n = 3$,

$$E_3 \sinh(-3) = -1 \quad \Rightarrow \quad E_3 = \frac{-1}{\sinh(-3)}.$$

For all other values of n, $E_n = 0$. By substituting these values into the expression found above for u_1, we obtain the formal solution to the first of our two problems given by

$$u_1(x,0) = \frac{\cos x \sinh(y-1)}{\sinh(-1)} - \frac{\cos 3x \sinh(3y-3)}{\sinh(-3)}. \tag{28}$$

To solve the second of our problems, we note that, except for the last two boundary conditions, it is similar to the problem solved in Example 1 on page 663 of the text. As in that example, using the separation of variables technique, we find that the ODE

$$X''(x) - KX(x) = 0, \qquad X'(0) = X'(\pi) = 0,$$

has solutions $X_n(x) = a_n \cos nx$, when $K = -n^2$, n = 1, 2, 3, ... By substituting these values for K into the ODE $Y''(y) + KY(y) = 0$, we again find that a family solutions to this differential equation is given by

$$Y_0(y) = A_0 + B_0 y,$$

$$Y_n(y) = C_n \sinh[n(y + D_n)], \qquad n = 1, 2, 3, \ldots .$$

At this point, the problem we are solving differs from the example. The boundary condition $u(x,0) = X(x)Y(0) = 0$, $0 \le x \le \pi$, implies that $Y(0) = 0$ (since we don't want the trivial solution). Therefore, applying this boundary condition to each of the solutions found above yields

$$Y_0(0) = A_0 + 0 = 0 \qquad \Rightarrow \qquad A_0 = 0,$$

$$Y_n(0) = C_n \sinh(nD_n) = 0 \qquad \Rightarrow \qquad D_n = 0,$$

where we have used the fact that $\sinh x = 0$ only when x = 0. By substituting these results into the solutions found above, we obtain

$$Y_0(y) = B_0 y,$$

$$Y_n(y) = C_n \sinh ny, \qquad n = 1, 2, 3, \ldots .$$

Combining these solutions with the solutions $X_n(x) = a_n \cos nx$ yields

$$u_{2,0}(x,y) = X_0(x)Y_0(y) = a_0 B_0 y \cos 0 = E_0 y,$$

$$u_{2,n}(x,y) = X_n(x)Y_n(y) = a_n C_n \cos nx \sinh ny = E_n \cos nx \sinh ny,$$

where $E_0 = a_0 B_0$ and $E_n = a_n C_n$. Thus, by the superposition principle, we find that a formal solution to the second problem is given by

$$u_2(x,y) = E_0 y + \sum_{n=1}^{\infty} E_n \cos nx \sinh ny .$$

By applying the last boundary condition of this second problem, namely $u(x,1) = \cos 2x$, to these solutions, we see that

$$u_2(x,1) = E_0 + \sum_{n=1}^{\infty} E_n \cos nx \sinh n = \cos 2x .$$

Therefore, when $n = 2$,

$$E_2 = \sinh 2 = 1 \quad \Rightarrow \quad E_2 = \frac{1}{\sinh 2},$$

and for all other values of n, $E_n = 0$. By substituting these coefficients into the solution $u_2(x,y)$ that we found above, we obtain the formal solution to this second problem

$$u_2(x,y) = \frac{\cos 2x \sinh 2y}{\sinh(2)}.$$

By the superposition principle (as noted at the beginning of this problem), a formal solution to the original partial differential equation is the sum of this solution and the solution given in equation (28). Thus, the solution that we seek is

$$u(x,y) = \frac{\cos x \sinh(y-1)}{\sinh(-1)} - \frac{\cos 3x \sinh(3y-3)}{\sinh(-3)} + \frac{\cos 2x \sinh 2y}{\sinh(2)}.$$

11. In this problem, the technique of separation of variables, as in Example 2 on page 666 of the text, leads to the two ODE's

$$r^2 R''(r) + rR'(r) - \lambda R(r) = 0, \quad \text{and} \quad T''(\theta) + \lambda T(\theta) = 0.$$

Again, as in Example 2, we require the solution $u(r, \theta)$ to be continuous on its domain. Therefore, $T(\theta)$ must again be periodic with period 2π. This implies that $T(-\pi) = T(\pi)$ and $T'(-\pi) = T'(\pi)$. Thus, as in Example 2, a family of solutions for the second ODE above which satisfies these periodic boundary conditions is

$$T_0(\theta) = B \quad \text{and} \quad T_n(\theta) = A_n \cos n\theta + B_n \sin n\theta, \quad n = 1, 2, 3, \ldots.$$

In solving this problem, it was found that $\lambda = n^2$, $n = 0, 1, 2, 3, \ldots$. Again, as in Example 2, substituting these values for λ into the first ODE above leads to the solutions

$$R_0(r) = C + D\ln r \quad \text{and} \quad R_n(r) = C_n r^n + D_n r^{-n}, \quad n = 1, 2, 3, \ldots.$$

Here, however, we are not concerned with what happens when $r = 0$. By our assumption that $u(r,\theta) = R(r)T(\theta)$, we see that solutions of the PDE given in this problem will have the form

$$u_0(r,\theta) = B(C + D\ln r) \quad \text{and} \quad u_n(r,\theta) = (C_n r^n + D_n r^{-n})(A_n \cos n\theta + B_n \sin n\theta),$$

where $n = 1, 2, 3, \ldots$. Thus, by the superposition principle, we see that a formal solution to this Dirichlet problem is given by

$$u(r,\theta) = BC + BD\ln r + \sum_{n=1}^{\infty} (C_n r^n + D_n r^{-n})(A_n \cos n\theta + B_n \sin n\theta),$$

or

$$u(r,\theta) = a + b\ln r + \sum_{n=1}^{\infty} \left[(c_n r^n + e_n r^{-n})\cos n\theta + (d_n r^n + f_n r^{-n})\sin n\theta\right], \qquad (29)$$

where $a = BC$, $b = BD$, $c_n = C_n A_n$, $e_n = D_n A_n$, $d_n = C_n B_n$, and $f_n = D_n B_n$. To find these coefficients, we apply the boundary conditions $u(1,\theta) = \sin 4\theta - \cos\theta$, and $u(2,\theta) = \sin\theta$, $-\pi \le \theta \le \pi$. From the first boundary condition, we see that

$$u(1,\theta) = a + \sum_{n=1}^{\infty} \left[(c_n + e_n)\cos n\theta + (d_n + f_n)\sin n\theta\right] = \sin 4\theta - \cos\theta,$$

which implies that $a = 0$, $d_4 + f_4 = 1$, $c_1 + e_1 = -1$, and for all other values of n, $c_n + e_n = 0$ and $d_n + f_n = 0$. From the second boundary condition, we have

$$u(2,\theta) = a + b\ln 2 + \sum_{n=1}^{\infty} \left[(c_n 2^n + e_n 2^{-n})\cos n\theta + (d_n 2^n + f_n 2^{-n})\sin n\theta\right] = \sin\theta,$$

which implies that $a = 0$, $b = 0$, $\left(2d_1 + 2^{-1} f_1\right) = 1$, and for all other values of n, $\left(2^n c_n + 2^{-n} e_n\right) = 0$ and $\left(2^n d_n + 2^{-n} f_n\right) = 0$. By combining these results, we obtain $a = 0$, $b = 0$, the three systems of two equations in two unknowns given by

$$\begin{aligned} d_4 + f_4 &= 1, \\ 2^4 d_4 + 2^{-4} f_4 &= 0, \end{aligned} \qquad \text{and} \qquad \begin{aligned} c_1 + e_1 &= -1, \\ 2c_1 + 2^{-1} e_1 &= 0, \end{aligned} \qquad \text{and} \qquad \begin{aligned} d_1 + f_1 &= 0, \\ 2d_1 + 2^{-1} f_1 &= 1, \end{aligned}$$

(where the first equation in each system was derived from the first boundary condition and the second equation in each system was derived from the second boundary condition), and for all other values of n, $c_n = 0$, $e_n = 0$, $d_n = 0$, $f_n = 0$. Solving each system of equations simultaneously yields $d_4 = \frac{-1}{255}$, $f_4 = \frac{256}{255}$, $c_1 = \frac{1}{3}$, $e_1 = \frac{-4}{3}$, $d_1 = \frac{2}{3}$, and $f_1 = \frac{-2}{3}$. By substituting these values for the coefficients into equation (29) above, we find that a solution to this Dirichlet problem is given by

$$u(r,\theta) = \left(\frac{1}{3}r - \frac{4}{3}r^{-1}\right)\cos\theta + \left(\frac{2}{3}r - \frac{2}{3}r^{-1}\right)\sin\theta + \left(-\frac{1}{255}r^4 + \frac{256}{255}r^{-4}\right)\sin 4\theta.$$

15. Here, as in Example 2 on page 666 of the text, the technique of separation of variables leads to the two ODE's given by

$$r^2 R''(r) + rR'(r) - \lambda R(r) = 0, \quad \text{and} \quad T''(\theta) + \lambda T(\theta) = 0.$$

Since we want to avoid the trivial solution, the boundary condition $u(r,0) = R(r)T(0) = 0$ implies that $T(0) = 0$ and the boundary condition $u(r,\pi) = R(r)T(\pi) = 0$ implies that $T(\pi) = 0$. Therefore, we seek a nontrivial solution to the ODE

$$T''(\theta) + \lambda T(\theta) = 0, \quad \text{with} \quad T(0) = 0 \ \text{and} \ T(\pi) = 0. \tag{30}$$

To do this we will consider three cases for λ.

<u>Case 1: $\lambda = 0$.</u> This case leads to the differential equation $T''(\theta) = 0$, which has solutions $T(\theta) = A\theta + B$. Applying the first boundary condition yields $T(0) = 0 = B$. Thus, $T(\theta) = A\theta$. The second boundary condition implies that $T(\pi) = 0 = A\pi$. Hence, $A = 0$. Therefore, in this case we find only the trivial solution.

Case 2: $\lambda < 0$. In this case, the auxiliary equation associated with the linear differential equation given in equation (30) above is $r^2 + \lambda = 0$, which has the real roots $r = \pm\sqrt{-\lambda}$ (where $-\lambda > 0$). Thus, the solution to this differential equation has the form

$$T(\theta) = c_1 e^{\sqrt{-\lambda}\theta} + c_2 e^{-\sqrt{-\lambda}\theta}.$$

Applying the first boundary condition yields

$$T(0) = c_1 + c_2 \qquad \Rightarrow \qquad c_1 = -c_2 \qquad \Rightarrow \qquad T(\theta) = -c_2 e^{\sqrt{-\lambda}\theta} + c_2 e^{-\sqrt{-\lambda}\theta}.$$

From the second boundary condition, we obtain

$$T(\pi) = c_2\left(-e^{\sqrt{-\lambda}\pi} + e^{-\sqrt{-\lambda}\pi}\right) \qquad \Rightarrow \qquad c_2\left(e^{2\sqrt{-\lambda}\pi} - 1\right) = 0.$$

Since we are assuming that $\lambda > 0$, the only way that this last expression can be zero is for $c_2 = 0$. Thus, $c_1 = -c_2 = 0$ and we again obtain the trivial solution.

Case 3: $\lambda > 0$. In this case, the roots to the auxiliary equation associated with this differential equation are $r = \pm\sqrt{\lambda}i$. Therefore, the solution to the differential equation given in equation (30) is

$$T(\theta) = c_1 \sin\sqrt{\lambda}\theta + c_2 \cos\sqrt{\lambda}\theta.$$

From the boundary conditions, we see that

$$T(0) = 0 = c_2 \qquad \Rightarrow \qquad T(\theta) = c_1 \sin\sqrt{\lambda}\theta,$$

and

$$T(\pi) = 0 = c_1 \sin\sqrt{\lambda}\pi.$$

Since we do not want the trivial solution, this last boundary condition implies that $\sin\sqrt{\lambda}\pi = 0$. This will be true if $\sqrt{\lambda} = n$ or in other words if $\lambda = n^2$, $n = 1, 2, 3, \ldots$. With these values for λ, we find nontrivial solution for the differential equation given in equation (30) above to be

$$T_n(\theta) = B_n \sin n\theta, \quad n = 1, 2, 3, \ldots.$$

Substituting the values for λ that we have just found into the differential equation

$$r^2 R''(r) + rR'(r) - \lambda R(r) = 0,$$

yields the ODE

$$r^2 R''(r) + rR'(r) - n^2 R(r) = 0, \qquad n = 1, 2, 3, \ldots.$$

This is the same Cauchy-Euler equation that was solved in Example 2 on page 666 of the text. There it was found that the solutions have the form

$$R_n(r) = C_n r^n + D_n r^{-n}, \quad n = 1, 2, 3, \ldots.$$

Since we require that $u(r, \theta)$ to be bounded on its domain, we see that $u(r,\theta) = R(r)T(\theta)$ must be bounded about $r = 0$. This implies that $R(0)$ must be bounded. Therefore, $D_n = 0$ and so the solutions to this Cauchy-Euler equation are given by

$$R_n(r) = C_n r^n, \quad n = 1, 2, 3, \ldots .$$

Since we have assumed that $u(r,\theta) = R(r)T(\theta)$, we see that formal solutions to the original partial differential equation are

$$u_n(r,\theta) = B_n C_n r^n \sin n\theta = c_n r^n \sin n\theta ,$$

where $c_n = B_n C_n$. Therefore, by the superposition principle, we obtain the formal solutions to this Dirichlet problem

$$u(r,\theta) = \sum_{n=1}^{\infty} c_n r^n \sin n\theta .$$

The final boundary condition yields

$$u(1,\theta) = \sin 3\theta = \sum_{n=1}^{\infty} c_n \sin n\theta .$$

This implies that $c_3 = 1$ and for all other values of *n*, $c_n = 0$. Substituting these values for the coefficients into the formal solution found above, yields the solution to this Dirichlet problem on the half disk

$$u(r,\theta) = r^3 \sin 3\theta .$$

17. As in Example 2 on page 666 of the text, we solve this problem by separation of variables. There it was found that we must solve the two ordinary differential equations

$$r^2 R''(r) + rR'(r) - \lambda R(r) = 0 , \tag{31}$$

and

$$T''(\theta) + \lambda T(\theta) = 0 , \tag{32}$$

with $T(\pi) = T(n\pi)$ and $T'(\pi) = T'(n\pi)$.

In Example 2, we found that when $\lambda = n^2$, n = 0, 1, 2, 3, … , the linear differential equation given in equation (32) has nontrivial solutions of the form

$$T_n(\theta) = A_n \cos n\theta + B_n \sin n\theta , \quad n = 0, 1, 2, 3, \ldots ,$$

and equation (31) has solutions of the form

$$R_0(r) = C + D\ln r , \qquad \text{and} \qquad R_n(r) = C_n r^n + D_n r^{-n} , \qquad n = 1, 2, 3, \ldots .$$

(Note that here we are not concerned with what happens to $u(r, \theta)$ around $r = 0$.) Thus, since we are assuming that $u(r,\theta) = R(r)T(\theta)$, we see that solutions to the original partial differential equation will be given by

$$u_0(r,\theta) = A_0\left(C + D\ln r\right) = a_0 + b_0 \ln r ,$$

and

$$u_n(r,\theta)=\left(C_n r^n + D_n r^{-n}\right)\left(A_n \cos n\theta + B_n \sin n\theta\right)$$
$$=\left(a_n r^n + b_n r^{-n}\right)\cos n\theta + \left(c_n r^n + d_n r^{-n}\right)\sin n\theta ,$$

where $a_0 = A_0 C$, $b_0 = A_0 D$, $a_n = C_n A_n$, $b_n = D_n A_n$, $c_n = C_n B_n$, and $d_n = D_n B_n$. Thus, by the superposition principle, we see that a formal solution to the partial differential equation given in this problem will have the form

$$u(r,\theta) = a_0 + b_0 \ln r + \sum_{n=1}^{\infty}\left[(a_n r^n + b_n r^{-n})\cos n\theta + (c_n r^n + d_n r^{-n})\sin n\theta\right]. \tag{33}$$

By applying the first boundary condition, we obtain

$$u(1,\theta) = f(\theta) = a_0 + \sum_{n=1}^{\infty}\left[(a_n + b_n)\cos n\theta + (c_n + d_n)\sin n\theta\right],$$

where we have used the fact that $\ln 1 = 0$. Comparing this equation with equation (8) on page 618 of the text, we see that a_0, $(a_n + b_n)$, and $(c_n + d_n)$ are the Fourier coefficients of $f(\theta)$ (with $T = \pi$). Therefore, by equations (9) and (10) on that same page we see that

$$a_0 = \frac{1}{2\pi}\int_{-\pi}^{\pi} f(\theta)d\theta , \tag{34}$$

$$a_n + b_n = \frac{1}{\pi}\int_{-\pi}^{\pi} f(\theta)\cos n\theta \, d\theta ,$$

$$c_n + d_n = \frac{1}{\pi}\int_{-\pi}^{\pi} f(\theta)\sin n\theta \, d\theta , \qquad n = 1, 2, 3, \ldots .$$

To apply the last boundary condition, we must find $\dfrac{\partial u}{\partial r}$. Hence, we find

$$\frac{\partial u(r,\theta)}{\partial r} = \frac{b_0}{r} + \sum_{n=1}^{\infty}\left[(a_n n r^{n-1} - b_n n r^{-n-1})\cos n\theta + (c_n n r^{n-1} - d_n n r^{-n-1})\sin n\theta\right].$$

Applying the last boundary condition yields

$$\frac{\partial u(3,\theta)}{\partial r} = g(\theta) = \frac{b_0}{3} + \sum_{n=1}^{\infty}\left[(a_n n 3^{n-1} - b_n n 3^{-n-1})\cos n\theta + (c_n n 3^{n-1} - d_n n 3^{-n-1})\sin n\theta\right].$$

Again by comparing this to equation (8) on page 618 of the text, we see that $\dfrac{b_0}{3}$, $\left(n3^{n-1} a_n - n3^{-n-1} b_n\right)$, and $\left(n3^{n-1} c_n - n3^{-n-1} d_n\right)$ are the Fourier coefficients of $g(\theta)$ (with $T = \pi$). Thus, by equations (9) and (10) on that same page of the text, we see that

$$b_0 = \frac{3}{2\pi}\int_{-\pi}^{\pi} g(\theta)d\theta , \tag{35}$$

$$n3^{n-1}a_n - n3^{-n-1}b_n = \frac{1}{\pi}\int_{-\pi}^{\pi} g(\theta)\cos n\theta \, d\theta,$$

$$n3^{n-1}c_n - n3^{-n-1}d_n = \frac{1}{\pi}\int_{-\pi}^{\pi} g(\theta)\sin n\theta \, d\theta, \quad n = 1, 2, 3, \ldots .$$

Therefore, the formal solution to this partial differential equation will be given by equation (33) with the coefficients given by equations (34) and (35).

CHAPTER 11: Eigenvalue Problems and Sturm-Liouville Equations

EXERCISES 11.2: Eigenvalues and Eigenfunctions, page 695

1. The auxiliary equation for this problem is $r^2+2r+26=0$, which has roots $r=-1\pm 5i$. Hence a general solution to the differential equation $y''+2y'+26y=0$ is given by

$$y(x)=C_1 e^{-x}\cos 5x+C_2 e^{-x}\sin 5x .$$

We will now try to determine C_1 and C_2 so that the boundary conditions are satisfied. Setting $x=0$ and $x=\pi$, we find

$$y(0)=C_1=1, \qquad y(\pi)=-C_1 e^{-\pi}=-e^{-\pi} .$$

Both boundary conditions yield the same result, $C_1=1$. Hence, there is a one parameter family of solutions,

$$y(x)=e^{-x}\cos 5x+C_2 e^{-x}\sin 5x ,$$

where C_2 is arbitrary.

13. First note that the auxiliary equation for this problem is $r^2+\lambda=0$. To find eigenvalues which yield nontrivial solutions we will consider the three cases $\lambda<0$, $\lambda=0$, and $\lambda>0$.

Case 1, $\lambda<0$: In this case the roots to the auxiliary equation are $\pm\sqrt{-\lambda}$ (where we note that $-\lambda$ is a positive number). Therefore, a general solution to the differential equation $y''+\lambda y=0$ is given by

$$y(x)=C_1 e^{\sqrt{-\lambda}x}+C_2 e^{-\sqrt{-\lambda}x} .$$

By applying the first boundary condition, we obtain

$$y(0)=C_1+C_2=0 \qquad \Rightarrow \qquad C_2=-C_1 .$$

Thus

$$y(x)=C_1\left(e^{\sqrt{-\lambda}x}-e^{-\sqrt{-\lambda}x}\right).$$

In order to apply the second boundary conditions, we need to find $y'(x)$. Thus, we have

$$y'(x)=C_1\sqrt{-\lambda}\left(e^{\sqrt{-\lambda}x}+e^{-\sqrt{-\lambda}x}\right).$$

Thus

$$y'(1)=C_1\sqrt{-\lambda}\left(e^{\sqrt{-\lambda}}+e^{-\sqrt{-\lambda}}\right)=0 . \qquad (1)$$

Since $\sqrt{-\lambda} > 0$ and $e^{\sqrt{-\lambda}} + e^{-\sqrt{-\lambda}} \neq 0$, the only way that equation (1) can be true is for $C_1 = 0$. So in this case we have only the trivial solution. Thus, there are no eigenvalues for $\lambda < 0$.

Case 2, $\lambda = 0$: In this case we are solving the differential equation $y'' = 0$. This equation has a general solution given by

$$y(x) = C_1 + C_2 x \qquad \Rightarrow \qquad y'(x) = C_2 .$$

By applying the boundary conditions, we obtain

$$y(0) = C_1 = 0 \qquad \text{and} \qquad y'(1) = C_2 = 0 .$$

Thus $C_1 = C_2 = 0$, and zero is not an eigenvalue.

Case 3, $\lambda > 0$: In this case the roots to the associated auxiliary equation are $r = \pm\sqrt{\lambda}i$. Therefore, the general solution is given by

$$y(x) = C_1 \cos\left(\sqrt{\lambda}x\right) + C_2 \sin\left(\sqrt{\lambda}x\right).$$

By applying the first boundary condition, we obtain

$$y(0) = C_1 = 0 \qquad \Rightarrow \qquad y(x) = C_2 \sin\left(\sqrt{\lambda}x\right).$$

In order to apply the second boundary conditions we need to find $y'(x)$. Thus, we have

$$y'(x) = C_2 \sqrt{\lambda} \cos\left(\sqrt{\lambda}x\right),$$

and so

$$y'(1) = C_2 \sqrt{\lambda} \cos\left(\sqrt{\lambda}\right) = 0 .$$

Therefore, in order to obtain a solution other than the trivial solution, we must have

$$\cos\left(\sqrt{\lambda}\right) = 0 \qquad \Rightarrow \qquad \sqrt{\lambda} = \left(n + \frac{1}{2}\right)\pi , \qquad n = 0, 1, 2, \ldots$$

$$\Rightarrow \qquad \lambda_n = \left(n + \frac{1}{2}\right)^2 \pi^2 , \qquad n = 0, 1, 2, \ldots .$$

Corresponding to the eigenvalue λ_n, we have the corresponding eigenfunctions,

$$y_n(x) = C_n \sin\left[\left(n + \frac{1}{2}\right)\pi x\right], \qquad n = 0, 1, 2, \ldots ,$$

where C_n is an arbitrary nonzero constant.

19. The equation $(xy')' + \lambda x^{-1} y = 0$ can be rewritten as the Cauchy–Euler equation

$$x^2 y'' + xy' + \lambda y = 0, \qquad x > 0. \qquad (2)$$

Substituting $y = x^r$ gives $r^2 + \lambda = 0$ as the auxiliary equation for (2). Again we will consider the three cases $\lambda < 0$, $\lambda = 0$, and $\lambda > 0$.

Case 1, $\lambda < 0$: Let $\lambda = -\mu^2$, for $\mu > 0$. The roots of the auxiliary equation are $r = \pm\mu$ and so a general solution to (2) is

$$y(x) = C_1 x^{\mu} + C_2 x^{-\mu}.$$

We first find $y'(x)$.

$$y'(x) = C_1 \mu x^{\mu-1} - C_2 \mu x^{-\mu-1} = \mu\left(C_1 x^{\mu-1} - C_2 x^{-\mu-1}\right).$$

Substituting into the first boundary condition gives

$$y'(1) = \mu\left(C_1 - C_2\right) = 0.$$

Since $\mu > 0$,

$$C_1 - C_2 = 0 \qquad \Rightarrow \qquad C_1 = C_2.$$

Substituting this into the second condition yields

$$y\left(e^{\pi}\right) = C_1\left(e^{\mu\pi} + e^{-\mu\pi}\right) = 0. \tag{3}$$

Since $e^{\mu\pi} + e^{-\mu\pi} \neq 0$ the only way that equation (3) can be true is for $C_1 = 0$. So in this case we have only the trivial solution. Thus, there are no eigenvalue for $\lambda < 0$.

Case 2, $\lambda = 0$: In this case we are solving the differential equation $\left(xy'\right)' = 0$. This equation can be solved as follows:

$$xy' = C_1 \qquad \Rightarrow \qquad y' = \frac{C_1}{x}$$

$$y(x) = C_2 + C_1 \ln x.$$

By applying the boundary conditions, we obtain

$$y'(1) = C_1 = 0 \qquad \text{and} \qquad y\left(e^{\pi}\right) = C_2 + C_1 \ln e^{\pi} = C_2 + \pi C_1 = 0.$$

Solving these equations simultaneously yields $C_2 = C_1 = 0$. Thus, we again find only the trivial solution. Therefore, $\lambda = 0$ is not an eigenvalue.

Case 3, $\lambda > 0$: Let $\lambda = \mu^2$, for $\mu > 0$. The roots of the auxiliary equation are $r = \pm\mu i$ and so a general solution (3) is

$$y(x) = C_1 \cos(\mu \ln x) + C_2 \sin(\mu \ln x).$$

We next find y'.

$$y'(x) = -C_1\left(\frac{\mu}{x}\right)\sin(\mu \ln x) + C_2\left(\frac{\mu}{x}\right)\cos(\mu \ln x).$$

By applying the first boundary condition, we obtain

$$y'(1) = \mu C_2 = 0 \qquad \Rightarrow \qquad C_2 = 0 .$$

Applying the second boundary condition, we obtain

$$y(e^{\pi}) = C_1 \cos(\mu \ln e^{\pi}) = C_1 \cos(\mu\pi) = 0 .$$

Therefore, in order to obtain a solution other than the trivial solution, we must have

$$\cos(\mu\pi) = 0 \qquad \Rightarrow \qquad \mu\pi = \frac{\pi}{2} + \pi n, \qquad n = 0, 1, 2, \ldots$$

$$\Rightarrow \qquad \mu = \frac{1}{2} + n, \qquad n = 0, 1, 2, \ldots$$

$$\Rightarrow \qquad \lambda_n = \left(n + \frac{1}{2}\right)^2, \qquad n = 0, 1, 2, \ldots .$$

Corresponding to the eigenvalues, λ_n's, we have the eigenfunctions.

$$y_n(x) = C_n \cos\left[\left(n + \frac{1}{2}\right)\ln x\right], \qquad n = 0, 1, 2, \ldots ,$$

where C_n is an arbitrary nonzero constant.

25. As in Problem 13, the auxiliary equation for this problem is $r^2 + \lambda = 0$. To find eigenvalues which yield nontrivial solutions we will consider the three cases $\lambda < 0$, $\lambda = 0$, and $\lambda > 0$.

Case 1, $\lambda < 0$: The roots of the auxiliary equation are $r = \pm\sqrt{-\lambda}$ and so a general solution to the differential equation $y'' + \lambda y = 0$ is given by

$$y(x) = C_1 e^{\sqrt{-\lambda}x} + C_2 e^{-\sqrt{-\lambda}x}$$

By applying the first boundary condition we obtain

$$y(0) = C_1 + C_2 = 0 \qquad \Rightarrow \qquad C_2 = -C_1 .$$

Thus

$$y(x) = C_1\left(e^{\sqrt{-\lambda}x} - e^{-\sqrt{-\lambda}x}\right).$$

Applying the second boundary conditions yields

$$y(1+\lambda^2) = C_1\left(e^{\sqrt{-\lambda}(1+\lambda^2)} - e^{-\sqrt{-\lambda}(1+\lambda^2)}\right) = 0 .$$

Multiplying by $e^{\sqrt{-\lambda}(1+\lambda^2)}$ yields

$$C_1\left(e^{2\sqrt{-\lambda}(1+\lambda^2)} - 1\right) = 0 .$$

Now either $C_1 = 0$ or

$$e^{2\sqrt{-\lambda}(1+\lambda^2)} = 1 \qquad \Rightarrow \qquad \sqrt{-\lambda}\left(1+\lambda^2\right) = 0$$

$$\Rightarrow \quad \sqrt{-\lambda} = 0 \qquad \Rightarrow \qquad \lambda = 0\,.$$

Since $\lambda < 0$, we must have $C_1 = 0$ and hence there are no eigenvalues for $\lambda < 0$.

Case 2, $\lambda = 0$: In this case we are solving the differential equation $y'' = 0$. This equation has a general solution given by

$$y(x) = C_1 + C_2 x\,.$$

By applying the boundary conditions, we obtain

$$y(0) = C_1 = 0 \qquad \text{and} \qquad y\left(1+\lambda^2\right) = C_1 + C_2\left(1+\lambda^2\right) = 0\,.$$

Solving these equations simultaneously yields $C_1 = C_2 = 0$. Thus, we find that is $\lambda = 0$ not an eigenvalue.

Case 3, $\lambda > 0$: The roots of the auxiliary equation are $r = \pm\sqrt{\lambda}i$ and so a general solution is

$$y(x) = C_1 \cos\left(\sqrt{\lambda}x\right) + C_2 \sin\left(\sqrt{\lambda}x\right).$$

Substituting in the first boundary condition yields

$$y(0) = C_1 \cos\left(\sqrt{\lambda}\cdot 0\right) + C_2 \sin\left(\sqrt{\lambda}\cdot 0\right) = C_1 = 0\,.$$

By applying the second boundary condition to $y(x) = C_2 \sin\left(\sqrt{\lambda}x\right)$, we obtain

$$y\left(1+\lambda^2\right) = C_2 \sin\left(\sqrt{\lambda}\left[1+\lambda^2\right]\right) = 0\,.$$

Therefore, in order to obtain a solution other than the trivial solution, we must have

$$\sin\left(\sqrt{\lambda}\left[1+\lambda^2\right]\right) = 0 \qquad \Rightarrow \qquad \sqrt{\lambda}\left(1+\lambda^2\right) = n\pi, \qquad n = 1, 2, 3, \ldots\,.$$

Hence choose the eigenvalues λ_n, $n = 1, 2, 3, \ldots$, such that $\sqrt{\lambda_n}\left(1+\lambda_n^2\right) = n\pi$; and the corresponding eigenfunctions are $y_n(x) = C_n \sin\left(\sqrt{\lambda_n}x\right)$, $n = 1, 2, 3, \ldots$, where the C_n's are arbitrary nonzero constants.

33. **(a)** We assume the $u(x,t) = X(x)T(t)$. Now $u_{tt} = X(x)T''(t)$, $u_{xx} = X''(x)T(t)$, and $u_x = X'(x)T(t)$. Substituting these functions into $u_{tt} = u_{xx} + 2u_x$, we obtain

$$X(x)T''(t) = X''(x)T(t) + 2X'(x)T(t)\,.$$

Separating variables yields

$$\frac{X''(x) + 2X'(x)}{X(x)} = -\lambda = \frac{T''(t)}{T(t)}, \tag{4}$$

where λ is some constant. The first equation in (4) gives

$$X''(x) + 2X'(x) + \lambda X(x) = 0 .$$

Let's now consider the boundary conditions. From $u(0,t) = 0$ and $u(\pi,t) = 0$, $t > 0$, we conclude that

$$X(0)T(t) = 0 \qquad \text{and} \qquad X(\pi)T(t) = 0, \qquad t > 0 .$$

Hence either $T(t) = 0$ for all $t > 0$, which implies $u(x,t) \equiv 0$, or

$$X(0) = X(\pi) = 0 .$$

Ignoring the trivial solution $u(x,t) \equiv 0$, we obtain the boundary value problem

$$X''(x) + 2X'(x) + \lambda X(x) = 0, \qquad X(0) = X(\pi) = 0 .$$

(b) The auxiliary equation for this problem is $r^2 + 2r + \lambda = 0$, which has roots $r = -1 \pm \sqrt{1-\lambda}$. To find eigenvalues which yield nontrivial solutions, we will consider the three cases $1 - \lambda < 0$, $1 - \lambda = 0$, and $1 - \lambda > 0$.

Case 1, $1 - \lambda < 0$ $(\lambda > 1)$: Let $\mu = \sqrt{-(1-\lambda)} = \sqrt{\lambda - 1}$. In this case the roots to the auxiliary equation are $-1 \pm \mu i$ (where μ is a positive number). Therefore, a general solution to the differential equation is given by

$$X(x) = C_1 e^{-x} \cos \mu x + C_2 e^{-x} \sin \mu x .$$

By applying the boundary conditions, we obtain

$$X(0) = C_1 = 0 \qquad \text{and} \qquad X(\pi) = e^{-\pi}\left(C_1 \cos \mu\pi + C_2 \sin \mu\pi\right) = 0 .$$

Solving these equations simultaneously yields $C_1 = 0$ and $C_2 \sin \mu\pi = 0$. Therefore, in order to obtain a solution other than the trivial solution, we must have

$$\sin \mu\pi = 0 \qquad \Rightarrow \qquad \mu\pi = n\pi \qquad \Rightarrow \qquad \mu = n, \qquad n = 1, 2, 3, \ldots .$$

Since $\mu = \sqrt{\lambda - 1}$

$$\sqrt{\lambda - 1} = n \qquad \Rightarrow \qquad \lambda - 1 = n^2 \qquad \Rightarrow \qquad \lambda = n^2 + 1, \qquad n = 1, 2, 3, \ldots .$$

Thus the eigenvalues are given by

$$\lambda_n = n^2 + 1, \qquad n = 1, 2, 3, \ldots .$$

Corresponding to the eigenvalue λ_n, we obtain the solutions

$$X_n(x) = C_n e^{-x} \sin nx , \qquad n = 1, 2, 3, \ldots ,$$

where C_n is arbitrary.

Case 2, $1 - \lambda = 0$ $(\lambda = 1)$: In this case the associated auxiliary equation has double root $r = -1$. Therefore, the general solution is given by

$$X(x) = C_1 e^{-x} + C_2 x e^{-x} .$$

By applying the boundary conditions we obtain

$$X(0) = C_1 = 0 \qquad \text{and} \qquad X(\pi) = e^{-\pi}(C_1 + C_2\pi) = 0.$$

Solving these equations simultaneously yields $C_1 = C_2 = 0$. So in this case we have only the trivial solution. Thus, $\lambda = 1$ is not an eigenvalue.

Case 3, $1 - \lambda > 0$ $(\lambda < 1)$: Let $\mu = \sqrt{1-\lambda}$. In this case the roots to the auxiliary equation are $-1 \pm \mu$ (where μ is a positive number). Therefore, a general solution to the differential equation is given by

$$X(x) = C_1 e^{(-1-\mu)x} + C_2 e^{(-1+\mu)x}.$$

By applying the first boundary condition we find

$$X(0) = C_1 + C_2 = 0 \qquad \Rightarrow \qquad C_2 = -C_1.$$

So we can express X as

$$X(x) = C_1\left(e^{(-1-\mu)x} - e^{(-1+\mu)x}\right).$$

Thus the second condition gives us

$$X(\pi) = C_1\left(e^{(-1-\mu)\pi} - e^{(-1+\mu)\pi}\right) = 0.$$

Since $e^{(-1-\mu)\pi} - e^{(-1+\mu)\pi} \neq 0$, $C_1 = 0$, and again in this case we have only the trivial solution. Thus, there are no eigenvalues for $\lambda < 1$.

Therefore, the eigenvalues are $\lambda_n = n^2 + 1$, $n = 1, 2, 3, \ldots$, with corresponding eigenfunctions $X_n(x) = C_n e^{-x} \sin nx$, $n = 1, 2, 3, \ldots$, where C_n is an arbitrary nonzero constant.

EXERCISES 11.3: Regular Sturm–Liouville Boundary Value Problems, page 706

3. Here $A_2 = x(1-x)$ and $A_1 = -2x$. Using formula (4) on page 697 of the text, we find

$$\mu(x) = \frac{1}{x(1-x)} e^{\int A_1(x)/A_2(x)\,dx} = \frac{1}{x(1-x)} e^{\int -2x/x(1-x)\,dx}$$

$$= \frac{1}{x(1-x)} e^{-2\int 1/(1-x)\,dx} = \frac{1}{x(1-x)} e^{2\ln(1-x)}$$

$$= \frac{1}{x(1-x)}(1-x)^2 = \frac{1-x}{x}.$$

Multiplying the original equation by $\mu(x) = \dfrac{1-x}{x}$, we get

$$(1-x)^2 y''(x) - 2(1-x)y'(x) + \lambda \frac{1-x}{x} y(x) = 0$$

$$\Rightarrow \qquad \left((1-x)^2 y'(x)\right)' + \lambda \frac{1-x}{x} y(x) = 0.$$

9. Here we consider the linear differential operator $L[y] := y'' + \lambda y$; $y(0) = -y(\pi)$, $y'(0) = -y'(\pi)$. We must show that $(u, L[v]) = (L[u], v)$, where $u(x)$ and $v(x)$ are any functions in the domain of L. Now

$$(u, L[v]) = \int_0^{\pi} u(v'' + \lambda v)\,dx = \int_0^{\pi} uv''\,dx + \int_0^{\pi} \lambda uv\,dx$$

and

$$(L[u], v) = \int_0^{\pi} (u'' + \lambda u)v\,dx = \int_0^{\pi} u''v\,dx + \int_0^{\pi} \lambda uv\,dx .$$

Hence it suffices to show that $\int_0^{\pi} uv''\,dx = \int_0^{\pi} u''v\,dx$. To do this we start with $\int_0^{\pi} u''v\,dx$ and integrate by parts twice. Doing this we obtain

$$\int_0^{\pi} u''v\,dx = u'v\Big|_0^{\pi} - uv'\Big|_0^{\pi} + \int_0^{\pi} uv''\,dx .$$

Hence, we just need to show $u'v\Big|_0^{\pi} - uv'\Big|_0^{\pi} = 0$. Expanding gives

$$u'v\Big|_0^{\pi} - uv'\Big|_0^{\pi} = u'(\pi)v(\pi) - u'(0)v(0) - u(\pi)v'(\pi) + u(0)v'(0).$$

Since u is in the domain of L, we have $u(0) = -u(\pi)$ and $u'(0) = -u'(\pi)$. Hence,

$$u'v\Big|_0^{\pi} - uv'\Big|_0^{\pi} = u'(\pi)\left[v(\pi) + v(0)\right] - u(\pi)\left[v'(\pi) + v'(0)\right].$$

But v also lies in the domain of L and hence $v(0) = -v(\pi)$ and $v'(0) = -v'(\pi)$. This makes the expressions in the brackets zero and we have $u'v\Big|_0^{\pi} - uv'\Big|_0^{\pi} = 0$.

Therefore, $L[y]$ is selfadjoint.

17. In Problem 13 of Section 11.2, we found the eigenvalues to be

$$\lambda_n = \left(n + \frac{1}{2}\right)^2 \pi^2, \qquad n = 0, 1, 2, \ldots$$

with the corresponding eigenfunctions

$$y_n(x) = C_n \sin\left[\left(n + \frac{1}{2}\right)\pi x\right], \qquad n = 0, 1, 2, \ldots,$$

where C_n is an arbitrary nonzero constant.

(a) We need only to choose the C_n so that $\int_0^1 C_n^2 \sin^2\left[\left(n+\frac{1}{2}\right)\pi x\right] dx = 1$. We compute

$$\int_0^1 C_n^2 \sin^2\left[\left(n+\frac{1}{2}\right)\pi x\right] dx = C_n^2 \int_0^1 \left(\frac{1}{2}-\frac{1}{2}\cos[(2n+1)\pi x]\right) dx$$

$$= \frac{1}{2}C_n^2 \int_0^1 (1-\cos[(2n+1)\pi x])\, dx$$

$$= \frac{1}{2}C_n^2 \left(x - \frac{1}{(2n+1)\pi}\sin[(2n+1)\pi x]\right)\Bigg|_0^1 = \frac{1}{2}C_n^2 .$$

Hence, we can take $C_n = \sqrt{2}$ which gives

$$\left\{\sqrt{2}\sin\left[\left(n+\frac{1}{2}\right)\pi x\right]\right\}_{n=0}^{\infty},$$

as an orthonormal system of eigenfunctions.

(b) To obtain the eigenfunction expansion for $f(x) = x$, we use formula (25) on page 703 of the text. Thus,

$$c_n = \int_0^1 x\sqrt{2}\sin\left[\left(n+\frac{1}{2}\right)\pi x\right] dx .$$

Using integration by parts with $u = \sqrt{2}x$ and $dv = \sin\left[\left(n+\frac{1}{2}\right)\pi x\right] dx$, we find

$$c_n = \frac{-\sqrt{2}x\cos\left[\left(n+\frac{1}{2}\right)\pi x\right]}{\left(n+\frac{1}{2}\right)\pi}\Bigg|_0^1 + \int_0^1 \frac{\sqrt{2}\cos\left[\left(n+\frac{1}{2}\right)\pi x\right] dx}{\left(n+\frac{1}{2}\right)\pi}$$

$$= \frac{-\sqrt{2}\cos\left[\left(n+\frac{1}{2}\right)\pi\right]}{\left(n+\frac{1}{2}\right)\pi} + \frac{\sqrt{2}\sin\left[\left(n+\frac{1}{2}\right)\pi x\right]}{\left(n+\frac{1}{2}\right)^2\pi^2}\Bigg|_0^1$$

$$= 0 + \frac{\sqrt{2}\sin\left[\left(n+\frac{1}{2}\right)\pi\right]}{\left(n+\frac{1}{2}\right)^2 \pi^2} = \frac{(-1)^n \sqrt{2}}{\left(n+\frac{1}{2}\right)^2 \pi^2}.$$

Therefore

$$x = \sum_{n=0}^{\infty} c_n \sqrt{2} \sin\left[\left(n+\frac{1}{2}\right)\pi x\right] = \sum_{n=0}^{\infty} \frac{2\,(-1)^n}{\left(n+\frac{1}{2}\right)^2 \pi^2} \sin\left[\left(n+\frac{1}{2}\right)\pi x\right]$$

$$= \frac{8}{\pi^2} \sum_{n=0}^{\infty} \frac{(-1)^n}{(2n+1)^2} \sin\left[\left(n+\frac{1}{2}\right)\pi x\right].$$

23. In Problem 19 of Section 11.2, we found the eigenvalues

$$\lambda_n = \left(n+\frac{1}{2}\right)^2, \qquad n = 0, 1, 2, \ldots,$$

with the corresponding eigenfunctions

$$y_n(x) = C_n \cos\left[\left(n+\frac{1}{2}\right)\ln x\right], \qquad n = 0, 1, 2, \ldots,$$

where C_n is an arbitrary nonzero constant.

(a) We need only to choose the C_n so that

$$\int_1^{e^\pi} C_n^2 \cos^2\left[\left(n+\frac{1}{2}\right)\ln x\right] \frac{1}{x}\, dx = 1.$$

To compute, we let $u = \ln x$ and so $du = x^{-1} dx$. Substituting, we find

$$\int_1^{e^\pi} C_n^2 \frac{\cos^2\left[\left(n+\frac{1}{2}\right)\ln x\right]}{x}\, dx = C_n^2 \int_0^{\pi} \cos^2\left[\left(n+\frac{1}{2}\right)u\right] du$$

$$= C_n^2 \int_0^{\pi} \left(\frac{1}{2} + \frac{1}{2}\cos[(2n+1)u]\right) du$$

$$= \frac{1}{2} C_n^2 \left(u + \frac{1}{2n+1}\sin[(2n+1)u]\right)\Bigg|_0^{\pi} = \frac{\pi}{2} C_n^2.$$

Hence, we can take $C_n = \sqrt{\dfrac{2}{\pi}}$. Which gives

$$\left\{\sqrt{\frac{2}{\pi}}\cos\left[\left(n+\frac{1}{2}\right)\ln x\right]\right\}_{n=0}^{\infty},$$

as an orthonormal system of eigenfunctions.

(b) To obtain the eigenfunction expansion for $f(x)=x$, we use formula (25) on page 703 of the text. Thus, with $w(x)=x^{-1}$, we have

$$c_n = \int_1^{e^{\pi}} x\sqrt{\frac{2}{\pi}}\cos\left[\left(n+\frac{1}{2}\right)\ln x\right]x^{-1}\,dx .$$

Let $u=\ln x$ so $du = x^{-1}dx$, we have

$$c_n = \sqrt{\frac{2}{\pi}}\int_0^{\pi} e^u \cos\left[\left(n+\frac{1}{2}\right)u\right]du$$

$$= \sqrt{\frac{2}{\pi}}\left\{\frac{e^u\cos\left[\left(n+\frac{1}{2}\right)u\right]+e^u\left(n+\frac{1}{2}\right)\sin\left[\left(n+\frac{1}{2}\right)u\right]}{1+\left(n+\frac{1}{2}\right)^2}\right\}\Bigg|_0^{\pi}$$

$$= \sqrt{\frac{2}{\pi}}\left\{\frac{e^{\pi}\left(n+\frac{1}{2}\right)\sin\left[\left(n+\frac{1}{2}\right)\pi\right]-1}{1+\left(n+\frac{1}{2}\right)^2}\right\} = \sqrt{\frac{2}{\pi}}\left\{\frac{(-1)^n e^{\pi}\left(n+\frac{1}{2}\right)-1}{1+\left(n+\frac{1}{2}\right)^2}\right\}.$$

Therefore,

$$x = \sum_{n=0}^{\infty} c_n\sqrt{\frac{2}{\pi}}\cos\left[\left(n+\frac{1}{2}\right)\ln x\right]$$

$$= \frac{2}{\pi}\sum_{n=0}^{\infty}\frac{(-1)^n e^{\pi}\left(n+\frac{1}{2}\right)-1}{1+\left(n+\frac{1}{2}\right)^2}\cos\left[\left(n+\frac{1}{2}\right)\ln x\right].$$

EXERCISES 11.4: Nonhomogeneous Boundary Value Problems and the Fredholm Alternative, page 716

3. Here our differential operator is given by

$$L[y]=\left(1+x^2\right)y''+2xy'+y.$$

Substituting into the formula (3) page 708 of the text, we obtain

$$L^+[y]=\left[\left(1+x^2\right)y\right]''-(2xy)'+y$$

$$=\left(2xy+\left(1+x^2\right)y'\right)'-2y-2xy'+y$$

$$=2y+2xy'+2xy'+\left(1+x^2\right)y''-2y-2xy'+y$$

$$=\left(1+x^2\right)y''+2xy'+y.$$

7. Here our differential operator is given by

$$L[y]=y''-2y'+10y; \qquad y(0)=y(\pi)=0.$$

Hence

$$L^+[v]=v''+2v'+10v.$$

To find the $D\left(L^+\right)$, we must have

$$P(u,v)(x)\Big|_0^\pi=0; \tag{5}$$

for all u in $D(L)$ and v in $D\left(L^+\right)$. Using formula (9) page 709 of the text for $P(u,v)$ with $A_1=-2$ and $A_2=1$, we find

$$P(u,v)=-2uv-uv'+u'v.$$

Evaluating at π and 0, condition (5) becomes

$$-2u(\pi)v(\pi)-u(\pi)v'(\pi)+u'(\pi)v(\pi)+2u(0)v(0)+u(0)v'(0)-u'(0)v(0)=0.$$

Since u in $D(L)$, we know that $u(0)=u(\pi)=0$. Thus the above equation becomes

$$u'(\pi)v(\pi)-u'(0)v(0)=0.$$

Since $u'(\pi)$ and $u'(0)$ can take on any value, we must have $v(0)=v(\pi)=0$ for this equation to hold for all u in $D(L)$. Hence $D\left(L^+\right)$ consists of all function v having continuous second derivatives on $[0, \pi]$ and satisfying the boundary condition

$$v(0)=v(\pi)=0.$$

11. Here our differential operator is given by

$$L[y] = y'' + 6y' + 10y; \qquad y'(0) = y'(\pi) = 0.$$

Hence

$$L^{+}[v] = v'' - 6v' + 10v.$$

To find the $D(L^{+})$, we must have

$$P(u,v)(x)\Big|_0^{\pi} = 0 \tag{6}$$

for all u in $D(L)$ and v in $D(L^{+})$. Again using formula (9) page 709 of the text for $P(u,v)$ with $A_1 = 6$ and $A_2 = 1$, we find

$$P(u,v) = 6uv - uv' + u'v.$$

Evaluating at π and 0, condition (6) becomes

$$6u(\pi)v(\pi) - u(\pi)v'(\pi) + u'(\pi)v(\pi) - 6u(0)v(0) + u(0)v'(0) - u'(0)v(0) = 0.$$

Applying the boundary conditions $u'(0) = u'(\pi) = 0$ to the above equation yields

$$6u(\pi)v(\pi) - u(\pi)v'(\pi) - 6u(0)v(0) + u(0)v'(0) = 0$$

$$\Rightarrow \qquad u(\pi)[6v(\pi) - v'(\pi)] - u(0)[6v(0) - v'(0)] = 0.$$

Since $u(\pi)$ and $u(0)$ can take on any value, we must have $6v(\pi) - v'(\pi) = 0$ and $6v(0) - v'(0) = 0$ in order for the equation to hold for all u in $D(L)$. Therefore, the adjoint boundary value problem is

$$L^{+}[v] = v'' - 6v' + 10v; \qquad 6v(\pi) = v'(\pi) \text{ and } 6v(0) = v'(0).$$

17. In Problem 7 we found the adjoint boundary value problem

$$L^{+}[v] = v'' + 2v' + 10v; \qquad v(0) = v(\pi) = 0. \tag{7}$$

The auxiliary equation for (7) is $r^2 + 2r + 10 = 0$, which has roots $r = -1 \pm 3i$. Hence a general solution to the differential equation in (7) is given by

$$y(x) = C_1 e^{-x}\cos 3x + C_2 e^{-x}\sin 3x.$$

Using the boundary conditions in (7) to determine C_1 and C_2, we find

$$y(0) = C_1 = 0 \qquad \text{and} \qquad y(\pi) = -C_1 e^{-\pi} = 0.$$

Thus $C_1 = 0$ and C_2 is arbitrary. Therefore, every solution to the adjoint problem (7) has the form

$$y(x) = C_2 e^{-x}\sin 3x.$$

It follows from the Fredholm alternative that if h is continuous, then the nonhomogeneous problem has a solution if and only if

$$\int_0^{\pi} h(x)\,\mathrm{e}^{-x}\sin 3x\,dx = 0\ .$$

21. In Problem 11 we found the adjoint boundary value problem

$$L^{+}[v] = v'' - 6v' + 10v\,; \qquad 6v(\pi) = v'(\pi) \quad \text{and} \quad 6v(0) = v'(0)\,. \qquad (8)$$

The auxiliary equation for (8) is $r^2 - 6r + 10 = 0$, which has roots $r = 3 \pm i$. Hence a general solution to the differential equation in (8) is given by

$$y(x) = C_1 \mathrm{e}^{3x}\cos x + C_2 \mathrm{e}^{3x}\sin x\,.$$

To apply the boundary conditions in (8), we first determine $y'(x)$

$$y'(x) = 3C_1 \mathrm{e}^{3x}\cos x - C_1 \mathrm{e}^{3x}\sin x + 3C_2 \mathrm{e}^{3x}\sin x + C_2 \mathrm{e}^{3x}\cos x\,.$$

Applying the first condition, we have

$$-6C_1 \mathrm{e}^{3\pi} = -3C_1 \mathrm{e}^{3\pi} - C_2 \mathrm{e}^{3\pi} \qquad \Rightarrow \qquad 3C_1 = C_2.$$

Applying the second condition, we have

$$6C_1 = 3C_1 + C_2 \qquad \Rightarrow \qquad 3C_1 = C_2.$$

Thus $C_2 = 3C_1$ where C_1 is arbitrary. Therefore, every solution to the adjoint problem (8) has the form

$$y(x) = C_1 \mathrm{e}^{3x}(\cos x + 3\sin x).$$

It follows from the Fredholm alternative that if h is continuous, then the nonhomogeneous problem has a solution if and only if

$$\int_0^{\pi} h(x)\,\mathrm{e}^{3x}(\cos x + 3\sin x)\,dx = 0\,.$$

EXERCISES 11.5: Solution by Eigenfunction Expansion, page 722

3. In Example 1 on page 720 of the text we noted that the boundary value problem

$$y'' + \lambda y = 0\,; \qquad y(0) = 0, \qquad y(\pi) = 0,$$

has eigenvalues $\lambda_n = n^2$, $n = 1, 2, 3, \ldots$, with corresponding eigenfunctions

$$\phi_n(x) = \sin nx\,, \qquad n = 1,\ 2,\ 3,\ \ldots\,.$$

Here $r(x) \equiv 1$, so we need to determine coefficients γ_n such that

$$f(x) = \frac{f(x)}{r(x)} = \sum_{n=1}^{\infty} \gamma_n \sin nx = \sin 2x + \sin 8x\,.$$

Clearly $\gamma_2 = \gamma_8 = 1$ and the remaining γ_n's are zero. Since $\mu = 4 = \lambda_2$ and $\gamma_2 = 1 \neq 0$ there is no solution to this problem.

5. In equation (18) on page 690 of the text we noted that the boundary value problem

$$y'' + \lambda y = 0; \qquad y'(0) = 0, \qquad y'(\pi) = 0,$$

has eigenvalues $\lambda_n = n^2$, $n = 0, 1, 2, \ldots$, with corresponding eigenfunctions

$$\phi_n(x) = \cos nx, \qquad n = 0, 1, 2, \ldots .$$

Here $r(x) \equiv 1$, so we need to determine coefficients γ_n such that

$$f(x) = \frac{f(x)}{r(x)} = \sum_{n=1}^{\infty} \gamma_n \cos nx = \cos 4x + \cos 7x .$$

Clearly $\gamma_4 = \gamma_7 = 1$ and the remaining γ_n's are zero. Since $\mu = 1 = \lambda_1$ and $\gamma_1 = 0$,

$$(\mu - \lambda_1)c_1 - \gamma_1 = 0$$

is satisfied for any value of c_1. Calculating c_4 and c_7, we get

$$c_4 = \frac{\gamma_4}{\mu - \lambda_4} = \frac{1}{1-16} = -\frac{1}{15}$$

and

$$c_7 = \frac{\gamma_7}{\mu - \lambda_7} = \frac{1}{1-49} = -\frac{1}{48}.$$

Hence a one parameter family of solutions is

$$\phi(x) = \sum_{n=1}^{\infty} c_n \phi_n(x) = c_1 \cos x - \frac{1}{15}\cos 4x - \frac{1}{48}\cos 7x,$$

where c_1 is arbitrary.

9. We first find the eigenvalues and corresponding eigenfunctions for this problem. Note that the auxiliary equation for this problem is $r^2 + \lambda = 0$. To find eigenvalues which yield nontrivial solutions we will consider the three cases $\lambda < 0$, $\lambda = 0$, and $\lambda > 0$.

Case 1, $\lambda < 0$: Let $\mu = \sqrt{-\lambda}$, then the roots to the auxiliary equation are $\pm\mu$ and a general solution to the differential equation is given by

$$y(x) = C_1 \sinh \mu x + C_2 \cosh \mu x .$$

Since

$$y'(x) = \mu C_1 \cosh \mu x + \mu C_2 \sinh \mu x ,$$

by applying the boundary conditions we obtain

$$y'(0) = \mu C_1 = 0 \quad \text{and} \quad y(\pi) = C_1 \sinh \mu\pi + C_2 \cosh \mu\pi = 0 .$$

Hence $C_1 = 0$ and $y(\pi) = C_2 \cosh \mu\pi = 0$. Therefore $C_2 = 0$ and we find only the trivial solution.

Case 2, $\lambda = 0$: In this case the differential equation becomes $y'' = 0$. This equation has a general solution given by

$$y(x) = C_1 + C_2 x .$$

Since $y'(x) = C_2$, by applying the boundary conditions we obtain

$$y'(0) = C_2 = 0 \quad \text{and} \quad y(\pi) = C_1 + C_2\pi = 0 .$$

Solving these equations simultaneously yields $C_1 = C_2 = 0$. Thus, we again find only the trivial solution.

Case 3, $\lambda > 0$: Let $\lambda = \mu^2$, for $\mu > 0$. The roots of the auxiliary equation are $r = \pm\mu i$ and so a general solution is $y(x) = C_1 \cos \mu x + C_2 \sin \mu x$. Since

$$y'(x) = -\mu C_1 \sin \mu x + \mu C_2 \cos \mu x ,$$

using the first boundary condition we find

$$y'(0) = -\mu C_1 \sin(\mu \cdot 0) + \mu C_2 \cos(\mu \cdot 0) = 0 \quad \Rightarrow \quad \mu C_2 = 0 \quad \Rightarrow \quad C_2 = 0 .$$

Thus substituting into the second boundary condition yields

$$y(\pi) = C_1 \cos \mu\pi = 0 .$$

Therefore, in order to obtain a solution other than the trivial solution, we must have

$$\cos \mu\pi = 0 \quad \Rightarrow \quad \mu = n + \frac{1}{2}, \quad n = 0, 1, 2, \ldots$$

$$\Rightarrow \quad \sqrt{\lambda} = n + \frac{1}{2}, \quad n = 0, 1, 2, \ldots .$$

Hence choose $\lambda_n = \left(n + \frac{1}{2}\right)^2$, $n = 0, 1, 2, \ldots$, and $y_n(x) = C_n \cos\left[\left(n + \frac{1}{2}\right)x\right]$, where the C_n's are arbitrary nonzero constants.

Next we need to choose the c_n so that $\int_0^{\pi} c_n^2 \cos^2\left[\left(n + \frac{1}{2}\right)x\right] dx = 1$. Computing we find

$$\int_0^{\pi} c_n^2 \cos^2\left[\left(n + \frac{1}{2}\right)x\right] dx = \frac{1}{2} c_n^2 \int_0^{\pi} \left(1 + \cos[(2n+1)x]\right) dx$$

$$= \frac{1}{2} c_n^2 \left(x + \frac{1}{2n+1} \sin[(2n+1)x] \right) \Bigg|_0^{\pi} = \frac{\pi}{2} c_n^2 .$$

An orthonormal system of eigenfunctions is given when we take $c_n = \sqrt{\frac{2}{\pi}}$,

$$\left\{ \sqrt{\frac{2}{\pi}} \cos\left[\left(n+\frac{1}{2}\right)x\right] \right\}_{n=0}^{\infty} .$$

Now f has the eigenfunction expansion

$$\sum_{n=0}^{\infty} \gamma_n \sqrt{\frac{2}{\pi}} \cos\left[\left(n+\frac{1}{2}\right)x\right],$$

where

$$\gamma_n = \sqrt{\frac{2}{\pi}} \int_0^{\pi} f(x) \cos\left[\left(n+\frac{1}{2}\right)x\right] dx .$$

Therefore, for γ_n as described above,

$$f(x) = \sum_{n=0}^{\infty} \frac{\gamma_n}{1-\lambda_n} \sqrt{\frac{2}{\pi}} \cos\left[\left(n+\frac{1}{2}\right)x\right] = \sum_{n=0}^{\infty} \frac{\gamma_n}{1-\left(n+\frac{1}{2}\right)^2} \sqrt{\frac{2}{\pi}} \cos\left[\left(n+\frac{1}{2}\right)x\right].$$

EXERCISES 11.6: Green's Functions, page 730

3. A general solution to the homogeneous problem $y''=0$ $y_h(x) = Ax + B$, so $z_1(x)$ and $z_2(x)$ must be of this form. To get $z_1(x)$ we want to choose A and B so that $z_1(0) = B = 0$. Since A is arbitrary, we can set it equal to 1 and $z_1(x) = x$. Next to get $z_2(x)$ we need to choose A and B so that $z_2(\pi) + z_2'(\pi) = A\pi + B + A = 0$. Thus $B = -(1+\pi)A$. Taking $A = 1$, we get

$$z_2(x) = x - 1 - \pi .$$

Now compute

$$C = p(x)W[z_1, z_2](x) = (1)[(x)(1) - (1)(x - 1 - \pi)] = 1 + \pi .$$

Thus, the Green's function is

$$G(x,s) = \begin{cases} -z_1(s)z_2(x)/C, & 0 \le s \le x, \\ -z_1(x)z_2(s)/C, & x \le s \le \pi. \end{cases} = \begin{cases} \dfrac{-s(x-1-\pi)}{1+\pi}, & 0 \le s \le x, \\ \dfrac{-x(s-1-\pi)}{1+\pi}, & x \le s \le \pi. \end{cases}$$

13. In Problem 3 we found the Green's function for this boundary value problem. When $f(x)=x$, the solution is given by equation (16) on page 708 of the text. Substituting for $f(x)$ and $G(x,s)$, we find

$$y(x)=\int_a^b G(x,s)f(s)\,ds=\int_0^x G(x,s)s\,ds+\int_x^\pi G(x,s)s\,ds$$

$$=\int_0^x \frac{-s^2(x-1-\pi)}{1+\pi}\,ds+\int_x^\pi \frac{-xs(s-1-\pi)}{1+\pi}\,ds$$

$$=\frac{x-1-\pi}{1+\pi}\left(\frac{-s^3}{3}\right)\Bigg|_0^x+\frac{-x}{1+\pi}\left(\frac{s^3}{3}-\frac{(1+\pi)s^2}{2}\right)\Bigg|_x^\pi$$

$$=\frac{x-1-\pi}{1+\pi}\left(\frac{-x^3}{3}\right)-\frac{x}{1+\pi}\left(\frac{\pi^3}{3}-\frac{(1+\pi)\pi^2}{2}\right)+\frac{x}{1+\pi}\left(\frac{x^3}{3}-\frac{(1+\pi)x^2}{2}\right)$$

$$=\frac{-x^4}{3(1+\pi)}+\frac{x^3}{3}-\frac{\pi^3 x}{3(1+\pi)}+\frac{\pi^2 x}{2}+\frac{x^4}{3(1+\pi)}-\frac{x^3}{2}$$

$$=\frac{-x^3}{6}+\left(\frac{\pi^2}{2}-\frac{\pi^3}{3(1+\pi)}\right)x=\frac{-x^3}{6}+\left(\frac{3\pi^2+\pi^3}{6(1+\pi)}\right)x\,.$$

17. A general solution to the homogeneous problem $y''-y=0$ is $y_h(x)=c_1e^x+c_2e^{-x}$, so $z_1(x)$ and $z_2(x)$ must be of this form. To get $z_1(x)$ we want to *choose* c_1 and c_2 so that

$$z_1(0)=c_1+c_2=0,$$

let $c_1=1$, then $c_2=-1$. So $z_1(x)=e^x-e^{-x}$. Likewise to get a $z_2(x)$, *choose* c_1 and c_2 so that

$$z_2(1)=c_1e+c_2e^{-1}=0\,.$$

If we let $c_1=1$, then $c_2=-e^2$. Hence $z_2(x)=e^x-e^2e^{-x}=e^x-e^{2-x}$. We now compute

$$C=p(x)W[z_1,z_2](x)=(1)\left[\left(e^x-e^{-x}\right)\left(e^x+e^{2-x}\right)-\left(e^x+e^{-x}\right)\left(e^x-e^{2-x}\right)\right]=2e^2-2\,.$$

Thus, the Green's function is

$$G(x,s)=\begin{cases}-z_1(s)z_2(x)/C, & 0\le s\le x,\\[2ex] -z_1(x)z_2(s)/C, & x\le s\le 1.\end{cases}=\begin{cases}\dfrac{\left(e^s-e^{-s}\right)\left(e^x-e^{2-x}\right)}{2-2e^2}, & 0\le s\le x,\\[2ex] \dfrac{\left(e^x-e^{-x}\right)\left(e^s-e^{2-s}\right)}{2-2e^2}, & x\le s\le 1.\end{cases}$$

Here $f(x)=-x$. Using Green's functions to solve the boundary value problem, we find

$$y(x)=\int_a^b G(x,s)f(s)\,ds=\int_0^x G(x,s)(-s)\,ds+\int_x^1 G(x,s)(-s)\,ds$$

$$=-\int_0^x \frac{s\left(e^s-e^{-s}\right)\left(e^x-e^{2-x}\right)}{2-2e^2}\,ds-\int_x^1 \frac{s\left(e^x-e^{-x}\right)\left(e^s-e^{2-s}\right)}{2-2e^2}\,ds$$

$$=-\left(\frac{e^x-e^{2-x}}{2-2e^2}\right)\int_0^x\left(se^s-se^{-s}\right)ds-\left(\frac{e^x-e^{-x}}{2-2e^2}\right)\int_x^1\left(se^s-se^{2-s}\right)ds$$

$$=-\left(\frac{e^x-e^{2-x}}{2-2e^2}\right)\left[se^s-e^s+se^{-s}+e^{-s}\right]\Bigg|_0^x-\left(\frac{e^x-e^{-x}}{2-2e^2}\right)\left[se^s-e^s+se^{2-s}+e^{2-s}\right]\Bigg|_x^1$$

$$=-\left(\frac{e^x-e^{2-x}}{2-2e^2}\right)\left[xe^x-e^x+xe^{-x}+e^{-x}\right]-\left(\frac{e^x-e^{-x}}{2-2e^2}\right)\left[2e-\left(xe^x-e^x+xe^{2-x}+e^{2-x}\right)\right.$$

$$=\frac{-\left(e^x-e^{2-x}\right)\left(xe^x-e^x+xe^{-x}+e^{-x}\right)-\left(e^x-e^{-x}\right)\left(2e-xe^x+e^x-xe^{2-x}-e^{2-x}\right)}{2-2e^2}$$

$$=\frac{-2x+2xe^2-2e^{1+x}+2e^{1-x}}{2-2e^2}=-x+\frac{e^{1+x}-e^{1-x}}{e^2-1}.$$

25. Substitution $y=x^r$ into the corresponding homogeneous Cauchy-Euler equation

$$x^2y''-2xy'+2y=0,$$

we obtain the auxiliary equation

$$r(r-1)-2r+2=0 \qquad \text{or} \qquad r^2-3r+2=(r-1)(r-2)=0.$$

Hence a general solution to the corresponding homogeneous equation is $y_h(x)=c_1x+c_2x^2$. To get $z_1(x)$ we want to choose c_1 and c_2 so that $z_1(1)=c_1+c_2=0$. Let $c_1=1$, then $c_2=-1$ and $z_1(x)=x-x^2$. Using $z_2(2)$ to choose c_1 and c_2, we obtain $z_2(2)=2c_1+4c_2=0$. Hence, we let $c_1=2$, then $c_2=-1$ and $z_2(x)=2x-x^2$. Now compute (see the formula for $K(x,s)$ in Problem 22, page 730)

$$C(s)=A_2(s)W\left[z_1,z_2\right](s)=\left(s^2\right)\left[\left(s-s^2\right)(2-2s)-(1-2s)\left(2s-s^2\right)\right]$$

$$=\left(s^2\right)\left(2s-4s^2+2s^3-2s+5s^2-2s^3\right)=s^4.$$

$$K(x,s)=\begin{cases}-z_1(s)z_2(x)/C(s), & 1\le s\le x,\\ -z_1(x)z_2(s)/C(s), & x\le s\le 2,\end{cases}$$

$$=\begin{cases}\dfrac{-(s-s^2)(2x-x^2)}{s^4}=-x(2-x)(s^{-3}-s^{-2}), & 1\le s\le x,\\[2ex] \dfrac{-(x-x^2)(2s-s^2)}{s^4}=-x(1-x)(2s^{-3}-s^{-2}), & x\le s\le 2.\end{cases}$$

Hence, a solution to the boundary value problem with $f(x) = -x$ is

$$y(x)=\int_a^b K(x,s)f(s)\,ds=\int_1^x K(x,s)(-s)\,ds+\int_x^2 K(x,s)(-s)\,ds$$

$$=(2x-x^2)\int_1^x(s^{-2}-s^{-1})ds+(x-x^2)\int_x^2(2s^{-2}-s^{-1})ds$$

$$=(2x-x^2)(-s^{-1}-\ln s)\Big|_1^x+(x-x^2)(-2s^{-1}-\ln s)\Big|_x^2$$

$$=(2x-x^2)\left[(-x^{-1}-\ln x)-(-1)\right]+(x-x^2)\left[(-1-\ln 2)-(-2x^{-1}-\ln x)\right]$$

$$=x^2\ln 2-x\ln 2-x\ln x\,.$$

29. Let $f(x)=\delta(x-s)$. Let $H(x,s)$ be the solution to

$$\frac{\partial^4 H(x,s)}{\partial x^4}=-\delta(x-s)$$

that satisfies the given boundary conditions, the jump condition

$$\lim_{x\to s^+}\frac{\partial^3 H(x,s)}{\partial x^3}-\lim_{x\to s^-}\frac{\partial^3 H(x,s)}{\partial x^3}=-1\,,$$

and H, $\dfrac{\partial H}{\partial x}$, $\dfrac{\partial^2 H}{\partial x^2}$ are continuous on the square $[0,\pi]\times[0,\pi]$. We begin by integrating to obtain

$$\frac{\partial^3 H(x,s)}{\partial x^3}=-u(x-s)+C_1\,,$$

where u is the unit step function and C_1 is a constant. (Recall in Section 7.8 we observed that $u'(t-a)=\delta(t-a)$, at least formally.) $\dfrac{\partial^3 H}{\partial x^3}$ is not continuous along the line $x=s$, but it does satisfy the jump condition

$$\lim_{x\to s^+}\frac{\partial^3 H(x,s)}{\partial x^3}-\lim_{x\to s^-}\frac{\partial^3 H(x,s)}{\partial x^3}=\lim_{x\to s^+}\left[-u(x-s)+C_1\right]-\lim_{x\to s^-}\left[-u(x-s)+C_1\right]$$

$$=(-1+C_1)-C_1=-1\,.$$

We want $H(x,s)$ to satisfy the boundary condition $y'''(\pi)=0$. So we solve

$$\frac{\partial^3 H(\pi, s)}{\partial x^3} = -u(\pi - s) + C_1 = -1 + C_1 = 0$$

to obtain $C_1 = 1$. Thus $\dfrac{\partial^3 H(x, s)}{\partial x^3} = 1 - u(x - s)$.

We now integrate again with respect to x to obtain

$$\frac{\partial^2 H(x, s)}{\partial x^2} = x - u(x - s)(x - s) + C_2 .$$

(The reader should verify this is the antiderivative for $x \neq s$ by differentiating it.) We selected this particular form of the antiderivative because we need $\dfrac{\partial^2 H}{\partial x^2}$ to be continuous on $[0, \pi] \times [0, \pi]$. (The jump of $u(x - s)$ when $x = s$ is canceled by the vanishing of this term by the factor $(x - s)$.) Since

$$\lim_{x \to s} \frac{\partial^2 H(x, s)}{\partial x^2} = s + C_2 ,$$

we can define

$$\frac{\partial^2 H(s, s)}{\partial x^2} = s + C_2 ,$$

and we now have a continuous function. Next, we want $y''(\pi) = 0$. Solving we find

$$0 = \frac{\partial^2 H(\pi, s)}{\partial x^2} = \pi - u(\pi - s)(\pi - s) + C_2 = \pi - (\pi - s) + C_2 = s + C_2 .$$

Thus, we find that $C_2 = -s$. Now,

$$\frac{\partial^2 H(x, s)}{\partial x^2} = (x - s) - u(x - s)(x - s) .$$

We integrate with respect to x again to get

$$\frac{\partial H(x, s)}{\partial x} = \frac{x^2}{2} - sx - u(x - s)\frac{(x - s)^2}{2} + C_3 ,$$

which is continuous on $[0, \pi] \times [0, \pi]$. We now want the boundary condition $y'(0) = 0$ satisfied. Solving, we obtain

$$0 = \frac{\partial H(0, s)}{\partial x} = -u(0 - s)\frac{s^2}{2} + C_3 = C_3 .$$

Hence,

$$\frac{\partial H(x, s)}{\partial x} = \frac{x^2}{2} - sx - u(x - s)\frac{(x - s)^2}{2} .$$

Integrating once more with respect to x, we have

$$H(x,s)=\frac{x^3}{6}-\frac{sx^2}{2}-u(x-s)\frac{(x-s)^3}{6}+C_4 .$$

Now $H(x,s)$ is continuous on $[0,\pi]\times[0,\pi]$. We want $H(x,s)$ to satisfy the boundary condition $y(0)=0$. Solving, we find

$$0=H(0,s)=-u(0-s)\frac{(0-s)^3}{6}+C_4=C_4 .$$

Hence,

$$H(x,s)=\frac{x^3}{6}-\frac{sx^2}{2}-u(x-s)\frac{(x-s)^3}{6},$$

which we can rewrite in the form

$$H(x,s)=\begin{cases}\dfrac{s^2(s-3x)}{6}, & 0\le s\le x,\\[2ex] \dfrac{x^2(x-3s)}{6}, & x\le s\le \pi.\end{cases}$$

EXERCISES 11.7: Singular Sturm-Liouville Boundary Value Problems, page 739

1. This is a typical singular Sturm-Liouville boundary value problem. Condition (ii) of Lemma 1 on page 734 of the main text holds since

$$\lim_{x\to 0^+} p(x)=\lim_{x\to 0^+} x=0$$

and $y(x)$, $y'(x)$ remain bounded as $x\to 0^+$. Because

$$\lim_{x\to 1^-} p(x)=p(1)=1$$

and $y(1)=0$, the analogue of condition (i) of Lemma 1 holds at the right endpoint. Hence L is selfadjoint.

The equation is Bessel's equation of order 2. On pages 736 and 737 of the text, we observed that the solution to this boundary value are given by

$$y_n(x)=c_nJ_2(\alpha_{2n}x),$$

where $\sqrt{\mu_n}=\alpha_{2n}$ is the increasing sequence of real zeros of J_2, that is, $J_2(\alpha_{2n})=0$.

Now to find an eigenfunction expansion for the given nonhomogeneous equation we compute the eigenfunction expansion for $f(x)/x$ (see page 720):

$$\frac{f}{x} \sim \sum_{n=1}^{\infty} a_n J_2(\alpha_{2n}\, x),$$

where

$$a_n = \frac{\int_0^1 f(x) J_2(\alpha_{2n}\, x)\, dx}{\int_0^1 x J_2^2(\alpha_{2n}\, x)\, dx}, \qquad n = 1, 2, 3, \ldots.$$

Therefore,

$$y(x) = \sum_{n=1}^{\infty} \frac{a_n}{\mu - \alpha_{2n}^2} J_2(\alpha_{2n}\, x).$$

3. Again, this is a typical singular Sturm-Liouville boundary value problem. L is selfadjoint since condition (ii) of Lemma 1 on page 734 of the main text holds at the left endpoint and the analogue of condition (i) holds at the right endpoint.

This is Bessel's equation of order 0. As we observed on page 736 of the text, $J_0\left(\sqrt{\mu x}\right)$ satisfies the boundary conditions at the origin. At the right endpoint, we want $J_0'\left(\sqrt{\mu}\right) = 0$. Now it follows from equation (32) on page 512 of the text, that the zeros of J_0' and J_1 are the same. So if we let $\sqrt{\mu_n} = \alpha_{1n}$, the increasing sequence of zeros of J_1, then $J_0'(\alpha_{1n}) = 0$. Hence, the eigenfunctions are given by

$$y_n(x) = J_0(\alpha_{1n} x), \qquad n = 1, 2, 3, \ldots.$$

To find an eigenfunction expansion for the solution to the nonhomogeneous equation, we first expand $f(x)/x$ (see page 720):

$$\frac{f}{x} \sim \sum_{n=1}^{\infty} b_n J_0(\alpha_{1n}\, x),$$

where

$$b_n = \frac{\int_0^1 f(x)\, J_0(\alpha_{1n}\, x)\, dx}{\int_0^1 x J_0^2(\alpha_{1n}\, x)\, dx}, \qquad n = 1, 2, 3, \ldots.$$

Therefore,

$$y(x) = \sum_{n=1}^{\infty} \frac{b_n}{\mu - \alpha_{1n}^2} J_0(\alpha_{1n}\, x).$$

11. **(a)** Let $\phi(x)$ be an eigenfunction for $\frac{d}{dx}\left[x \frac{dy}{dx}\right] - \frac{v^2}{x} y + \lambda x y = 0$. Therefore,

$$\frac{d}{dx}[x\phi'(x)] - \frac{v^2}{x}\phi(x) + \lambda x\phi(x) = 0$$

$$\Rightarrow \qquad \phi'(x) + x\phi''(x) - \frac{v^2}{x}\phi(x) + \lambda x\phi(x) = 0 .$$

Multiplying both side by $\phi(x)$ and integrating both sides from 0 to 1, we obtain

$$\int_0^1 \phi(x)\phi'(x)\,dx + \int_0^1 x\phi(x)\phi''(x)\,dx - \int_0^1 \frac{v^2}{x}[\phi(x)]^2\,dx + \int_0^1 \lambda x[\phi(x)]^2\,dx = 0 . \qquad (9)$$

Now integrating by parts with $u = \phi(x)\phi'(x)$ and $dv = dx$, we have

$$\int_0^1 \phi(x)\phi'(x)\,dx = x\phi(x)\phi'(x)\Big|_0^1 - \int_0^1 x[\phi'(x)\phi'(x) + \phi(x)\phi''(x)]\,dx$$

$$= x\phi(x)\phi'(x)\Big|_0^1 - \int_0^1 x[\phi'(x)]^2\,dx - \int_0^1 x\phi(x)\phi''(x)\,dx .$$

Since $\phi(1) = 0$, we have

$$x\phi(x)\phi'(x)\Big|_0^1 = 0 ,$$

and

$$\int_0^1 \phi(x)\phi'(x)\,dx = -\int_0^1 x[\phi'(x)]^2\,dx - \int_0^1 x\phi(x)\phi''(x)\,dx .$$

Thus equation (9) reduces to

$$-\int_0^1 x[\phi'(x)]^2\,dx - \int_0^1 x\phi(x)\phi''(x)\,dx + \int_0^1 x\phi(x)\phi''(x)\,dx$$

$$- v^2 \int_0^1 x^{-1}[\phi(x)]^2\,dx + \lambda \int_0^1 x[\phi(x)]^2\,dx = 0$$

$$\Rightarrow \qquad -\int_0^1 x[\phi'(x)]^2\,dx - v^2 \int_0^1 x^{-1}[\phi(x)]^2\,dx + \lambda \int_0^1 x[\phi(x)]^2\,dx = 0 . \qquad (10)$$

(b) First note that each integrand in (10) is nonnegative on the interval (0,1), hence each integral is nonnegative. Moreover, since $\phi(x)$ is an eigenfunction, it is a continuous function which is not the zero function. Hence, the second and third integrals are strictly positive. Thus, if $v > 0$, then λ must be positive in order for the left-hand side of (10) to sum to zero.

(c) If $v = 0$, then only the first and third terms remain on the left hand side of equation (10). Since the first integral need only be nonnegative, we only need λ to be nonnegative in order for equation (10) to be satisfied.

To show $\lambda = 0$ is not an eigenvalue, we solve Bessel's equation with $v = 0$, that is, we solve

$$xy'' + y' = 0,$$

which is the same as the Cauchy–Euler equation

$$x^2 y'' + xy' = 0.$$

Solving this Cauchy–Euler equation, we find a general solution

$$y(x) = c_1 + c_2 \ln x.$$

Since $\lim_{x \to 0^+} y(x)$ is undefined if $c_2 \neq 0$, we take $c_2 = 0$. Now $y(x) = c_1$ satisfies the boundary condition (17) in the text. The right end-point boundary condition (18) is $y(1) = 0$. So we want $0 = y(1) = c_1$. Hence the only solution to Bessel's equation of order 0 with that satisfies the boundary conditions (17) and (18) is the trivial solution. Hence $\lambda = 0$ is not an eigenvalue.

EXERCISES 11.8: Oscillation and Comparison Theory, page 749

5. To apply the Sturm fundamental theorem to

$$y'' + \left(1 - e^x\right)y = 0, \qquad 0 < x < \infty, \tag{11}$$

we must find a $q(x)$ and a function $\phi(x)$ such that $q(x) \geq 1 - e^x$, $0 < x < \infty$, and $\phi(x)$ is a solution to

$$y'' + q(x)y = 0, \qquad 0 < x < \infty. \tag{12}$$

Because, for $x > 0$, $1 - e^x < 0$, we choose $q(x) \equiv 0$. Hence equation (12) becomes $y'' = 0$. The function $\phi(x) = x + 4$ is a nontrivial solution to this differential equation. Since $\phi(x) = x + 4$ does not have a zero for $x > 0$, any nontrivial solution to (11) can have at most one zero in $0 < x < \infty$. To use the Sturm fundamental theorem to show that any nontrivial solution to

$$y'' + \left(1 - e^x\right)y = 0, \qquad -\infty < x < 0, \tag{13}$$

has infinitely many zeros we must find a $q(x)$ and a function $\phi(x)$ such that $q(x) \leq 1 - e^x$, $x < 0$, and $\phi(x)$ is a solution to

$$y'' + q(x)y = 0, \qquad -\infty < x < 0.$$

Because $1 - e^{-1} \approx 0.632$, we choose $q(x) \equiv \dfrac{1}{4}$ and only consider the interval $(-\infty, -1)$. Hence, we obtain

$$y'' + \frac{1}{4}y = 0,$$

which has nontrivial solution $\phi(x) = \sin\left(\frac{x}{2}\right)$. Now the function $\phi(x)$ has infinitely many zeros in $(-\infty,-1)$ and between any two consecutive zeros of $\phi(x)$ any nontrivial solution to (13) must have a zero; hence any nontrivial solution to (13) will have infinitely many zeros in $(-\infty,-1)$.

9. First express

$$y'' + x^{-2}y' + (4 - e^{-x})y = 0,$$

in Strum–Liouville form by multiplying by the integrating factor $e^{-1/x}$

$$e^{-1/x}y'' + e^{-1/x}x^{-2}y' + e^{-1/x}(4 - e^{-x})y = 0$$

$$\Rightarrow \quad \left(e^{-1/x}y'\right)' + e^{-1/x}(4 - e^{-x})y = 0.$$

Now when x gets large, we have

$$\sqrt{\frac{p}{q}} \approx \sqrt{\frac{e^{-1/\text{large}}}{e^{-1/\text{large}}\left(4 - e^{-\text{large}}\right)}} \approx \sqrt{\frac{1}{(1)(4 - \text{small})}} \approx \sqrt{\frac{1}{4}} = \frac{1}{2}.$$

Hence, the distance between consecutive zeros is approximately $\frac{1}{2}\pi$.

11. We apply Corollary 5 with $p(x) = 1 + x$, $q(x) = e^{-x}$, and $r(x) = 1$ to a nontrivial solution on the interval [0,5]. On this interval we have $p_M = 6$, $p_m = 1$, $q_M = 1$, $q_m = e^{-5}$, and $r_M = r_m = 1$. Therefore, for $\phi(x)$ a nontrivial solution where

$$\lambda > \max\left\{\frac{-q_M}{r_M}, \frac{-q_m}{r_m}, 0\right\} = 0,$$

the distance between consecutive zeros of $\phi(x)$ is bounded between

$$\pi\sqrt{\frac{p_m}{q_M + \lambda r_M}} = \pi\sqrt{\frac{1}{1+\lambda}},$$

and

$$\pi\sqrt{\frac{p_M}{q_m + \lambda r_m}} = \pi\sqrt{\frac{6}{e^{-5} + \lambda}}.$$

CHAPTER 12: Stability of Autonomous Systems

EXERCISES 12.2: Linear Systems in the Plane, page 777

3. The characteristic equation for this system is $r^2 + 2r + 10 = 0$, which has roots $r = -1 \pm 3i$. Since the real part of each root is negative, the trajectories approach the origin, and the origin is an asymptotically stable spiral point.

7. The critical point is the solution to the system

$$-4x + 2y + 8 = 0,$$
$$x - 2y + 1 = 0.$$

Solving this system, we obtain the critical point (3, 2). Now we use the change of variables

$$x = u + 3, \quad y = v + 2,$$

to translate the critical point (3, 2) to the origin (0, 0). Substituting into the system of this problem and simplifying, we obtain a system of differential equations in u and v:

$$\frac{du}{dt} = \frac{dx}{dt} = -4(u+3) + 2(v+2) + 8 = -4u + 2v,$$

$$\frac{dv}{dt} = \frac{dy}{dt} = (u+3) - 2(v+2) + 1 = u - 2v.$$

The characteristic equation for this system is $r^2 + 6r + 6 = 0$, which has roots $r = -3 \pm \sqrt{3}$. Since both roots are distinct and negative, the origin is an asymptotically stable improper node of the new system. Therefore, the critical point (3, 2) is an asymptotically stable improper node of the original system.

9. The critical point is the solution to the system

$$2x + y + 9 = 0,$$
$$-5x - 2y - 22 = 0.$$

Solving this system, we obtain the critical point (–4, –1). Now we use the change of variables

$$x = u - 4, \quad y = v - 1,$$

to translate the critical point (–4, –1) to the origin (0, 0). Substituting into the system of this problem and simplifying we obtain a system of differential equations in u and v:

$$\frac{du}{dt} = \frac{dx}{dt} = 2(u-4) + (v-1) + 9 = 2u + v,$$

$$\frac{dv}{dt} = \frac{dy}{dt} = -5(u-4) - 2(v-1) - 22 = -5u - 2v.$$

The characteristic equation for this system is $r^2+1=0$, which has roots $r=\pm i$. Since both roots are distinct and pure imaginary, the origin is a stable center of the new system. Therefore, the critical point (–4, –1) is a stable center of the original system.

15. The characteristic equation for this system is $r^2+r-12=0$, which has roots $r=-4$ and $r=3$. Since the roots are real and have opposite signs, the origin is an unstable saddle point. To sketch the phase plane diagram, we must first determine two lines passing through the origin that correspond to the transformed axes. To find the transformed axes, we make the substitution $y=mx$ into

$$\frac{dy}{dx}=\frac{dy/dt}{dx/dt}=\frac{5x-2y}{x+2y}$$

to obtain

$$m=\frac{5x-2mx}{x+2mx}.$$

Solving for m yields

$$m(x+2mx)=5x-2mx \quad\Rightarrow\quad 2m^2+3m-5=0$$

$$\Rightarrow\quad (2m+5)(m-1)=0 \quad\Rightarrow\quad m=-\frac{5}{2} \quad\text{or}\quad m=1.$$

So $m=-\frac{5}{2}$ or $m=1$. Hence, the two axes are $y=-\frac{5}{2}x$ and $y=x$. On the line $y=x$ one finds

$$\frac{dx}{dt}=3x,$$

so the trajectories move away from the origin. On the line $y=-\frac{5}{2}x$ one finds

$$\frac{dy}{dt}=-4y,$$

so the trajectories move towards the origin. A phase plane diagram is given in Figure B.56 in the answers of the text.

19. The characteristic equation for this system is $(r+2)(r+2)=0$ which has roots $r=-2,-2$. Since the roots are equal, real, and negative, the origin is an asymptotically stable point. To sketch the phase plane diagram, we determine the slope of the two lines passing through the origin that correspond to the transformed axes by substituting $y=mx$ into

$$\frac{dy}{dx}=\frac{dy/dt}{dx/dt}=\frac{-2y}{-2x+y},$$

to obtain

$$m = \frac{-2mx}{-2x + mx}.$$

Solving for m yields

$$m(-2x + mx) = -2mx \qquad \Rightarrow \qquad m^2 = 0 \qquad \Rightarrow \qquad m = 0.$$

Since there is only one line ($y = 0$) through the origin that is a trajectory, the origin is an improper node. A phase plane diagram is given in Figure B.58 in the answers of the text.

EXERCISES 12.3: Almost Linear Systems, page 788

5. This system is almost linear since $ad - bc = (1)(-1) - (5)(-1) \neq 0$, and $F(x, y) = G(x, y) = -y^2 = 0$ involve only high order terms in y. Since the characteristic equation for this system is $r^2 + 4 = 0$ which has pure imaginary roots $r = \pm 2i$, the origin is either a center or a spiral point and the stability is indeterminant.

7. To see that this system is almost linear, we first express e^{x+y}, $\cos x$, and $\cos y$ using their respective Maclaurin series. Hence, the system

$$\frac{dx}{dt} = e^{x+y} - \cos x,$$

$$\frac{dy}{dt} = \cos y + x - 1,$$

becomes

$$\frac{dx}{dt} = \left[1 + (x + y) + \frac{(x + y)^2}{2!} + \cdots\right] - \left[1 - \frac{x^2}{2!} + \cdots\right] = x + y + (\text{higher orders}) = x + y + F(x, y),$$

$$\frac{dy}{dt} = \left[1 - \frac{y^2}{2!} + \cdots\right] + x - 1 = x + (\text{higher orders}) = x + G(x, y).$$

This system is almost linear since $ad - bc = (1)(0) - (1)(1) \neq 0$, and $F(x, y)$, $G(x, y)$ each only involve higher order forms in x and y. The characteristic equation for this system is $r^2 - r - 1 = 0$ which has roots $r = \dfrac{1 \pm \sqrt{5}}{2}$. Since these roots are real and have different signs the origin is an unstable saddle point.

9. The critical points for this system are the solutions to the pair of equations

$$16 - xy = 0,$$
$$x - y^3 = 0.$$

Solving the second equation for x in terms of y and substituting this into the first equation we obtain

$$16 - y^4 = 0$$

which has solutions $y = \pm 2$. Hence the critical points are (8, 2) and (–8, –2).

We consider the critical point (8, 2). Using the change of variables $x = u + 8$ and $y = v + 2$, we obtain the system

$$\frac{du}{dt} = 16 - (u+8)(v+2),$$

$$\frac{dv}{dt} = (u+8) - (v+2)^3,$$

which simplifies to the almost linear system

$$\frac{du}{dt} = -2u - 8v - uv,$$

$$\frac{dv}{dt} = u - 12v - 6v^2 - v^3.$$

The characteristic equation for this system is $r^2 + 14r + 32 = 0$, which has the distinct negative roots $r = -7 \pm \sqrt{17}$. Hence (8, 2) is an improper node which is asymptotically stable.

Next we consider the critical point (–8, –2). Using the change of variables $x = u - 8$ and $y = v - 2$, we obtain the system

$$\frac{du}{dt} = 16 - (u-8)(v-2),$$

$$\frac{dv}{dt} = (u-8) - (v-2)^3,$$

which simplifies to the almost linear system

$$\frac{du}{dt} = 2u + 8v - uv,$$

$$\frac{dv}{dt} = u - 12v + 6v^2 - v^3.$$

The characteristic equation for this system is $r^2 + 10r - 32 = 0$ which has roots $r = -5 \pm \sqrt{57}$. Since these roots are real and have different signs, (–8, –2) is an unstable saddle point.

13. The critical points for this system are the solutions to the pair of equations

$$1 - xy = 0,$$
$$x - y^3 = 0.$$

Solving the second equation for x in terms of y and substituting this into the first equation we obtain

$$1 - y^4 = 0$$

which has solutions $y = \pm 1$. Hence the critical points are (1, 1) and (–1, –1).

We consider the critical point (1, 1). Using the change of variables $x = u + 1$ and $y = v + 1$, we obtain the almost linear system

$$\frac{du}{dt} = 1 - (u+1)(v+1) = -u - v - uv,$$

$$\frac{dv}{dt} = (u+1) - (v+1)^3 = u - 3v - 3v^2 - v^3.$$

The characteristic equation for this system is $r^2 + 4r + 4 = 0$ which has the equal negative roots $r = -2$. Hence (1, 1) is an improper or proper node or spiral point which is asymptotically stable.

Next we consider the critical point (–1, –1). Using the change of variables $x = u - 1$ and $y = v - 1$, we obtain the almost linear system

$$\frac{du}{dt} = 1 - (u-1)(v-1) = u + v - uv,$$

$$\frac{dv}{dt} = (u-1) - (v-1)^3 = u - 3v + 3v^2 - v^3.$$

The characteristic equation for this system is $r^2 + 2r - 4 = 0$ which has roots $r = -1 \pm \sqrt{5}$. Since these roots are real and have different signs, (–1, –1) is an unstable saddle point. A phase plane diagram is given in Figure B.59 in the answers of the text.

21. <u>Case 1: $h = 0$.</u> The critical points for this system are the solutions to the pair of equations

$$x(1 - 4x - y) = 0,$$
$$y(1 - 2y - 5x) = 0.$$

To solve this system, we first let $x = 0$, then

$$y(1 - 2y) = 0.$$

So $y = 0$ or $y = \frac{1}{2}$. When $y = 0$, we must have

$$x(1 - 4x) = 0.$$

So $x = 0$ or $x = \frac{1}{4}$. And if $x \neq 0$ and $y \neq 0$, we have the system

$$1-4x-y=0,$$
$$1-2y-5x=0,$$

which has the solution $x=\frac{1}{3}$, $y=-\frac{1}{3}$. Hence the critical points are (0, 0), $\left(0,\frac{1}{2}\right)$, $\left(\frac{1}{4},0\right)$, and $\left(\frac{1}{3},-\frac{1}{3}\right)$.

At the critical point (0, 0), the characteristic equation for is $r^2-2r+1=0$, which has equal positive roots $r=1$. Hence (0, 0) is an improper or proper node or spiral point which is unstable. From Figure B.61 in the text, we see that (0, 0) is an improper node.

Next we consider the critical point $\left(0,\frac{1}{2}\right)$. Using the change of variables $x=u$ and $y=v+\frac{1}{2}$, we obtain the almost linear system

$$\frac{du}{dt}=u\left(1-4u-v-\frac{1}{2}\right)=\frac{1}{2}u-4u^2-uv,$$

$$\frac{dv}{dt}=\left(v+\frac{1}{2}\right)(1-2v-1-5u)=-\frac{5}{2}u-v-2v^2-5uv.$$

The characteristic equation for this system is $r^2+\frac{1}{2}r-\frac{1}{2}=0$, which has roots $r=\frac{1}{2}$ and $r=-1$. Since these roots are real and have different signs, $\left(0,\frac{1}{2}\right)$ is an unstable saddle point.

Now consider the critical point $\left(\frac{1}{4},0\right)$. Using the change of variables $x=u+\frac{1}{4}$ and $y=v$, we obtain the almost linear system

$$\frac{du}{dt}=\left(u+\frac{1}{4}\right)(1-4u-1-v)=-u-\frac{1}{4}v-4u^2-uv,$$

$$\frac{dv}{dt}=v\left(1-2v-5u-\frac{5}{4}\right)=-\frac{1}{4}v-2v^2-5uv.$$

The characteristic equation for this system is $r^2+\frac{5}{4}r+\frac{1}{4}=0$, which has roots $r=-\frac{1}{4}$ and $r=-1$. Since these roots are distinct and negative, $\left(\frac{1}{4},0\right)$ is an improper node which is asymptotically stable.

At the critical point $\left(\frac{1}{3},-\frac{1}{3}\right)$, we use the change of variables $x=u+\frac{1}{3}$ and $y=v-\frac{1}{3}$ to obtain the almost linear system

$$\frac{du}{dt}=\left(u+\frac{1}{3}\right)\left(1-4u-\frac{4}{3}-v+\frac{1}{3}\right)=-\frac{4}{3}u-\frac{1}{3}v-4u^2-uv\,,$$

$$\frac{dv}{dt}=\left(v-\frac{1}{3}\right)\left(1-2v+\frac{2}{3}-5u-\frac{5}{3}\right)=\frac{5}{3}u+\frac{2}{3}v-2v^2-5uv\,.$$

The characteristic equation for this system is $r^2+\frac{2}{3}r-\frac{1}{3}=0$ which has roots $r=\frac{1}{3}$ and $r=-1$. Again since these roots are real and have different signs, $\left(\frac{1}{3},-\frac{1}{3}\right)$ is an unstable saddle point, but not of interest since $y<0$. Species x survives while species y dies off. A phase plane diagram is given in Figure B.61 in the answers of the text.

<u>Case 2: $h=\frac{1}{32}$.</u> The critical points for this system are the solutions to the pair of equations

$$x(1-4x-y)-\frac{1}{32}=0,$$

$$y(1-2y-5x)=0.$$

To solve this system, we first set $y=0$ and solve

$$x(1-4x)-\frac{1}{32}=0,$$

which has solutions $x=\frac{2\pm\sqrt{2}}{16}$. And if $y\neq 0$, we have

$$1-2y-5x=0\,.$$

So

$$y=\frac{1}{2}-\frac{5}{2}x\,.$$

Substituting, we obtain

$$x\left[1-4x-\left(\frac{1}{2}-\frac{5}{2}x\right)\right]-\frac{1}{32}=0.$$

Simplifying, we obtain

$$-\frac{3}{2}x^2+\frac{1}{2}x-\frac{1}{32}=0.$$

which has the solution $x = \frac{1}{4}$ or $x = \frac{1}{12}$. When $x = \frac{1}{4}$, we have

$$y = \frac{1}{2} - \frac{5}{2}\left(\frac{1}{4}\right) = -\frac{1}{8}.$$

And when $x = \frac{1}{12}$, we have

$$y = \frac{1}{2} - \frac{5}{2}\left(\frac{1}{12}\right) = \frac{7}{24}.$$

Hence the critical points are $\left(\frac{2-\sqrt{2}}{16}, 0\right)$, $\left(\frac{2+\sqrt{2}}{16}, 0\right)$, $\left(\frac{1}{4}, -\frac{1}{8}\right)$, and $\left(\frac{1}{12}, \frac{7}{24}\right)$.

At the critical point $\left(\frac{2-\sqrt{2}}{16}, 0\right)$, we use the change of variables $x = u + \frac{2-\sqrt{2}}{16}$ and $y = v$ to obtain the almost linear system

$$\frac{du}{dt} = \left(u + \frac{2-\sqrt{2}}{16}\right)\left(1 - 4u - \frac{2-\sqrt{2}}{4} - v\right) - \frac{1}{32} = \frac{\sqrt{2}}{2}u - \frac{2-\sqrt{2}}{16}v - 4u^2 - uv,$$

$$\frac{dv}{dt} = v\left(1 - 2v - 5u - \frac{10-5\sqrt{2}}{16}\right) = \frac{6+5\sqrt{2}}{16}v - 2v^2 - 5uv.$$

The characteristic equation for this system is $\left(r - \frac{\sqrt{2}}{2}\right)\left(r - \frac{6+5\sqrt{2}}{16}\right) = 0$, which has distinct positive roots. Hence $\left(\frac{2-\sqrt{2}}{16}, 0\right)$ is an unstable improper node.

Now consider the critical point $\left(\frac{2+\sqrt{2}}{16}, 0\right)$ where we use the change of variables $x = u + \frac{2+\sqrt{2}}{16}$ and $y = v$ to obtain the almost linear system

$$\frac{du}{dt} = \left(u + \frac{2+\sqrt{2}}{16}\right)\left(1 - 4u - \frac{2+\sqrt{2}}{4} - v\right) - \frac{1}{32} = -\frac{\sqrt{2}}{2}u - \frac{2+\sqrt{2}}{16}v - 4u^2 - uv,$$

$$\frac{dv}{dt} = v\left(1 - 2v - 5u - \frac{10+5\sqrt{2}}{16}\right) = \frac{6-5\sqrt{2}}{16}v - 2v^2 - 5uv.$$

The characteristic equation for this system is $\left(r+\frac{\sqrt{2}}{2}\right)\left(r-\frac{6-5\sqrt{2}}{16}\right)=0$, which has distinct negative roots. Hence $\left(\frac{2+\sqrt{2}}{16},0\right)$ is an asymptotically stable improper node.

When the critical point is $\left(\frac{1}{12},\frac{7}{24}\right)$, the change of variables $x=u+\frac{1}{12}$ and $y=v+\frac{7}{24}$ leads to the almost linear system

$$\frac{du}{dt}=\left(u+\frac{1}{12}\right)\left(1-4u-\frac{1}{3}-v-\frac{7}{24}\right)-\frac{1}{32}=\frac{1}{24}u-\frac{1}{12}v-4u^2-uv,$$

$$\frac{dv}{dt}=\left(v+\frac{7}{24}\right)\left(1-2v-\frac{7}{12}-5u-\frac{5}{12}\right)=-\frac{35}{24}u-\frac{7}{12}v-2v^2-5uv.$$

The characteristic equation for this system is $r^2+\frac{13}{24}r-\frac{7}{48}=0$, which has roots $r=\frac{-13\pm\sqrt{505}}{48}$. Since these roots have opposite signs, $\left(\frac{1}{12},\frac{7}{24}\right)$ is an unstable saddle point.

And when the critical point is $\left(\frac{1}{4},-\frac{1}{8}\right)$, the change of variables $x=u+\frac{1}{4}$ and $y=v-\frac{1}{8}$ leads to the almost linear system

$$\frac{du}{dt}=\left(u+\frac{1}{4}\right)\left(1-4u-1-v+\frac{1}{8}\right)-\frac{1}{32}=-\frac{7}{8}u-\frac{1}{4}v-4u^2-uv,$$

$$\frac{dv}{dt}=\left(v-\frac{1}{8}\right)\left(1-2v+\frac{1}{4}-5u-\frac{5}{4}\right)=\frac{5}{8}u+\frac{1}{4}v-2v^2-5uv.$$

The characteristic equation for this system is $r^2+\frac{5}{8}r-\frac{1}{16}=0$, which has roots $r=\frac{-5\pm\sqrt{41}}{16}$. Since these roots have opposite signs, $\left(\frac{1}{4},-\frac{1}{8}\right)$ is an unstable saddle point. But since $y<0$, this point is not of interest.

Hence, this is competitive exclusion; one species survives while the other dies off. A phase plane diagram is given in Figure B.62 in the answers of the text.

<u>Case 3: $h=\frac{5}{32}$</u>. The critical points for this system are the solutions to the pair of equations

$$x(1-4x-y)-\frac{5}{32}=0,$$
$$y(1-2y-5x)=0.$$

To solve this system, we first set $y=0$ and solve

$$x(1-4x)-\frac{5}{32}=0,$$

which has complex solutions. If $y\neq 0$ then we must have

$$1-2y-5x=0 \qquad \Rightarrow \qquad y=\frac{1}{2}-\frac{5}{2}x.$$

Substituting we obtain

$$x\left[1-4x-\left(\frac{1}{2}-\frac{5}{2}x\right)\right]-\frac{5}{32}=0.$$

Simplifying, we obtain

$$-\frac{3}{2}x^2+\frac{1}{2}x-\frac{5}{32}=0,$$

which also has only complex solutions. Hence there are no critical points. The phase plane diagram shows that species y survives while the x dies off. A phase plane diagram is given in Figure B.63 in the answers of the text.

EXERCISES 12.4: Energy Methods, page 798

3. Here $g(x)=\frac{x^2}{x-1}=x+1+\frac{1}{x-1}$. By integrating $g(x)$, we obtain the potential function

$$G(x)=\frac{x^2}{2}+x+\ln|x-1|+C,$$

and so

$$E(x,v)=\frac{v^2}{2}+\frac{x^2}{2}+x+\ln|x-1|+C.$$

Since $E(0,0)=0$ implies $C=0$, let $E(x,v)=\frac{v^2}{2}+\frac{x^2}{2}+x+\ln|x-1|$. Now, since we are interested in E near the origin, we let $|x-1|=1-x$ (because for x near 0, $|x-1|=1-x$). Therefore,

$$E(x,v)=\frac{v^2}{2}+\frac{x^2}{2}+x+\ln(1-x).$$

9. Here we have $g(x) = 2x^2 + x - 1$ and hence the potential function $G(x) = \frac{2x^3}{3} + \frac{x^2}{2} - x$. The local maxima and minima of $G(x)$ occur when $G'(x) = g(x) = 2x^2 + x - 1 = 0$. Thus the phase plane diagram has critical points at (–1, 0) and $\left(\frac{1}{2}, 0\right)$. Since $G(x)$ has a strict local minimum at $x = \frac{1}{2}$, the critical point $\left(\frac{1}{2}, 0\right)$ is a center. Furthermore, since $x = -1$ is strict local maximum, the critical point (–1, 0) is a saddle point. A sketch of the potential plane and phase plane diagram is given in Figure B.65 in the answers of the text.

11. Here we have $g(x) = \frac{x}{x-2} = 1 + \frac{2}{x-2}$ so the potential function is

$$G(x) = x + 2\ln|x-2| = x + 2\ln(2-x),$$

for x near zero. The local maxima and minima of $G(x)$ occur when $G'(x) = g(x) = \frac{x}{x-2} = 0$. Thus the phase plane diagram has critical points at (0, 0). Furthermore we note that $x = 2$ is not in the domain of $g(x)$ nor of $G(x)$. Now $G(x)$ has a strict local maximum at $x = 0$, hence the critical point (0, 0) is a saddle point. A sketch of the potential plane and phase plane diagram for $x < 2$ is given in Figure B.66 in the answers of the text.

13. We first observe that $vh(x,v) = v^2 > 0$ for $v \neq 0$. Hence, the energy is continually decreasing along a trajectory. The level curves for the energy function $E(x,v) = \frac{v^2}{2} + \frac{x^2}{2} - \frac{x^4}{4}$ are just the integral curves for Example 2(a) and are sketch in Figure 12.22 on page 794 of the text. The critical points for this damped system are the same as in the example and moreover, they are of the same type. The resulting phase plane is given in Figure B.67 on page A-69 in the answers of the text.

EXERCISES 12.5: Lyapunov's Direct Method, page 806

3. We compute $\dot{V}(x,y)$ with $V(x,y) = x^2 + y^2$.

$$\dot{V}(x,y) = V_x(x,y)f(x,y) + V_y(x,y)g(x,y)$$

$$= 2x\left(y^2 + xy^2 - x^3\right) + 2y\left(-xy + x^2y - y^3\right)$$

$$= 4x^2y^2 - 2x^4 - 2y^4 = -2\left(x^2 - y^2\right)^2.$$

According to Theorem 3, since $\dot{V}$ is negative semidefinite, V is positive definite function, and

(0, 0) is an isolated critical point of the system, the origin is stable.

5. The origin is an isolated critical point for the system. Using the hint, we compute $\dot{V}(x,y)$ with $V(x,y)=x^2-y^2$. Computing, we obtain

$$\dot{V}(x,y)=V_x(x,y)f(x,y)+V_y(x,y)g(x,y)$$

$$=2x(2x^3)-2y(2x^2y-y^3)=4x^4-4x^2y^2+2y^4=2x^4+2(x^2-y^2)^2,$$

which is positive definite. Now $V(0,0)=0$, and in every disk centered at the origin, V is positive at some point, (namely those points where $|x|>|y|$). Therefore by Theorem 4 the origin is unstable.

7. We compute $\dot{V}(x,y)$ with $V(x,y)=ax^4+by^2$.

$$\dot{V}(x,y)=V_x(x,y)f(x,y)+V_y(x,y)g(x,y)$$

$$=4ax^3(2y-x^3)+2by(-x^3-y^5)$$

$$=8ax^3y-4ax^6-2bx^3y-2by^6$$

To eliminate the x^3y term, we let $a=1$ and $b=4$, then

$$\dot{V}(x,y)=-4x^6-8y^6,$$

and we get that $\dot{V}$ is negative definite. Since V is positive definite and the origin is an isolated critical point, according to Theorem 3, the origin is asymptotically stable.

11. Here we set

$$y=\frac{dx}{dt}.$$

Then, we obtain the system

$$y=\frac{dx}{dt},$$

$$\frac{d^2x}{dt^2}=\frac{dy}{dt}=-(1-y^2)y-x.$$

Clearly, the zero solution is a solution to this system. To apply Lyapunov's direct method, we try the positive definite function $V(x,y)=ax^2+by^2$ and compute $\dot{V}(x,y)$.

$$\dot{V}(x,y)=V_x(x,y)f(x,y)+V_y(x,y)g(x,y)$$

$$= 2ax(y) + 2by\left(-\left(1-y^2\right)y - x\right) = 2axy - 2by^2 + 2by^4 - 2bxy .$$

To eliminate the xy terms, we choose $a = b = 1$, then

$$\dot{V}(x,y) = -2y^2\left(1-y^2\right).$$

Hence $\dot{V}(x,y)$ is negative semidefinite for $|y| < 1$, so by Theorem 3, the origin is stable.

EXERCISES 12.6: Limit Cycles and Periodic Solutions, page 815

5. We compute $r\dfrac{dr}{dt}$:

$$r\frac{dr}{dt} = x\frac{dx}{dt} + y\frac{dy}{dt} = x\left[x - y + x\left(r^3 - 4r^2 + 5r - 3\right)\right] + y\left[x + y + y\left(r^3 - 4r^2 + 5r - 3\right)\right.$$

$$= x^2 - xy + x^2\left(r^3 - 4r^2 + 5r - 3\right) + xy + y^2 + y^2\left(r^3 - 4r^2 + 5r - 3\right)$$

$$= r^2 + r^2\left(r^3 - 4r^2 + 5r - 3\right).$$

Hence

$$\frac{dr}{dt} = r + r\left(r^3 - 4r^2 + 5r - 3\right) = r\left(r^3 - 4r^2 + 5r - 2\right) = r(r-1)^2(r-2) .$$

Now $\dfrac{dr}{dt} = 0$ when $r = 0$, 1, 2. The critical point is represented by $r = 0$ and when $r = 1$ or 2, we have limit cycles of radius 1 and 2. When r lies in (0, 1) we have $\dfrac{dr}{dt} < 0$, so a trajectory in this region spirals into the origin. Therefore, the origin is an asymptotically stable spiral point. Now when r lies in (1, 2) we again have $\dfrac{dr}{dt} < 0$, so a trajectory in this region spirals into the limit cycle $r = 1$. This tells us that $r = 1$ is a semistable limit cycle. Finally, when $r > 2$, $\dfrac{dr}{dt} > 0$, so a trajectory in this region spirals away from the limit cycle $r = 2$. Hence, $r = 2$ is an unstable limit cycle.

To find the direction of the trajectories, we compute $r^2\dfrac{d\theta}{dt}$.

$$r^2\frac{d\theta}{dt} = x\frac{dy}{dt} - y\frac{dx}{dt} = x\left[x + y + y\left(r^3 - 4r^2 + 5r - 3\right)\right] - y\left[x - y + x\left(r^3 - 4r^2 + 5r - 3\right)\right]$$

$$= x^2 + xy + xy\left(r^3 - 4r^2 + 5r - 3\right) - xy + y^2 - xy\left(r^3 - 4r^2 + 5r - 3\right)$$

$$= x^2 + y^2 = r^2.$$

Hence $\frac{d\theta}{dt} = 1$ which tells us that the trajectories revolve counterclockwise about the origin. A phase plane diagram is given in Figure B.74 in the answers of the text.

11. We compute $r\frac{dr}{dt}$:

$$r\frac{dr}{dt} = x\frac{dx}{dt} + y\frac{dy}{dt} = x[y + x\sin(1/r)] + y[-x + y\sin(1/r)]$$

$$= xy + x^2\sin(1/r) - xy + y^2\sin(1/r)$$

$$= (x^2 + y^2)\sin(1/r) = r^2\sin(1/r).$$

Hence, $\frac{dr}{dt} = r\sin(1/r)$, and $\frac{dr}{dt} = 0$ when $r = \frac{1}{n\pi}$, $n = 1, 2, \ldots$. Consequently, the origin ($r = 0$) is not an isolated critical point. Observe that $\frac{dr}{dt} > 0$ for $\frac{1}{(2n+1)\pi} < r < \frac{1}{2n\pi}$ and $\frac{dr}{dt} < 0$ for $\frac{1}{2n\pi} < r < \frac{1}{(2n+1)\pi}$.

Thus, trajectories spiral into the limit cycles $r = \frac{1}{2n\pi}$ and away from the limit cycles $r = \frac{1}{(2n+1)\pi}$. To determine the direction of the spiral, we compute $r^2\frac{d\theta}{dt}$.

$$r^2\frac{d\theta}{dt} = x\frac{dy}{dt} - y\frac{dx}{dt} = x[-x + y\sin(1/r)] - y[y + x\sin(1/r)]$$

$$= -x^2 + xy\sin(1/r) - y^2 - xy\sin(1/r) = -x^2 - y^2 = -r^2.$$

Hence $\frac{d\theta}{dt} = -1$ which tells us that the trajectories revolve clockwise about the origin. A phase plane diagram is given in Figure B.68 in the answers of the text.

15. We compute $f_x + g_y$ in order to apply Theorem 6. Thus

$$f_x(x,y) + g_y(x,y) = (-8 + 3x^2) + (-7 + 3y^2) = 3(x^2 + y^2 - 5),$$

which is less than 0 for the given domain. Hence, by Theorem 6, there are no nonconstant periodic solutions in the disk $x^2 + y^2 < 5$.

19. It is easily seen that (0, 0) is a critical point, however, it is not easily shown that it is the only critical point for this system. Using the Lyapunov function $V(x,y)=2x^2+y^2$, we compute $\dot{V}(x,y)$. Thus

$$\begin{aligned}\dot{V}(x,y)&=V_x(x,y)\frac{dx}{dt}+V_y(x,y)\frac{dy}{dt}\\&=4x\left(2x-y-2x^3-3xy^2\right)+2y\left(2x+4y-4y^3-2x^2y\right)\\&=8x^2-8x^4-16x^2y^2+8y^2-8y^4=8\left(x^2+y^2\right)-8\left(x^2+y^2\right)^2.\end{aligned}$$

Therefore $\dot{V}(x,y)<0$ for $x^2+y^2>1$ and $\dot{V}(x,y)>0$ for $0<x^2+y^2<1$. Let C_1 be the curve $2x^2+y^2=\frac{1}{2}$, which lies inside $x^2+y^2=1$ and let C_2 be the curve $2x^2+y^2=3$, which lies outside $x^2+y^2=1$. Now $\dot{V}(x,y)>0$ on C_1 and $\dot{V}(x,y)<0$ on C_2. Hence, we let *R* be the region between the curves C_1 and C_2. Now, any trajectory that enters *R* is contained in *R*. So by Theorem 7, the system has a nonconstant periodic solution in *R*.

25. To apply Theorem 8, we check to see that all five conditions hold. Here we have $g(x)=x$ and $f(x)=x^2\left(x^2-1\right)$. Clearly $f(x)$ is even, hence condition **(a)** holds. Now

$$F(x)=\int_0^x s^2(s^2-1)\,ds=\frac{x^5}{5}-\frac{x^3}{3}.$$

Hence $F(x)<0$ for $0<x<\sqrt{\frac{5}{3}}$ and $F(x)>0$ for $x>\sqrt{\frac{5}{3}}$. Therefore condition **(b)** holds. Furthermore, condition **(c)** holds since $F(x)\to+\infty$ as $x\to+\infty$, monotonically for $x>\sqrt{\frac{5}{3}}$. As stated above $g(x)=x$ is an odd function with $g(x)>0$ for $x>0$, thus condition **(d)** holds. Finally, since

$$G(x)=\int_0^x s\,ds=\frac{x^2}{2},$$

we clearly have $G(x)\to+\infty$ as $x\to+\infty$, hence condition **(e)** holds. It follows from Theorem 8, that the Lienard equation has a unique nonconstant periodic solution .

EXERCISES 12.7: Stability of Higher–Dimensional Systems, page 822

5. From the characteristic equation

$$-(r-1)\left(r^2+1\right)=0,$$

we find that the eigenvalues are 1, ± *i*. Since at least one eigenvalue, 1, has a positive real part, the zero solution is unstable.

9. The characteristic equation is

$$\left(r^2+1\right)\left(r^2+1\right)=0,$$

which has eigenvalues ± *i*, ± *i*. Next we determine the eigenspace for the eigenvalue *i*. Computing we find

$$\begin{vmatrix} i & -1 & -1 & 0 \\ 1 & i & 0 & -1 \\ 0 & 0 & i & -1 \\ 0 & 0 & 1 & i \end{vmatrix} \quad \Rightarrow \quad \begin{vmatrix} 1 & i & 0 & 0 \\ 0 & 0 & 1 & 0 \\ 0 & 0 & 0 & 1 \\ 0 & 0 & 0 & 0 \end{vmatrix}.$$

Hence the eigenspace is degenerate and by problem 8c on page 822 of the text, the zero solution is unstable. Note: it can be shown that the eigenspace for the eigenvalue – *i* is also degenerate.

13. To find the fundamental matrix for this system we first recall the Taylor series e^x, $\sin x$, and $\cos x$. These are

$$e^x = 1 + x + \frac{x^2}{2!} + \frac{x^3}{3!} + \cdots,$$

$$\sin x = x - \frac{x^3}{3!} + \frac{x^5}{5!} - \cdots,$$

$$\cos x = 1 - \frac{x^2}{2!} + \frac{x^4}{4!} - \cdots.$$

Hence

$$\frac{dx_1}{dt} = \left(1 - x_1 + \frac{x_1^2}{2!} - \cdots\right) + \left(1 - \frac{x_2^2}{2!} + \cdots\right) - 2 = -x_1 + \left(\frac{x_1^2}{2!} - \cdots\right) + \left(-\frac{x_2^2}{2!} + \cdots\right),$$

$$\frac{dx_2}{dt} = -x_2 + \left(x_3 - \frac{x_3^2}{3!} + \cdots\right) = -x_2 + x_3 + \left(-\frac{x_3^2}{3!} + \cdots\right),$$

$$\frac{dx_3}{dt} = 1 - \left[1 + (x_2 + x_3) + \frac{(x_2 + x_3)^2}{2!} + \cdots\right] = -x_2 - x_3 - \left[\frac{(x_2 + x_3)^2}{2!} + \cdots\right].$$

Thus,

$$\mathbf{A} = \begin{bmatrix} -1 & 0 & 0 \\ 0 & -1 & 1 \\ 0 & -1 & -1 \end{bmatrix}.$$

Calculating the eigenvalues, we have

$$|\mathbf{A} - r\mathbf{I}| = \begin{vmatrix} -1-r & 0 & 0 \\ 0 & -1-r & 1 \\ 0 & -1 & -1-r \end{vmatrix} = 0.$$

Hence, the characteristic equation is $-(r+1)(r^2+2r+2)=0$. Therefore, the eigenvalues are -1, $-1 \pm i$. Since the real part of each is negative, the zero solution is asymptotically stable.

15. Solving for the critical points, we must have

$$\begin{aligned} -x_1 + 1 &= 0 \\ -2x_1 - x_2 + 2x_3 - 4 &= 0 \\ -3x_1 - 2x_2 - x_3 + 1 &= 0. \end{aligned}$$

Solving this system, we find the only solution is $(1, -2, 2)$. We now use the change of variables

$$x_1 = u + 1,$$

$$x_2 = v - 2,$$

$$x_3 = w + 2,$$

to translate the critical point to the origin. Substituting, we obtain the system

$$\frac{du}{dt} = -u,$$

$$\frac{dv}{dt} = -2u - v + 2w,$$

$$\frac{dw}{dt} = -3u - 2v - w.$$

Here $\mathbf{A}$ is given by

$$\begin{bmatrix} -1 & 0 & 0 \\ -2 & -1 & 2 \\ -3 & -2 & -1 \end{bmatrix}.$$

Finding the characteristic equation, we have $-(r+1)(r^2+2r+5)=0$. Hence the eigenvalues are $-1,\ -1\pm 2i$. Since each eigenvalue has a negative real part, the critical point $(1,-2,2)$ is asymptotically stable.

CHAPTER 13: Existence and Uniqueness Theory

EXERCISES 13.1: Introduction: Successive Approximations, page 836

9. To start the method of successive substitutions, we observe that

$$g(x)=\left(\frac{5-x}{3}\right)^{1/4}.$$

Therefore, according to equation (7) on page 831 of the text, we can find the next approximation from the previous one by using the recurrence relation

$$x_{n+1}=g(x_n)=\left(\frac{5-x_n}{3}\right)^{1/4}.$$

We start the procedure at the point $x_0=1$. Thus, using a hand calculator, we obtain

$$x_1=\left(\frac{5-x_0}{3}\right)^{1/4}=\left(\frac{5-1}{3}\right)^{1/4}=\left(\frac{4}{3}\right)^{1/4}\approx 1.0745699,$$

$$x_2=\left(\frac{5-x_1}{3}\right)^{1/4}=\left(\frac{5-1.0745699}{3}\right)^{1/4}\approx 1.0695264,$$

$$x_3=\left(\frac{5-x_2}{3}\right)^{1/4}=\left(\frac{5-1.0695264}{3}\right)^{1/4}\approx 1.0698697.$$

By continuing this process, we fill in Table 13-A below. Notice that, to the accuracy of the calculator used, $x_6=x_7$. Hence, we stopped the procedure after seven steps to obtain the approximation given by

$$x \approx 1.0698479.$$

Table 13-A. Approximations for a solution of $x=\left(\frac{5-x}{3}\right)^{1/4}$.

$x_0=1$	$x_4=1.0698464$
$x_1=1.0745699$	$x_5=1.0698480$
$x_2=1.0695264$	$x_6=1.0698479$
$x_3=1.0698697$	$x_7=1.0698479$

15. We first write this differential equation as an integral equation by integrating both sides from $x_0 = 0$ to x. Therefore, using the fact that $y(0) = 0$, we obtain

$$\int_0^x y'(t)\,dt = y(x) - y(0) = \int_0^x \left[y(t) - e^t\right]dt$$

$$\Rightarrow \qquad y(x) = \int_0^x \left[y(t) - e^t\right]dt.$$

Hence, by equation (15) on page 835 of the text, we observe that the Picard iterations are given by

$$y_{n+1}(x) = \int_0^x \left[y_n(t) - e^t\right]dt.$$

Thus, starting with $y_0(x) = y(0) = 0$, we calculate

$$y_1(x) = \int_0^x \left[y_0(t) - e^t\right]dt = -\int_0^x e^t\,dt = 1 - e^x,$$

$$y_2(x) = \int_0^x \left[y_1(t) - e^t\right]dt = \int_0^x \left[\left(1 - e^t\right) - e^t\right]dt = \int_0^x \left[1 - 2e^t\right]dt = 2 + x - 2e^x.$$

19. The functions $y = \dfrac{x^2+1}{2}$ and $y = x$ are graphed on the same coordinate axes in Figure 13-A. By examining this figure, we see that these two graphs intersect only at (1, 1). We can find this point by solving the equation

$$x = \frac{x^2+1}{2},$$

for x. Thus, we have

$$2x = x^2 + 1$$

$$\Rightarrow \qquad x^2 - 2x + 1 = 0$$

$$\Rightarrow \qquad (x-1)^2 = 0.$$

Since $y = x$, the only intersection point is (1, 1).

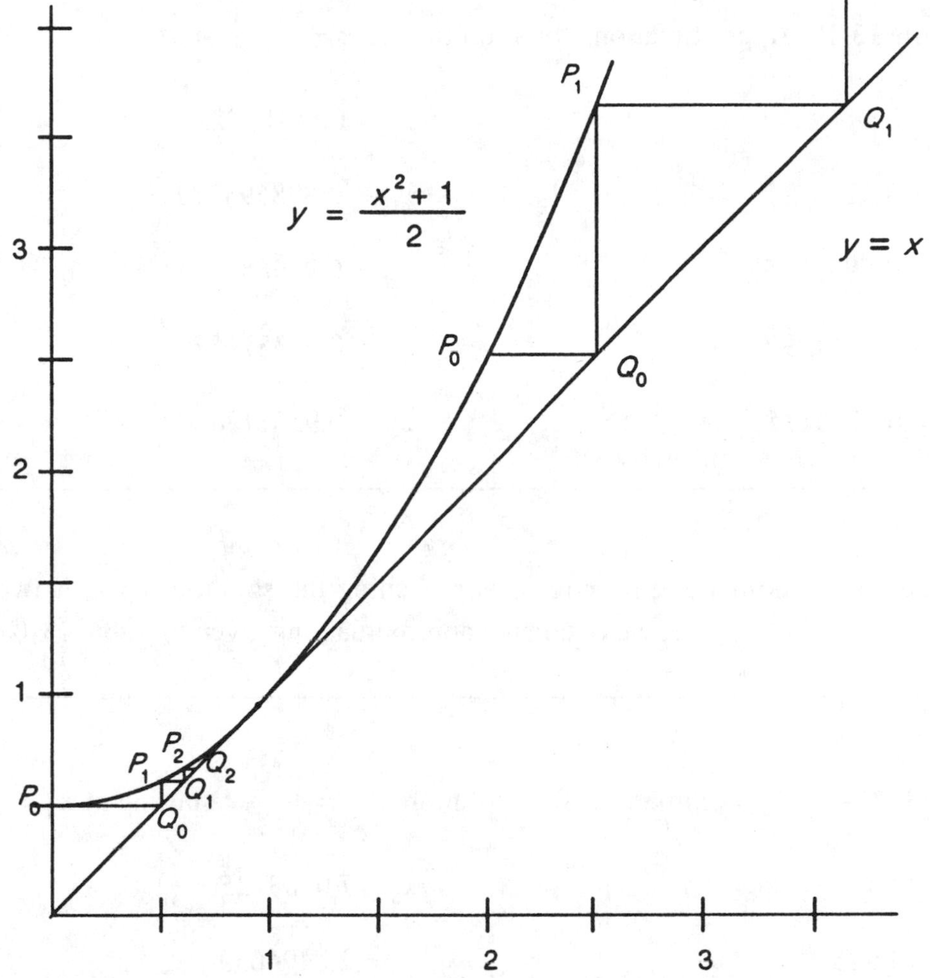

Figure 13-A. The method of successive substitution for the equation $x = \dfrac{x^2+1}{2}$.

To approximate the solution to the equation $x = \dfrac{x^2+1}{2}$ using the method of successive substitutions, we use the recurrence relation

$$x_{n+1} = \frac{x_n^{\ 2}+1}{2}.$$

Starting this method at $x_0 = 0$, we obtain the approximations given in Table 13-B below.

Table 13-B. Approximations for a solution of $x = \dfrac{x^2+1}{2}$ starting at $x_0 = 0$.

$x_1 = 0.5$	$x_{10} = 0.86109821$
$x_2 = 0.625$	$x_{15} = 0.898598372$
$x_3 = 0.6953125$	$x_{20} = 0.919887445$
$x_4 = 0.7417297$	$x_{30} = 0.943371575$
$x_5 = 0.7750815$	$x_{40} = 0.95611749$

These approximations do appear to be approaching the solution $x = 1$. However, if we start the process at the point $x_0 = 2$, we obtain the approximations given in Table 13-C.

Table 13-C. Approximations for a solution of $x = \dfrac{x^2+1}{2}$ starting at $x_0 = 2$.

$x_1 = 2.5$	$x_3 = 7.0703125$
$x_2 = 3.625$	$x_4 = 25.494659$

We observe that these approximations are getting larger and so do not seem to approach a fixed point. This also appears to be the case if we examine the pictorial representation for the method of successive substitutions given in Figure 13-A. By plugging $x_0 = 0$ into the function $y = \dfrac{x^2+1}{2}$, we find P_0 to be the point $\left(0, \dfrac{1}{2}\right)$. Then by moving parallel to the *x*-axis from the point P_0 to the line $y = x$, we observe that Q_0 is the point $\left(\dfrac{1}{2}, \dfrac{1}{2}\right)$. Next, by moving parallel to the *y*-axis from the point Q_0 to the curve $y = \dfrac{x^2+1}{2}$, we find that P_1 is the point $\left(\dfrac{1}{2}, \dfrac{5}{8}\right)$. Continuing this process moves us slowly in a step fashion to the point (1, 1). However, if we start this process at $x_0 = 2$, we observe that this method moves us through larger and larger steps away from the point of intersection (1, 1).

Note that for this equation, the movement of the method of successive substitutions is to the right. This is because the term $\frac{x_n{}^2+1}{2}$, in the recurrence relation, is increasing for $x > 0$. Thus, *the sequence of approximations* $\{x_n$ *of is an increasing sequence.* Starting at a nonnegative point less than 1 moves us to the fixed point at $x = 1$, but starting at a point larger that 1 moves us to ever increasing values for our approximations and, therefore, away from the fixed point.

EXERCISES 13.2: Picard's Existence and Uniqueness Theorem, page 844

3. In order to determine whether this sequence of functions converges uniformly, we must find

$$\lim_{n\to\infty}\|y_n - y\|.$$

Therefore, we first compute

$$\|y_n - y\| = \max_{x\in[0,1]}\left|\frac{nx}{1+n^2x^2}\right| = \max_{x\in[0,1]}\frac{nx}{1+n^2x^2},$$

where we have removed the absolute value signs because the term $\frac{nx}{1+n^2x^2}$ is nonnegative when $x \in [0,1]$. We will use calculus methods to obtain this maximum value. Thus, we differentiate the function $y_n = \frac{nx}{1+n^2x^2}$ to obtain

$$y_n'(x) = \frac{n\left(1-n^2x^2\right)}{\left(1+n^2x^2\right)^2}.$$

Setting $y_n'(x)$ equal to zero and solving yields

$$n\left(1-n^2x^2\right) = 0 \quad\Rightarrow\quad n^2x^2 = 1 \quad\Rightarrow\quad x = \pm\frac{1}{n}.$$

Since we are interested in the values of x on the interval [0, 1], we will only examine the critical point $x = \frac{1}{n}$. By the first derivative test, we observe that the function $y_n = \frac{nx}{1+n^2x^2}$ has a maximum value at the point $x = \frac{1}{n}$. Hence, the term $\frac{nx}{1+n^2x^2}$ will have a maximum value for $x \in [0,\ 1]$ when $x = n^{-1}$. Therefore, we have

$$\|y_n - y\| = \max_{x\in[0,1]}\frac{nx}{1+n^2x^2} = \frac{n\left(n^{-1}\right)}{1+n^2\left(n^{-2}\right)} = \frac{1}{2},$$

and so

$$\lim_{n\to\infty}\|y_n - y\| = \lim_{n\to\infty}\frac{1}{2} = \frac{1}{2} \neq 0 .$$

Thus, the given sequence of functions does *not* converge uniformly to the function $y(x) \equiv 0$ on the interval [0, 1].

This sequence of functions does, however, converge pointwise to the function $y(x) \equiv 0$ on the interval [0, 1]. To see this, notice that for any fixed $x \in [0,1]$ we have

$$\lim_{n\to\infty}[y_n(x) - y(x)] = \lim_{n\to\infty}\frac{nx}{1+n^2x^2} = \lim_{n\to\infty}\frac{1}{2nx} = 0,$$

where we have found this limit by using L'Hopital's rule. At the point $x = 0$, we observe that

$$\lim_{n\to\infty}[y_n(0) - y(0)] = \lim_{n\to\infty}\frac{0}{1} = 0 .$$

Thus, we have pointwise convergence but *not* uniform convergence.

5. We know (as was stated on page 455 of the text) that for all x such that $|x|<1$ the geometric series, $\sum_{k=0}^{\infty} x^k$, converges to the function $f(x) = \frac{1}{1-x}$. Thus, for all $x \in \left[0, \frac{1}{2}\right]$, we have

$$\frac{1}{1-x} = 1 + x + x^2 + \cdots = \sum_{k=0}^{\infty} x^k .$$

Therefore, we see that

$$\|y_n - y\| = \max_{x\in[0,1/2]}|y_n - y| = \max_{x\in[0,1/2]}\left|\sum_{k=0}^{n} x^k - \sum_{k=0}^{\infty} x^k\right| = \max_{x\in[0,1/2]}\sum_{k=n+1}^{\infty} x^k = \sum_{k=n+1}^{\infty}\left(\frac{1}{2}\right)^k .$$

Thus, we have

$$\lim_{n\to\infty}\|y_n - y\| = \lim_{n\to\infty}\left[\sum_{k=n+1}^{\infty}\left(\frac{1}{2}\right)^k\right].$$

Since $\sum_{k=n+1}^{\infty} (1/2)^k$ is the tail end of a convergent series, its limit must be zero. Hence, we have

$$\lim_{n\to\infty}\|y_n - y\| = \lim_{n\to\infty}\left[\sum_{k=n+1}^{\infty}\left(\frac{1}{2}\right)^k\right] = 0 .$$

Therefore, the given sequence of functions converges uniformly to the function $y(x) = \frac{1}{1-x}$ on the interval $\left[0, \frac{1}{2}\right]$.

9. We need to find an $h > 0$ such that $h < \min\left(h_1, \frac{\alpha_1}{M}, \frac{1}{L}\right)$. We are given that

$$R_1 = \{(x,y) : |x-1| \le 1, |y| \le 1\} = \{(x,y) : 0 \le x \le 2, -1 \le y \le 1\},$$

and so $h_1 = 1$ and $\alpha_1 = 1$. Thus, we must find values for M and L.

In order to find M, notice that, as was stated on page 840 of the text, we require that M satisfy the condition

$$|f(x,y)| = |y^2 - x| \le M,$$

for all (x, y) in R_1. To find this upper bound for $|f(x,y)|$, we must find the maximum and the minimum values of $f(x,y)$ on R_1. (Since $f(x,y)$ is a continuous function on the closed and bounded region R_1, it will have a maximum and a minimum there.) We will use calculus methods to find this maximum and this minimum. Since the first partial derivatives of $f(x,y)$, given by

$$f_x(x,y) = -1 \qquad \text{and} \qquad f_y(x,y) = 2y,$$

are never both zero, the maximum and minimum must occur on the boundary of R_1. Notice that R_1 is bounded on the left by the line $x = 0$, on the right by the line $x = 2$, on the top by the line $y = 1$, and on the bottom by the line $y = -1$. Therefore, we will examine the behavior of $f(x,y)$ (and, thus, of $|f(x,y)|$) on each of these lines.

Case 1: On the left side of R_1 where $x = 0$, the function $f(x,y)$ becomes the function in the single variable y, given by

$$f(0,y) = F_1(y) = y^2 - 0 = y^2, \qquad y \in [-1,1].$$

This function has a maximum at $y = \pm 1$ and a minimum at $y = 0$. Thus, on the left side of R_1 we see that f reaches a maximum value of $f(0,\pm 1) = 1$ and a minimum value of $f(0,0) = 0$.

Case 2: On the right side of R_1 where $x = 1$, the function $f(x,y)$ becomes the function in the single variable y, given by

$$f(1,y) = F_2(y) = y^2 - 1, \quad y \in [-1,1].$$

This function also has a maximum at $y = \pm 1$ and a minimum at $y = 0$. Thus, on the right side of R_1, the function $f(x,y)$ reaches a maximum value of $f(1,\pm 1) = 0$ and a minimum value of $f(1,0) = -1$.

Case 3: On the top and bottom of R_1 where $y = \pm 1$, the function $f(x,y)$ becomes the function given by

$$f(x,\pm 1) = F_3(x) = (\pm 1)^2 - x = 1 - x, \qquad x \in [0,1].$$

This function also has a maximum at $x = 0$ and a minimum at $x = 1$. Thus, on both the top and bottom of the region R_1, the function $f(x,y)$ reaches a maximum value of $f(0,\pm 1) = 1$ and a minimum value of $f(1,\pm 1) = 0$.

From the above cases we see that the maximum value of $f(x,y)$ is 1 and the minimum value is – 1 on the boundary of R_1. Thus, we have $|f(x,y)| \le 1$ for all (x,y) in R_1. Hence, we choose $M = 1$. To find L, we observe that L is an upper bound for

$$\left|\frac{\partial f}{\partial y}\right| = |2y| = 2|y| ,$$

on R_1. Since $y \in [-1,1]$ in this region, we have $|y| \le 1$. Hence, we see that

$$\left|\frac{\partial f}{\partial y}\right| = |2y| = 2|y| \le 2 ,$$

for all (x,y) in R_1. Therefore, we choose $L = 2$.

Now we can choose $h \ge 0$ such that

$$h < \min\left(h_1, \frac{\alpha_1}{M}, \frac{1}{L}\right) = \min\left(1, \frac{1}{1}, \frac{1}{2}\right) = \frac{1}{2}.$$

Thus, Theorem 3 guarantees that the given initial value problem will have a unique solution on the interval $[1-h, 1+h]$, where $0 < h < \frac{1}{2}$.

11. We are given that the recurrence relation for these approximations is $y_{n+1} = T[y_n]$. Using the definition of $T[y_n]$, we have

$$y_{n+1}(x) = x^3 - x + 1 + \int_0^x (u-x) y_n(u)\, du .$$

Thus, starting these approximations with $y_0(x) = x^3 - x + 1$, we obtain

$$y_1(x) = x^3 - x + 1 + \int_0^x (u-x) y_0(u)\, du = x^3 - x + 1 + \int_0^x (u-x)(u^3 - u + 1)\, du$$

$$= x^3 - x + 1 + \int_0^x (u^4 - u^2 + u - xu^3 + xu - x)\, du$$

$$= x^3 - x + 1 + \frac{x^5}{5} - \frac{x^3}{3} + \frac{x^2}{2} - \frac{x^5}{4} + \frac{x^3}{2} - x^2 .$$

By simplifying, we obtain

$$y_1(x) = -\frac{1}{20}x^5 + \frac{7}{6}x^3 - \frac{1}{2}x^2 - x + 1 .$$

Substituting this result into the recurrence relation yields

$$y_2(x) = x^3 - x + 1 + \int_0^x (u - x) y_1(u)\, du$$

$$= x^3 - x + 1 + \int_0^x (u - x)\left[-\frac{1}{20}u^5 + \frac{7}{6}u^3 - \frac{1}{2}u^2 - u + 1\right] du$$

$$= x^3 - x + 1 + \left[-\frac{1}{140}x^7 + \frac{7}{30}x^5 - \frac{1}{8}x^4 - \frac{1}{3}x^3 + \frac{1}{2}x^2\right.$$

$$\left. + \frac{1}{120}x^7 - \frac{7}{24}x^5 + \frac{1}{6}x^4 + \frac{1}{2}x^3 + \frac{1}{2}x^2\right].$$

When simplified, this yields

$$y_2(x) = \frac{1}{840}x^7 - \frac{7}{120}x^5 + \frac{1}{24}x^4 + \frac{7}{6}x^3 - \frac{1}{2}x^2 - x + 1.$$

EXERCISES 13.3: Existence of Solutions of Linear Equations, page 850

3. By comparing this problem to the problem given in (14) on page 849 of the text, we see that in this case

$$p_1(t) = -\ln t, \qquad p_2(t) = 0, \qquad p_3(t) = \tan t, \qquad \text{and} \qquad g(t) = e^{2t}.$$

We also observe that $t_0 = 1$. Thus, we must find an interval containing $t_0 = 1$ on which all of the functions $p_1(t)$, $p_2(t)$, $p_3(t)$, and $g(t)$ are simultaneously continuous. Therefore, we note that $p_2(t)$ and $g(t)$ are continuous everywhere; $p_1(t)$ is continuous on the interval $(0, \infty)$; and the interval which contains $t_0 = 1$ on which $p_3(t)$ is continuous is $\left(-\frac{\pi}{2}, \frac{\pi}{2}\right)$. Hence, these four functions are simultaneously continuous on the interval $\left(0, \frac{\pi}{2}\right)$ and this interval contains the point $t_0 = 1$. Therefore, Theorem 7 given on page 849 of the text guarantees that we will have a unique solution to this initial value problem on the whole interval $\left(0, \frac{\pi}{2}\right)$.

EXERCISES 13.4: Continuous Dependence of Solutions, page 856

3. To apply Theorem 9, we first determine the constant L for $f(x,y)=\mathrm{e}^{\cos y}x^2$. To do this, we observe that

$$\frac{\partial f}{\partial y}(x,y)=-\mathrm{e}^{\cos y}\sin y\,.$$

Now on any rectangle R_0, we have

$$\left|\frac{\partial f}{\partial y}\right|=\left|-\mathrm{e}^{\cos y}\sin y\right|\le \mathrm{e}\,.$$

Thus, since $h=1$, we have by Theorem 9,

$$|\phi(x,y_0)-\phi(x,\tilde{y}_0)|\le|y_0-\tilde{y}_0|\mathrm{e}^{\mathrm{e}}\,.$$

Since we are given that $|y_0-\tilde{y}_0|\le 10^{-2}$, we obtain the result

$$|\phi(x,y_0)-\phi(x,\tilde{y}_0)|\le 10^{-2}\mathrm{e}^{\mathrm{e}}\,.$$

9. We can use inequality (18) in Theorem 10 to obtain the bound, but first must determine the constant L and the constant ε. Here $f(x,y)=\sin x+\left(1+y^2\right)^{-1}$ and $F(x,y)=x+1-y^2$. Now

$$\left|\frac{\partial f}{\partial y}\right|=\left|\frac{2y}{\left(1+y^2\right)^2}\right|,$$

and

$$\left|\frac{\partial F}{\partial y}\right|=|2y|\le 2\,.$$

To find an upper bound for $\left|\frac{\partial f}{\partial y}\right|$ on R_0, we maximize $\frac{2y}{\left(1+y^2\right)^2}$. Hence, we obtain

$$\left(\frac{2y}{\left(1+y^2\right)^2}\right)'=\frac{2\left(1+y^2\right)^2-2y\cdot 2\left(1+y^2\right)2y}{\left(1+y^2\right)^4}=\frac{2\left(1+y^2\right)-8y^2}{\left(1+y^2\right)^3}=\frac{2-6y^2}{\left(1+y^2\right)^3}\,.$$

Setting this equal to zero and solving for y, we obtain

$$\frac{2-6y^2}{\left(1+y^2\right)^3}=0 \quad\Rightarrow\quad 2-6y^2=0 \quad\Rightarrow\quad y=\pm\frac{1}{\sqrt{3}}\,.$$

Since $\frac{2y}{\left(1+y^2\right)^2}$ is odd, we need only use $y=\frac{1}{\sqrt{3}}$. Thus

$$\left|\frac{\partial f}{\partial y}\right| \le \frac{\frac{2}{\sqrt{3}}}{\left(1+\frac{1}{3}\right)^2} = \frac{3\sqrt{3}}{8},$$

and so $L = \frac{3\sqrt{3}}{8}$. To obtain ε we seek an upper bound for

$$|f(x,y) - F(x,y)| = \left|\sin x + \frac{1}{1+y^2} - x - 1 + y^2\right| = \left|\sin x - x + \frac{1}{1+y^2} - (1-y)^2\right|$$

$$\le |\sin x - x| + \left|\frac{1}{1+y^2} - (1-y^2)\right|.$$

Using Taylor's Theorem with remainder we have $\sin x = x - \frac{\cos(\xi)x^3}{3!}$, where $0 \le \xi \le x$. Thus for $-1 \le x \le 1$ we obtain

$$|\sin x - x| = \left|x - \frac{\cos(\xi)x^3}{3!} - x\right| \le \frac{\cos(\xi)x^3}{6} \le \frac{1}{6}.$$

Applying Taylor's Theorem with remainder to $g(y) = \frac{1}{1+y^2} - 1 + y^2$, we obtain

$$g(y) = (1+y^2)^{-1} - 1 + y^2,$$

$$g'(y) = -2y(1+y^2)^{-2} + 2y,$$

$$g''(y) = -2(1+y^2)^{-2} + 2(1+y^2)^{-3}(2y)^2 + 2,$$

$$g'''(y) = 4(1+y^2)^{-3}2y - 6(1+y^2)^{-4}(2y)^3 + 2(1+y^2)^{-3}8y.$$

Since $g(0) = g'(0) = g''(0) = 0$, we have $\frac{1}{1+y^2} - (1-y^2) = \frac{g'''(\xi)}{3!}$, where $0 \le \xi \le y$. Thus, we obtain

$$\left|\frac{1}{1+y^2} - (1-y^2)\right| = \left|\frac{g'''(\xi)}{3!}\right| \le \frac{8+48+16}{6} = \frac{72}{6}.$$

Hence

$$|f(x,y) - F(x,y)| \le \frac{1}{6} + \frac{72}{6} = \frac{73}{6}.$$

It now follows from inequality (18) in Theorem 10 that

$$|\phi(x)-\psi(x)|\le\frac{73}{6}e^{3\sqrt{3}/8},$$

for x in $[-1, 1]$.

MICHAEL A. DRUMMOND 737-6179